Signaling and Communication in Plants

Series Editors

František Baluška
Department of Plant Cell Biology, IZMB, University of Bonn, Kirschallee 1,
D-53115 Bonn, Germany

Jorge Vivanco
Center for Rhizosphere Biology, Colorado State University, 217 Shepardson Building,
Fort Collins, CO 80523-1173, USA

For further volumes:
http://www.springer.com/series/8094

Frans Tax • Birgit Kemmerling
Editors

Receptor-like Kinases in Plants

From Development to Defense

Springer

Editors
Frans Tax
University of Arizona
Department of Molecular and
Cellular Biology
1007 E. Lowell
Tucson, AZ 85721
USA
fetax@email.arizona.edu

Birgit Kemmerling
University of Tuebingen
Department of Plant
Biochemistry
Auf der Morgenstelle 5
72076 Tuebingen
Germany
birgit.kemmerling@zmbp.uni-tuebingen.de

ISSN 1867-9048 e-ISSN 1867-9056
ISBN 978-3-642-23043-1 e-ISBN 978-3-642-23044-8
DOI 10.1007/978-3-642-23044-8
Springer Heidelberg Dordrecht London New York

Library of Congress Control Number: 2011944013

Printed on acid-free paper

Springer is part of Springer Science+Business Media (www.springer.com)

Preface

In 1991, John Walker and Ren Zhang reported the identification of the first plant receptor protein kinase gene (1). Based on its predicted structure of an extracellular domain related to the Self-incompatibility (S)-locus secreted glycoprotein, and a cytoplasmic serine/threonine kinase domain, Walker and Zhang predicted that the discovery of these novel plant proteins "provides a unique opportunity to gain fresh insights into signal transduction in higher plants".

Twenty years after Walker and Zhang's initial findings, it is difficult to imagine plant biology without receptor kinases coming to mind. The next decade saw the emergence of additional receptor kinases through forward genetic screens and through molecular biology. As alluded to above, the S-locus in Brassica also encoded receptor kinases with extracellular domains related to the S-locus glycoproteins. Developmental functions for various receptor kinases included epidermal differentiation for the CRINKLY-4 gene in maize, morphogenesis for the ERECTA gene in Arabidopsis, and maintenance of stem cells in the shoot apical meristem for the CLAVATA-1 gene in Arabidopsis. A role for receptor kinases in recognition of pathogens was first revealed by the identification of Xa21 in rice. A big surprise came with the finding that the BRI1 receptor kinase was the receptor for the plant steroid hormone brassinosteroids.

The emerging genome sequence of Arabidopsis was also uncovering hundreds of receptor kinases, ultimately more than 600. In the report of the NSF-Sponsored Workshop: "New Directions in Plant Biological Research" in April of 1999 (http://www.arabidopsis.org/carnegie_rep.html), the authors asked: "What are the roles of the hundreds of these proteins? Their existence implies a massive network of cell–cell and environment–plant communication, via a series of ligands yet to be discovered. Understanding this network will give us an entirely new view of plant development, environmental response, and organismal integration." Analysis of other genomes, including rice, which has more than 900 receptor kinases, indicates that the large number of receptor kinases in Arabidopsis was not an anomaly.

This book focuses on the momentum created within the plant biology community since Walker and Zhang's initial discovery. Thanks to a combination of collaborative "omics" projects, as well as the deep research efforts of many labs, portions of this "massive network" are emerging. This book opens with a view of the evolution and conservation of receptor kinases in plants, focusing on the rapid expansion of this gene family. After the first chapter, the following seven chapters update the known functions of receptor kinases in various biological contexts, extending the initial discoveries mentioned above. The second half of the book focuses on the diverse ligands, signaling mechanisms, and regulation of receptor kinases. The authors of all of these chapters reveal the amazing results from the past 20 years, and hint at the discoveries that may come in the next 20 years.

Tucson, USA *Frans Tax*
Tübingen, Germany *Birgit Kemmerling*

Contents

Origin, Diversity, Expansion History, and Functional Evolution of the Plant Receptor-Like Kinase/*Pelle* Family

Melissa D. Lehti-Shiu, Cheng Zou, and Shin-Han Shiu

Abstract The RLK/*Pelle* gene family is one of the largest gene families in plants with several hundred to more than a thousand members, but only a few family members exist in animals. This unbalanced distribution indicates a rather dramatic expansion of this gene family in land plants. In this chapter we review what is known about the RLK/*Pelle* family's origin in eukaryotes, its domain content evolution, expansion patterns across plant and animal species, and the duplication mechanisms that contribute to its expansion. We conclude by summarizing current knowledge of plant RLK/*Pelle* functions for a discussion on the relative importance of neutral evolution and natural selection as the driving forces behind continuous expansion and innovation in this gene family.

1 Introduction

In 1990, John Walker and Ren Zhang reported the cloning of a maize protein kinase resembling animal receptor tyrosine kinases (Walker and Zhang 1990). This maize kinase, which contains a putative extracellular domain (ECD) delineated by a signal sequence and a hydrophobic transmembrane region, represents the prototypical Receptor-Like Kinase (RLK). In the following two decades, extensive genetic and phenotypic studies revealed diverse roles of RLKs, ranging from control of development to stress responses (Walker 1994; Braun and Walker 1996; Torii 2000; Shiu and Bleecker 2001b; Lease and Walker 2006; Morillo and Tax 2006). In the late 1990s, as expressed sequence tags and genomic sequence accumulated, more and more RLKs were found in the model plant *Arabidopsis thaliana*. A global computational analysis of *A. thaliana* kinases then established that RLKs were one of the largest gene families in plants (Shiu and Bleecker 2001a). Interestingly, RLKs were

M.D. Lehti-Shiu • C. Zou • S.-H. Shiu (✉)
Department of Plant Biology, Michigan State University, East Lansing, MI 48824, USA
e-mail: lehtishi@msu.edu; czou@caas.net.cn; shius@msu.edu

F. Tax and B. Kemmerling (eds.), *Receptor-like Kinases in Plants*,
Signaling and Communication in Plants 13,
DOI 10.1007/978-3-642-23044-8_1,

found to be most closely related to *Drosophila melanogaster Pelle* (Belvin and Anderson 1996) and mammalian Interleukin Receptor-Associated Kinases (IRAKs) (Cao et al. 1996; Flannery and Bowie 2010), which comprise a very small family of cytoplasmic kinases without ECDs or membrane spanning regions. Similar to *Pelle* and IRAKs, some plant members of the RLK/*Pelle* family also lack ECDs and transmembrane regions and are referred to as Receptor-Like Cytoplasmic Kinases (RLCKs).The phylogenetic relationships between RLKs and *Pelle*/IRAKs indicate that they likely are orthologs derived from an ancestral kinase present in the common ancestor of plants and animals. Thus, these related kinases are collectively classified as members of the RLK/*Pelle* family (Shiu and Bleecker 2001a).

In this chapter, we discuss the current knowledge of the evolutionary history of the RLK/*Pelle* family with a focus on the following areas. Our first focus is on the origin of the RLK/*Pelle* family, particularly on the relationship between the kinase domains of plant RLK/*Pelle*s and other kinase families. The second focus concerns

Table 1 Protein domains found in the extracellular regions of RLK/*Pelles*

Protein domain	Description	Reference
Bulb-type lectin domain (B-lectin)	Binds α-D-mannose	Hester et al. (1995)
Cyclases/Histidine kinases Associated Sensory Extracellular (CHASE)	Found in diverse transmembrane receptors and is predicted to bind low molecular weight ligands	Mougel and Zhulin (2001)
C-type lectin (C-LEC)	Binds carbohydrates	Sharon and Lis (2001)
Domain of Unknown Function26 (DUF26)	Contains four conserved cysteines and implicated in defense response	Miyakawa et al. (2007)
Epidermal Growth Factor (EGF)	Contains calcium binding motif and calcium binding may be necessary for protein–protein interactions	Handford et al. (1990)
Glycerophosphodiester Phosphodiesterase Domain (GDPD)	Found in enzymes that hydrolyze glycerophosphodiesters	Santelli et al. (2004)
Glycoside hydrolase, family 18 catalytic domain	Hydrolyzes chitin oligosaccharides	Perrakis et al. (1994)
Legume lectin B	Carbohydrate binding	Loris et al. (1998)
Lysin Motif (LysM)	Peptidoglycan, chitin-binding	Buist et al. (2008)
Leucine-Rich Repeat (LRR)	Protein–protein interactions	Kobe and Deisenhofer (1994)
Plasminogen/Apple/Nematode protein domain (PAN)	Protein–protein and protein–carbohydrate interactions	Tordai et al. (1999)
S-locus glycoprotein	Secreted proteins involved in self-incompatibility	Hinata et al. (1995)
Pathogenesis-Related Protein-1/Sperm-Coating Glycoprotein domain (SCP)	Found in pathogenesis-related proteins expressed during defense response	Szyperski et al. (1998)
Thaumatin	Antifungal and chitinase activity	Pan et al. (1999)
WAK	Domain found in Wall-associated Kinases, which bind cell wall components	Kohorn (2001)

how the RLK/*Pelle* family has diversified in plants. Plant RLKs with a clear receptor configuration possess a diverse array of extracellular regions implicated in interactions with proteins, polysaccharides, lipids, and other ligands (Shiu and Bleecker 2001b) (see Table 1, Chapters "Receptor Ligands in Development" and "Phosphorylation and RLK signaling"). In particular, we will address two questions regarding RLK diversification: (1) how RLKs acquired their extracellular regions and (2) how frequently these acquisitions have occurred. Third, we will focus on how the RLK/*Pelle* family expanded during plant evolution. The RLK/*Pelle* family is among the largest in plants, but there is only one *D. melanogaster Pelle* kinase and four human IRAKs (Belvin and Anderson 1996; Janssens and Beyaert 2003). Why is there such a large size difference between plant and animal RLK/*Pelle* members? Based on kinase phylogeny, the RLK/*Pelle* family can be subdivided into multiple subfamilies where members in each subfamily possess mostly similar ECDs (Shiu and Bleecker 2001a). How have these subfamilies expanded differentially among plant lineages and what are the implications for plant adaptation and evolution? Finally, we will summarize current knowledge of plant RLK/*Pelle* functions to pinpoint potential sources of selective pressure that drive continuous expansion and innovation in this gene family.

2 Origin of Receptor-Like/*Pelle* Kinases in Eukaryotes

Phylogenetic studies of representative kinase family members indicate that plant RLKs and animal *Pelle* kinases and IRAKs are each others' closest relative and that the RLK/*Pelle* family predates the divergence of the plant and animal lineages (Shiu and Bleecker 2001a). In addition, the kinase domains of RLK/*Pelles* are more closely related to animal receptor tyrosine kinases and Raf kinases than to any other kinase family (Shiu and Bleecker 2001a). Thus, RLK/*Pelle* and receptor tyrosine kinases likely have a monophyletic origin. Interestingly, aside from G-protein coupled receptors, receptor tyrosine kinases belong to the largest family of transmembrane receptors in animals (Hunter et al. 1992; van der Geer et al. 1994). Taking into consideration that plant RLKs and animal receptor tyrosine kinases have highly similar structural configurations and are likely monophyletic in origin, it is likely that their common ancestor was involved in the perception of extracellular stimuli. However, it is not clear if the ancestral kinase of RLK/*Pelle* and tyrosine kinases was a transmembrane receptor protein or a cytoplamic kinase that interacted with a receptor complex that perceived extracellular signals.

The RLK/*Pelle* family was originally established using plant and metazoan homologs only (Shiu and Bleecker 2001b). Subsequent studies with expanded taxonomic sampling have revealed that there are no clear RLK/*Pelle* homologs in fungi (Shiu and Bleecker 2003). For this review, we have also surveyed multiple nonplant eukaryotic genomes currently available in GenBank. Our findings confirm the absence of RLK/*Pelle* homologs in fungi and indicate that the taxonomic distribution of RLK/*Pelles* is surprisingly sparse among eukaryotes (Fig. 1).

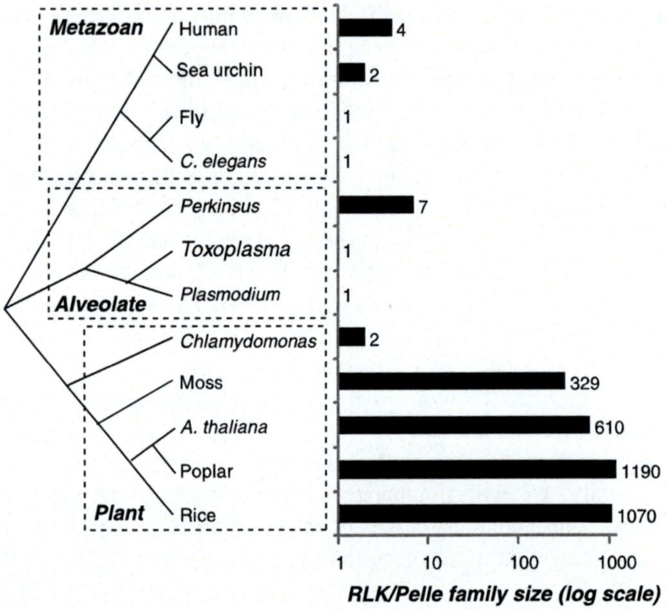

Fig. 1 Phylogenetic relationships between species harboring RLK/*Pelle* homologs and the RLK family size in each species

Although *Pelle* and IRAK related sequences can be found in multiple vertebrate and invertebrate animals, there are no clear homologs in *Monosiga brevicollis*, a unicellular choanoflagellate that is basal to all metazoans (King et al. 2008). In addition, among nonplant and nonmetazoan eukaryotes, RLK/*Pelle* homologs are only found in alveolata species including *Plasmodium*, *Toxoplasma*, and *Perkinsus*. The distribution of RLK/*Pelle* genes among eukaryotic species suggests an ancient origin before the divergence between the plant and animal lineages. The absence of RLK/*Pelle* in most eukaryotic lineages can be explained by either gene losses in multiple lineages or sequence divergence. The difficulty in using gene loss as an explanation is that multiple independent losses would have had to occur to account for the patterns we see (Fig. 1). On the other hand, given that the eukaryotes diverged >1–2 billion years ago (Hedges 2002), it is likely that divergent evolution of RLK/*Pelle* homologs in eukaryotes has resulted in a situation where little signal is left for generating reliable phylogenies. If this is the case, why can we still find clear RLK/*Pelle* homologs in apicomplexans and metazoans? One intriguing possibility is that these apicomplexan RLK/*Pelles* may be products of horizontal transfer from a secondary symbiotic event involving green algae (Kohler et al. 1997), although this remains to be substantiated. In the case of metazoan RLK/*Pelles*, we speculate that conserved ancestral functions between animal and plant RLK/*Pelles* may have limited the degree of sequence divergence (see Sects. 3.1, 3.2, and 5.1).

In Viridiplantae, RLK/*Pelle* family members are found in multiple green algal species and all land plants (Fig. 1). Among chlorophyte algae, although no RLK/*Pelle* gene is found in *Ostreococcus tauri*, a green alga that is regarded as the smallest eukaryote (Courties et al. 1994), two RLCK genes are present in *Chlamydomonas reinhardtii* (Lehti-Shiu et al. 2009). In the charophyte lineage, which among green algae shares the closest common ancestor with land plants, multiple RLK/*Pelle* genes are found in *Closterium ehrenbergii*, a unicellular species, and *Nitella axillaris*, a multicellular charophyte (Sasaki et al. 2007). In land plants, the number of RLK/*Pelle* members rises sharply, with 329, 1,070, 610, and 1,192 genes found in moss, rice, *A. thaliana*, and poplar, respectively (Lehti-Shiu et al. 2009). This, together with the small number of RLK/*Pelle* genes in animal species, indicates that this gene family has experienced dramatic expansion specifically in the land plant lineage. In addition, given the presence of RLKs in charophytes, the receptor configuration likely arose before the divergence of charophytes from land plants. However, the charophyte RLKs were identified from transcript sequences. Thus, a more thorough phylogenetic study using the entire repertoire of RLK/*Pelle* genes in a charophyte genome will be necessary to assess whether there is a single or multiple origins of receptor configurations involving members of this family.

Why have RLK/*Pelle*s undergone such dramatic expansion only in plants? It is clear that different receptor kinase families have expanded in different lineages. In a comparative analysis of tyrosine kinases in animal species ranging from *Caenorhabditis elegans* to human, the tyrosine kinase family was found to have expanded substantially over the course of metazoan evolution (Shiu and Li 2004). In plants the expanded receptor kinase family is the RLK/*Pelle* family, and brown algae and oomycetes each contain phylogenetically distinct families of receptor kinases (Cock et al. 2010). Therefore, it is likely that kinases in different families were paired with ECDs in different lineages, and expansion of distinct receptor kinases were selected for independently due to the adaptive advantage conferred by the ability to perceive extracellular signals.

3 Evolution of RLK/*Pelle* Domain Content

3.1 *RLK/*Pelle *Domain Content Diversity and the Creation of Receptor Chimera*

The RLK/*Pelle* family can be divided into several subfamilies based on phylogenetic relationships between kinase domains (Shiu and Bleecker 2001a), and RLK/*Pelle* genes with related kinase domains almost always have the same type of ECD (Shiu and Bleecker 2001a, b). The diversity of plant RLK extracellular regions is similar to the domain complexity seen in animal receptor kinases (Cock et al. 2002), and this diversity has two implications. The first is the importance of RLKs in perceiving a wide range of extracellular signals. Supporting this, multiple RLKs have been shown to directly bind to protein, lipid, polysaccharides, and other molecular ligands, both self and nonself (see Chapters "Receptor Ligands in Development" and "Phosphorylation and RLK signaling"). The second implication is that, because of the utility of transmembrane signaling, RLK/*Pelle* gene fusions were repeatedly selected for over the course of plant evolution. The fact that chimeric RLKs containing nonnative ECDs activate the same downstream signaling pathways as the native ECD (He et al. 2000; Albert et al. 2010; Brutus et al. 2010) illustrates how naturally occurring novel chimeric RLKs may have been created.

In plants, a large number of secreted or membrane spanning proteins have similar domain content as RLKs (Shiu and Bleecker 2003; Fritz-Laylin et al. 2005). These RLK-like proteins, collectively named Receptor-Like Proteins (RLPs), have a signal sequence and transmembrane domain but no intracellular kinase domain. Multiple RLPs have been shown to function with RLKs to regulate development and defense response. The RLKs *CLAVATA1* and *CORYNE* interact with the RLP *CLAVATA2* to regulate meristem development (Jeong et al. 1999; Muller et al. 2008). The RLP *TOO MANY MOUTHS* is proposed to form a complex with *ERECTA* family members to regulate stomatal patterning (Shpak et al. 2005). The rice LysM domain-containing RLP, *CEBiP*, and RLK, *OsCERK1*, transiently form a complex when treated with chitin oligosaccharides and activate defense response pathways (Shimizu et al. 2010). The similarities between RLPs and RLKs and their functional relationships are consistent with the possibility that RLKs with novel domain configurations may have been created through fusions between existing RLPs and RLKs/RLCKs. In addition, most RLPs, which are secreted or membrane spanning proteins, are likely integral components of extracellular signaling networks. Fusions between ancestral RLPs and RLK/*Pelle* kinases could therefore have led to novel signal transduction pathways by linking ligand perception to different downstream kinase targets. Alternatively, fusions may simply have occurred between RLPs and RLK/*Pelles* that were already components of the same signaling networks.

The interaction between receptors lacking kinase domains and cytoplasmic kinases is known in animal systems; *D. melanogaster Pelle* is a cytoplasmic kinase that is part of signaling networks involving Toll, a transmembrane receptor without a kinase domain, mediating both innate immunity and development (Hecht and Anderson 1993; Shelton and Wasserman 1993; Belvin and Anderson 1996). Similarly mammalian IRAKs are parts of innate immunity signaling networks involving Toll-Like Receptors (Flannery and Bowie 2010). Multiple plant RLK/*Pelle* members are involved in innate immunity (Boller and Felix 2009). Thus, the innate immunity function of some members in the RLK/*Pelle* family is likely an ancestral trait. Given that the receptor configuration must arise from a fusion between an RLP and an RLCK, it is plausible that these RLKs with innate immunity functions were originally RLPs and RLCKs that fused together later on.

3.2 Rate of Domain Gain and Loss

Domain content diversity among RLK/*Pelle* family members clearly points to repeated creation of new RLKs (Shiu and Bleecker 2001a). Comparative genomic studies of RLK/*Pelle* families also provide support for multiple domain gain and loss events during flowering plant evolution. For example, after the poplar lineage separated from the *A. thaliana* lineage 100–120 million years ago (Tuskan et al. 2006), a novel poplar and a novel *A. thaliana* RLK were created by the fusion of a glycosyl hydrolase 18 (chitinase domain; Perrakis et al. 1994) domain and a glycerophosphodiester phosphodiesterase domain (GDPD; Santelli et al. 2004), respectively. In the rice lineage after its separation from dicots, two novel RLKs were generated involving a Pathogenesis-Related Protein-1/Sperm-Coating Glycoprotein domain (Szyperski et al. 1998) and a CHASE domain (Anantharaman and Aravind 2001; Mougel and Zhulin 2001).

There are also multiple examples where existing RLK ECDs were replaced by protein domains that are novel to the RLK/*Pelle* family (Shiu et al. 2004; Lehti-Shiu et al. 2009). In addition, several RLK ECDs are associated with more than one RLK subfamily. In these cases, based on the phylogenetic relationships between subfamilies, these common ECDs were likely acquired independently (Shiu et al. 2004; Lehti-Shiu et al. 2009). It is possible that the signaling mechanisms between subfamilies where domain swapping took place were similar and thus the swapped RLKs play the same roles. Alternatively, it is possible the swapping allowed an alternative signaling route for perceiving the same kinds of stimuli, assuming that different RLK subfamilies have distinct downstream signaling networks. As we learn more about RLK/*Pelle* functions, comparative analyses across RLK/*Pelle* subfamilies will allow us to distinguish between these two possibilities. In many cases, the domains gained have been implicated in biotic stress response, suggesting the selective pressure for proper response to biotic agents may have contributed to their retention (Shiu et al. 2004; Lehti-Shiu et al. 2009). Consistent with this view, a recent study has demonstrated that a gain of function mutation in the Arabidopsis LRK10L kinase containing the novel GDPD domain leads to constitutive activation of defense responses (Bi et al. 2010).

In addition to domain gain, domain loss has occurred during the evolution of the RLK/*Pelle* family. In *A. thaliana*, 28 members of RLK subfamilies have likely lost their ECDs based on phylogenetic analysis, leaving only an intracellular kinase domain typical of RLCKs. It is also likely that RLK/*Pelle* genes have lost kinase domains, leading to the generation of RLPs. This is consistent with the observation that some RLPs are located in close proximity to RLKs in their chromosomal positions (Shiu and Bleecker 2003). It was recently found that a *Drosophila* protein, Tube, which lacks a kinase domain, is evolutionarily related to mammalian *IRAK4* (Towb et al. 2009). Although it lacks kinase activity, *Tube* still functions in Toll signaling pathways as a scaffold protein. It is conceivable that, similarly, some RLPs may be derived from RLKs through kinase domain losses but still function in the same signaling pathways as scaffold proteins similar to *Tube*. This can potentially be resolved through detailed phylogenomic analysis of RLP and RLK ECDs.

Although the diversity of RLKs is clearly the consequence of repeated innovation involving RLK/*Pelle* members, this does not mean that novel RLK configurations arose all that frequently. Among the RLK/*Pelle* subfamilies defined based on kinase phylogeny, 77% were established prior to the divergence of the vascular plant lineage from nonvascular plants ~450–700 million years ago (Hedges 2002; Lehti-Shiu et al. 2009). In addition, there are very few lineage-specific RLK/*Pelle* subfamilies. In a comparative analysis of the RLK/*Pelle* family in moss, rice, poplar, and *A. thaliana*, only two moss-specific and one poplar-specific subfamily were identified (Lehti-Shiu et al. 2009). Many RLK/*Pelle* subfamilies are also represented in liverwort (Sasaki et al. 2007). This suggests that most RLK/*Pelle* subfamilies were established very early in land plant evolution, well before the divergence of the land plant lineage ~450–700 million years ago. While domain fusions involving members of the RLK/*Pelle* family may have happened frequently early in land plant evolution, the rate of domain gain is very low. There have been only 12 instances of domain gain or swapping in three flowering plant lineages over the last 150 million years (Lehti-Shiu et al. 2009). The major reason for substantial differences in RLK/*Pelle* family size in plants is not due to the presence of many different RLKs with different domain content but due to differential expansion of existing subfamilies (Lehti-Shiu et al. 2009).

4 Expansion of the Plant RLK/*Pelle* Family

4.1 Dramatic Expansion in the Land Plant Lineage

As discussed in Sect. 2, there are zero or very few RLK/*Pelle* members in most eukaryotes except land plants, and in Viridiplantae, only two RLCKs are present in the chlorophyte alga *C. reinhardtii*. Thus, the small RLK/*Pelle* family sizes in animals and green algae are taken as evidence that the RLK/*Pelle* family likely was very small before the chlorophyte lineage diverged from land plants and related charophyte algae ~1 billion years ago (Hedges 2002). On the other hand, a charophyte alga, *C. ehrenbergii*, has at least 14 RLK/*Pelle*s, including an RLK (Sasaki et al. 2007). Note that charophyte algae belong to a polyphyletic group (Qiu 2008). *C. ehrenbergii* is in the family Zygnemophyceae that is not as close to land plants as those in the Charophyceae family such as *Chara* (Lewis and McCourt 2004). Thus, it will be of great interest to determine the abundance of RLK/*Pelle* genes in Charophyceae algae. The RLK/*Pelle* family expanded continuously over the course of land plant evolution; the early diverging bryophyte species, *Physcomitrella patens* (moss), has 329 RLK/*Pelle*s (Lehti-Shiu et al. 2009) and there are 610 RLK/*Pelle* genes in *A. thaliana* (Shiu and Bleecker 2001a), and over a thousand in rice (Shiu et al. 2004) and poplar (Lehti-Shiu et al. 2009). Given that there were likely very few RLK/*Pelle* genes in the green algal-land plant common ancestor, the ancestral RLK/*Pelle* genes have been amplified greatly in the land plant lineage, particularly in vascular plants.

Based on a comparative analysis of RLK/*Pelle* protein sequences from moss and three flowering plants, it is estimated that the RLK/*Pelle* family has expanded by ~24% in the moss lineage since its divergence from the flowering plant lineage (Lehti-Shiu et al. 2009). In contrast, the expansion of the RLK/*Pelle* family in flowering plant lineages ranges from 93% to 212%. Based on a global comparison of plant gene families, the kinase superfamily, in particular, has undergone a much higher degree of expansion than nearly all gene families (Hanada et al. 2008). Given that the number of other kinases is ~400 in algae and land plants, the much higher degree of expansion seen in the kinase superfamily is mostly a consequence of expansion of the RLK/*Pelle* family (Lehti-Shiu et al. 2009). The expansion seen in the RLK/*Pelle* family is not only a result of many gene duplication events involving this gene family but also because many duplicates were retained. Because most duplicate genes become pseudogenes rather quickly (Li 1983), the much higher degree of expansion in the RLK/*Pelle* family relative to most other families in plants is mostly a consequence of differential retention.

Differential expansion among land plants is also apparent at the RLK/*Pelle* subfamily level (Lehti-Shiu et al. 2009). While some subfamilies have remained relatively constant in size, others have undergone pronounced differential expansion. For example, there are two C-LEC subfamily members in moss and only one in rice, poplar, and *A. thaliana*. In contrast, the rice WAK subfamily, which has over 114 members, is six times larger than in *A. thaliana*. Strikingly the rice and poplar SD1 subfamilies are more than ten times larger than in *A. thaliana*. Similar numbers of RLK/*Pelle* subfamily members in different species do not necessarily indicate that family sizes were established prior to divergence of these lineages, however. For example, the DUF26 subfamily expanded after the divergence of vascular plants, and there are similar numbers of DUF26 RLKs in poplar, *A. thaliana,* and rice. However, based on the phylogenetic relationships of DUF26 subfamily members in these plant species, the similar DUF26 subfamily sizes in these different lineages are mainly due to parallel, lineage-specific expansion and frequent gene loss (Shiu et al. 2004; Lehti-Shiu et al. 2009).

*4.2 Duplication Mechanisms That Contribute to the RLK/*Pelle *Family Expansion*

Expansion of a gene family involves a combination of gene duplication and subsequent retention. Thus, the dramatic expansion of the RLK/*Pelle* family in the land plant lineage and substantial variation in differential expansion among RLK/*Pelle* subfamilies across species prompt two important questions. The first is how these RLK/*Pelle*s were duplicated. The second question is whether these

RLK/*Pelle* duplicates were retained preferentially. The latter question will be discussed in Sect. 5. Compared to other eukaryotes, plants are unique in that whole genome duplications occurred much more frequently (Lockton and Gaut 2005; Cui et al. 2006). Together with other duplication mechanisms, such as tandem duplication, segmental duplication, and replicative transposition, a very large number of duplicate genes, in all gene families, have been generated in the long evolutionary history of land plants. Our current understanding is that the RLK/*Pelle* family has undergone dramatic expansion in land plant species and that this expansion is mainly a consequence of high rates of duplicate retention. However, substantial variation in retention rates within the family is revealed upon closer examination of subfamilies and orthologous groups. Interestingly, many RLK/*Pelle* genes are tandemly duplicated. In fact, 33% of the RLK/*Pelle* family in *A. thaliana* is found in tandem repeats (Lehti-Shiu et al. 2009). In rice and poplar, which have nearly double the number of RLK/*Pelle* genes, the proportion of members in tandem repeats is even higher (50% and 39%, respectively; Lehti-Shiu et al. 2009). In several studies, protein coding genes containing kinase domains have been shown to be overrepresented in tandem repeats (Rizzon et al. 2006; Tuskan et al. 2006; Hanada et al. 2008). Consistent with these studies, RLK/*Pelle* genes in orthologous groups that have expanded tend to be tandem genes (Lehti-Shiu et al. 2009), indicating that tandem duplication contributed more significantly to lineage-specific expansion of the RLK/*Pelle* family than other duplication mechanisms, such as whole genome duplication, combined.

Unequal crossovers, or tandem duplication, can change the content of tandem clusters dramatically (for example, resulting in multiple linked copies in one homologous chromosome and none in the other) and quickly (taking place every generation) (Leister 2004; Reams and Neidle 2004). This provides a mechanistic explanation for why RLKs are among the most variable gene families among *A. thaliana* accessions (Clark et al. 2007; Cao et al. 2011). Unequal crossover is most effective between genomic regions that are highly similar either because they are derived from recent duplication or have experienced gene conversion. Thus, the contribution of tandem duplication to RLK/*Pelle* family expansion is mostly on relatively young RLK/*Pelle* duplicates. We should also emphasize that other duplication mechanisms likely contributed to expansion as well, just that there is an overabundance of relatively young RLK/*Pelle* genes derived from tandem duplication. Based on the duplicated blocks that were derived from the most recent whole genome duplication in the *A. thaliana* lineage ~25–40 million years ago (Blanc et al. 2003), 63 RLK/*Pelle* gene pairs were retained (Lehti-Shiu et al. 2009). During the same time period, there were likely many more tandem duplicate RLK/*Pelles* generated but lost. It is difficult to ascertain how the RLK/*Pelle* family was established early in land plant evolution because genomic features that allow us to discern duplication mechanisms quickly degenerated over time. In addition, it is not clear how mechanisms other than tandem and whole genome duplications have contributed to the RLK/*Pelle* family expansion. For example, retrogenes have contributed significantly to the gene content in animals (Pan and Zhang 2009) and in plants (Benovoy and Drouin 2006; Wang et al. 2006). Particularly in repeat-rich plant genomes, it will be important to evaluate how retrotransposition contributed to RLK/*Pelle* family expansion.

Based on results of comparative genomic studies (Shiu et al. 2004; Hanada et al. 2008; Lehti-Shiu et al. 2009), it is clear that the RLK/*Pelle* family, together with the F-box gene family (Hua et al. 2011), has undergone the most dramatic lineage-specific expansion in land plants. Does this dramatic lineage specific expansion reflect the adaptive consequence of RLK/*Pelle* duplication? Or is it possible that the success of this gene family merely reflects a rare case of neutral evolution that can be explained by the genome drift hypothesis (Nozawa and Nei 2007)? To facilitate our discussion on why RLK/*Pelle* duplicates may be retained, we will first provide a short introduction of RLK/*Pelle* functions and how these functions are correlated with mechanisms of duplication and patterns of expansion in this gene family. For more details on RLK/*Pelle* functions, the readers are referred to in-depth reviews (Morillo and Tax 2006; Berger 2009; Tor et al. 2009; Zhao 2009; Li 2010; Postel et al. 2010; Rowe and Bergmann 2010) and other chapters in this volume.

5 Why Has the RLK/*Pelle* Family Expanded in the Land Plant Lineage

*5.1 Functions of RLK/*Pelle* Duplicates*

While the functions of most of the RLKs in plants remain unknown, much progress in elucidating RLK signaling networks has been made in recent years. Like the *Pelle* and IRAK genes, several plant RLK/*Pelle* family members have been shown to function in innate immunity as pattern recognition receptors (Albert et al. 2010). Several additional RLK/*Pelle*s function in defense response, binding pathogen components directly or acting in downstream pathways (Dodds and Rathjen 2010). A smaller group of RLKs have also been implicated in abiotic stress response (Yang et al. 2010; Sivaguru et al. 2003; Osakabe et al. 2005). In addition to plant stress response, by far the most RLKs with known functions are involved in some aspect of plant development regulating, for example, meristem size, organ identity, and cell-type specificity (reviewed in De Smet et al. 2009). Although most RLK/*Pelle*s with known functions are either involved in stress response or development regulation, recent studies indicate some RLK/*Pelle*s have more than one defined role. For example, BAK1 functions in both innate immunity and brassinosteroid signaling pathways, forming complexes with FLS2 and the BRI1 brassinosteroid receptor (reviewed in Chinchilla et al. 2009).

Members of a subfamily sometimes, but not always have related functions. One particularly well-studied subfamily is the LRR-XI family, of which *CLAVATA1* (*CLV1*) is a member. In *A. thaliana CLV1* functions in the meristem to restrict proliferation and promote differentiation (Brand et al. 2000). The closely related *BAM* RLKs are expressed in different regions of the meristem and promote stem cell maintenance (DeYoung et al. 2006). The *CLV1* signaling pathway also appears to be conserved in monocots; mutations in the maize *CLV1* ((Taguchi-Shiobara et al. 2001; Bommert et al. 2005) and the rice *CLV1* (Suzaki et al. 2004) orthologs result in inflorescence meristem proliferation. However, not all LRR-XI family members function in meristems. In legumes, the orthologous LRR-XI subfamily members *NARK* from soybean, *HAR* and *Klavier* from *Lotus japonicus*, and *SUNN* from *Medicago truncatula* regulate nodule number but do not have meristem phenotypes when mutated (Krusell et al. 2002; Searle et al. 2003; Schnabel et al. 2005). In *A. thaliana,* the LRR RLK XI family members *PEPR1* and *PEPR2* recognize damage associated molecular patterns and activate innate immunity pathways (Yamaguchi et al. 2006; Krol et al. 2010; Yamaguchi et al. 2010). Clearly, different subfamily members have been recruited to play different roles during evolution. In addition, members from different subfamilies can have similar functions. For example, *ACR4*, a member of the *Crinkly4*-like subfamily, has a function analogous to *CLV1* in the root meristem, and despite having a different ECD, binds a similar ligand (Stahl et al. 2009).

5.2 *Co-option of RLK/*Pelle *Family Members for Plant–Microbe Interactions*

Co-option is the new use of existing genes, organs, or biological structures through natural selection, particularly, adaptive evolution (True and Carroll 2002). For duplicate genes, it has long been hypothesized that one copy retains the original role and the other copy may "neofunctionalize," i.e., take on novel functions (Ohno 1970). In this case, the duplicate copy can be retained if the novel function confers fitness advantage. Co-option of existing RLK functions has been shown to be important for the evolution of symbiosis. For example, given that nodulation is a legume-specific trait and that many LRR-XI members are involved in aspects of meristem development, the legume LRR-XI members discussed in the previous section are likely examples of co-option. Another example is the LysM subfamily whose members bind to chitin and are found in many plant species (Zhang et al. 2007; Buist et al. 2008). In legumes this family recognizes nod factors required for symbiosis, suggesting that nod recognition RLKs were recruited from existing chitin binding proteins (Shimizu et al. 2010). Nodulation recognition factors belonging to the LRR-I family were also co-opted to act in different nodulation pathways. SYMbiosis Receptor-like Kinase (*SYMRK*) and Nodulation Receptor Kinase (*NORK*) perceive lipo-chitooligosaccharide nodulation factors that are required for bacterial and fungal symbiosis (Endre et al. 2002; Stracke et al. 2002). *SYMRK* is required for symbiosis with mycorrhizal fungi as well as the formation of nitrogen fixing bacteria root nodule symbioses, which are restricted to legumes. It is thought that root nodule symbiosis likely evolved from the already existing arbuscular mycorrhizal symbiotic pathway (Soltis et al. 1995). Recently, it was found that *SYMRK* played an important role in the evolution of root nodule symbiosis. While many species contain *SYMRK* orthologs and form arbuscular mychorriza, only legumes that have longer versions of the SYMRK gene are capable of nitrogen fixation (Markmann et al. 2008).

Although co-option and neofunctionalization may explain the retention of RLK/*Pelle* duplicates involved in symbiosis, there is little evidence that other RLK/*Pelle* duplicates were retained due to the acquisition of novel functions. In addition to neofunctionalization, the retention of duplicate genes can be attributed to subfunctionalization (Force et al. 1999; Lynch et al. 2001) where the ancestral functions of an RLK/*Pelle* were partitioned among the duplicate copies. If the partitioned "subfunctions" were all important, then both duplicate copies must have been retained even though the presence of both copies did not confer selective advantage over the ancestral gene. As more knowledge of RLK/*Pelle* functions accumulates, we will have a better understanding of the relative importance of neofunctionalization and/or subfunctionalization on RLK/*Pelle* duplicate retention.

5.3 *Relationship Between RLK/*Pelle* Function and Duplication Mechanism*

The RLK/*Pelle* members can be classified into multiple subfamilies, and these subfamilies show dramatically different degrees of expansion over the course of land plant evolution. Because many RLK/*Pelle* genes have been implicated in biotic stress, mostly on the basis of expression studies (Wrzaczek et al. 2010; Chae et al. 2009; Lehti-Shiu et al. 2009; Postel et al. 2010), one hypothesis explaining RLK/*Pelle* family expansion is that the selective pressure to properly perceive rapidly changing environmental signals has been one of the driving forces behind expansion. This notion is consistent with the finding that the RLK/*Pelle* family has a significantly overrepresented number of biotic stress responsive genes compared to all *A. thaliana* genes (Lehti-Shiu et al. 2009) and that many RLK/*Pelles* are responsive to abiotic stress and hormone/chemical treatments (Chae et al. 2009). Furthermore, several RLK/*Pelle* subfamilies have significantly overrepresented numbers of genes that are responsive to particular stress conditions. For example, the DUF26, L-LEC, LRR-I, LRR-VIII-2, LRR-Xb, RLCK-VIIa, SD1, SD-2b, WAK, and WAK_LRK10L-1 subfamilies are enriched in genes that are responsive to a number of biotic stresses, and these families are also the ones that have undergone expansion (Lehti-Shiu et al. 2009).

There is evidence that lineage-specific expansion of stress-responsive RLKs has been driven mainly by tandem duplication. Among plant genes, duplicates generated by lineage-specific tandem duplication are more likely to function in stress response (Hanada et al. 2008). This is also true for tandem RLK/*Pelle* genes, which are significantly more likely to be responsive to biotic stress than nontandem genes (Lehti-Shiu et al. 2009). One explanation for the relationship between tandem duplication and stress responsiveness is that tandem duplications allow rapid changes in gene content over just a few generations. This can generate a higher degree of diversity than other duplication mechanisms such as whole genome duplication. In fact, the NBS-LRR family of resistance genes which recognize pathogen effectors and trigger disease resistance pathways have a similar pattern of expansion as RLK/*Pelle* genes, with varying numbers of genes found between species and even between accessions of the same species (Zhang et al. 2010).

The expansion of stress-responsive RLKs by tandem duplication is supported when looking at RLKs with known functions. RLKs with roles in development tend not to be in tandem clusters whereas those with roles in defense tend to be in tandem repeats (Shiu et al. 2004). Interesting exceptions to this are the LRR-XII FLS2 and EFR pattern recognition receptors in *A. thaliana*. Members of the LRR-XII subfamily have broadly conserved functions, with subfamily members in different species activating similar transduction pathways. FLS2 can be found in tomato and rice (Robatzek et al. 2007; Takai et al. 2008), and the signaling pathways induced by binding to LRR-XII receptors are conserved. Interestingly, while there are only ten LRR-XII subfamily members in *A. thaliana*, there are ten times as many members in rice and poplar. It was reasoned that the relatively small number of members in *A. thaliana* may be due to the fact that since PAMP receptors recognize conserved features of microbes, diversification of this subfamily is not selected for (Tang et al. 2010). Another explanation is that the differential expansion of LRR-XII in poplar and rice is the consequence of intense selective pressure, perhaps due to their involvement in a recent or ongoing arms race with yet to be identified biotic agents. Note that the previous two explanations assume natural selection in the form of adaptive evolution is the driving force behind RLK/*Pelle* retention. Another possibility, however, is that differences in numbers between species could be due to random "genomic drift" and that expansion does not necessarily imply an adaptive advantage (Nozawa and Nei 2007; Nei et al. 2008).

5.4 Is RLK/Pelle Family Expansion the Result of Neutral or Adaptive Evolution?

Based on our understanding of the differences and similarities between the RLK/*Pelle* families within and between land plant species, it is now possible to speculate on the contribution of neutral and adaptive evolution to the expansion of this gene family. Gene duplication is one type of mutation that is thought to occur randomly. Thus, in general, larger gene families tend to have more lineage-specific duplicates compared to smaller gene families in plants (Hanada et al. 2008; Zou et al. 2009). Nonetheless, duplicated genes are not necessarily retained. Thus, the expansion of a gene family is controlled by how often its members are duplicated and retained. The central question is whether retention is due to natural selection or some neutral processes that occur randomly. The latter possibility is the central tenet of the genome drift hypothesis, which postulates that the expansion and shrinkage of gene families can be explained by random duplication and inactivation events analogous to how genetic drift affects allelic frequencies (Nozawa and Nei 2007; Nei et al. 2008; Nei et al. 2010). The metazoan olfactory receptor (OR) family has been used as an important example of genome drift (Nozawa and Nei 2007). The OR family has three important features consistent with a family that may have experienced genome drift: (1) extensive lineage-specific differences between metazoan species, (2) highly variable within species, (3) a large number of OR pseudogenes littering the human genome. Among plant gene families, the F-box family is the most similar to the metazoan OR family in these three aspects (Hua et al. 2011). In the case of RLK/*Pelle*, there are some similarities but the pattern is rather complicated.

First of all, as detailed in Sect. 4.1, there is extensive lineage-specific expansion in the RLK/*Pelle* family among flowering plants. RLK/*Pelle* members also rank highly in the degree of intraspecific variation (Clark et al. 2007; Cao et al. 2011). However, although there are a large number of RLK/*Pelle* pseudogenes in *A. thaliana* and rice, the RLK/*Pelle* family in fact has significantly fewer pseudogenes compared to most other gene families (Zou et al. 2009). In contrast, the LRR-R genes and F-box genes both have significantly more pseudogenes than most plant gene families. Thus, neutral processes such as genome drift do not entirely account for the degree of RLK/*Pelle* family expansion. At least some of the duplicated copies were retained due to natural selection. Consistent with this speculation, the co-opted RLK/*Pelle*s involved in symbiotic associations are clearly examples of selection, in these cases through adaptive evolution. And results from comparative genomics provide ample examples of ancient RLK/*Pelle* duplicates that remain conserved across land plant species, arguing against retention solely due to neutral explanations (Shiu et al. 2004; Lehti-Shiu et al. 2009).

We should point out that some RLK/*Pelle* subfamilies evolve much like the OR family with extensive lineage-specific differences, substantial intraspecific variation, and a high pseudogene-to-functional gene ratio (Lehti-Shiu et al. 2009; Zou et al. 2009). Are these RLK/*Pelle*s still around due to genome drift? A significant number of these types of OR-like RLK/*Pelle*s are derived from tandem duplication and evolve quickly. In addition, multiple protein domains found in these RLK/*Pelle*s are involved in perceiving fungal or bacterial molecular patterns (see Sect. 3.2). Furthermore, an overrepresented number of RLK/*Pelle*s are differentially upregulated under biotic stress conditions. Taking all this information into consideration, the evolution of the RLK/*Pelle* family may be best explained by a mixture of drift and selection. It is conceivable that many RLK/*Pelle* duplicates are generated by chance. Over the course of land plant evolution, although many duplicates did not survive, some were selected for because of their ability to regulate novel developmental processes or mediate responses to symbiotic or pathogenic microbes. Many RLK/*Pelle*s with known functions are involved in developmental regulation and are clearly indispensible and conserved across species. In some cases the molecular patterns recognized by the selected RLK/*Pelle*s are not easily mutable, such as flagellin. In other cases, members of this gene family may engage in arms races with biotic agents much like the LRR-R genes. To determine if the above scenario is true or not, we need more information on how RLK/*Pelle* ortholog functions have evolved in multiple land plant species. We also need a better handle on the roles of rapidly evolving RLK/*Pelle*s, particularly those in tandem repeats.

6 Conclusions

The timing and extent of RLK/*Pelle* family expansion suggests that this family has played a significant role in the evolution of land plants. Closer examination of the patterns of RLK/*Pelle* subfamily expansion has revealed that much expansion is lineage specific and driven by both whole genome and tandem duplication. Based on rapidly accumulating information on the functions of RLK/*Pelle* genes, a clearer picture of the selection pressures driving expansion is emerging. Most duplicate genes become pseudogenes within a few million years. However, because larger families tend to generate more duplicates, and the RLK/*Pelle* family is one of the largest in plants, some RLK/*Pelle* duplicates may still be around simply because many of them are generated and there is not sufficient time for their pseudogenization. Thus, some of the observed expansion is likely a consequence of completely neutral processes that do not involve natural selection.

Nonetheless, many RLK/*Pelle* genes are known to have roles in biotic stress signaling and have likely been co-opted for symbiotic interactions. This suggests that the pressure to perceive and respond to rapidly evolving biotic factors is a likely driving force behind expansion. Furthermore, the fact that many RLK/*Pelle* duplicates have survived over tens to hundreds of millions of years indicates that natural selection plays an important role in RLK/*Pelle* expansion. One of the biggest challenges to understanding the mechanisms that drive the expansion of this family is in identifying the roles of fast evolving, tandem RLK/*Pelles*. In addition, although related RLK/*Pelle* sequences can function in similar pathways in different species, it is clear that RLK/*Pelle*s have been co-opted for new signaling roles. This makes extrapolations of functions based on studies from one species complicated. As more knowledge about RLK/*Pelle* functions becomes available, it can be combined with information about expansion, duplication mechanism, and conservation of family members across species to better understand the functional evolution of this large gene family.

References

Albert M, Jehle AK, Mueller K, Eisele C, Lipschis M, Felix G (2010) *Arabidopsis thaliana* pattern recognition receptors for bacterial elongation factor Tu and flagellin can be combined to form functional chimeric receptors. J Biol Chem 285:19035–19042

Anantharaman V, Aravind L (2001) The CHASE domain: a predicted ligand-binding module in plant cytokinin receptors and other eukaryotic and bacterial receptors. Trends Biochem Sci 26:579–582

Belvin MP, Anderson KV (1996) A conserved signaling pathway: the Drosophila Toll-Dorsal pathway. Annu Rev Cell Dev Biol 12:393–416

Benovoy D, Drouin G (2006) Processed pseudogenes, processed genes, and spontaneous mutations in the Arabidopsis genome. J Mol Evol 62:511–522

Berger F (2009) Reproductive biology: receptor-like kinases orchestrate love songs in plants. Curr Biol 19:R647–R649

Bi D, Cheng YT, Li X, Zhang Y (2010) Activation of plant immune responses by a gain-of-function mutation in an atypical receptor-like kinase. Plant Physiol 153:1771–1779

Blanc G, Hokamp K, Wolfe KH (2003) A recent polyploidy superimposed on older large-scale duplications in the Arabidopsis genome. Genome Res 13:137–144

Boller T, Felix G (2009) A renaissance of elicitors: perception of microbe-associated molecular patterns and danger signals by pattern-recognition receptors. Annu Rev Plant Biol 60:379–406

Bommert P, Lunde C, Nardmann J, Vollbrecht E, Running M, Jackson D, Hake S, Werr W (2005) Thick tassel dwarf1 encodes a putative maize ortholog of the Arabidopsis CLAVATA1 leucine-rich repeat receptor-like kinase. Development 132:1235–1245

Brand U, Fletcher JC, Hobe M, Meyerowitz EM, Simon R (2000) Dependence of stem cell fate in Arabidopsis on a feedback loop regulated by CLV3 activity. Science 289:617–619

Braun DM, Walker JC (1996) Plant transmembrane receptors: new pieces in the signaling puzzle. Trends Biochem Sci 21:70–73

Brutus A, Sicilia F, Macone A, Cervone F, De Lorenzo G (2010) A domain swap approach reveals a role of the plant wall-associated kinase 1 (WAK1) as a receptor of oligogalacturonides. Proc Natl Acad Sci USA 107:9452–9457

Buist G, Steen A, Kok J, Kuipers OP (2008) LysM, a widely distributed protein motif for binding to (peptido)glycans. Mol Microbiol 68:838–847

Cao Z, Henzel WJ, Gao X (1996) IRAK: a kinase associated with the interleukin-1 receptor. Science 271:1128–1131

Cao J, Schneeberger K, Ossowski S, Günther T, Bender S, Fitz J, Koenig D, Lanz C, Stegle O, Lippert C, Wang X, Ott F, Müller J, Alonso-Blanco C, Borgwardt K, Schmid KJ, Weigel D (2011) Whole-genome sequencing of multiple *Arabidopsis thaliana* populations. Nature Genet 43:956–963

Chae L, Sudat S, Dudoit S, Zhu T, Luan S (2009) Diverse transcriptional programs associated with environmental stress and hormones in the Arabidopsis receptor-like kinase gene family. Mol Plant 2:84–107

Chinchilla D, Shan L, He P, de Vries S, Kemmerling B (2009) One for all: the receptor-associated kinase BAK1. Trends Plant Sci 14:535–541

Clark RM, Schweikert G, Toomajian C, Ossowski S, Zeller G, Shinn P, Warthmann N, Hu TT, Fu G, Hinds DA, Chen H, Frazer KA, Huson DH, Scholkopf B, Nordborg M, Ratsch G, Ecker JR, Weigel D (2007) Common sequence polymorphisms shaping genetic diversity in *Arabidopsis thaliana*. Science 317:338–342

Cock JM, Vanoosthuyse V, Gaude T (2002) Receptor kinase signalling in plants and animals: distinct molecular systems with mechanistic similarities. Curr Opin Cell Biol 14:230–236

Cock JM, Sterck L, Rouze P, Scornet D, Allen AE, Amoutzias G, Anthouard V, Artiguenave F, Aury JM, Badger JH, Beszteri B, Billiau K, Bonnet E, Bothwell JH, Bowler C, Boyen C, Brownlee C, Carrano CJ, Charrier B, Cho GY, Coelho SM, Collen J, Corre E, Da Silva C, Delage L, Delaroque N, Dittami SM, Doulbeau S, Elias M, Farnham G, Gachon CM, Gschloess B, Heesch S, Jabbari K, Jubin C, Kawai H, Kimura K, Kloareg B, Kupper FC, Lang D, Le Bail A, Leblanc C, Lerouge P, Lohr M, Lopez PJ, Martens C, Maumus F, Michel G, Miranda-Saavedra D, Morales J, Moreau H, Motomura T, Nagasato C, Napoli CA, Nelson DR, Nyvall-Collen P, Peters AF, Pommier C, Potin P, Poulain J, Quesneville H, Read B, Rensing SA, RitterA RS, Samanta M, Samson G, Schroeder DC, Segurens B, Strittmatter M, Tonon T, Tregear JW, Valentin K, von Dassow P, Yamagishi T, Van de Peer Y, Wincker P (2010) The Ectocarpus genome and the independent evolution of multicellularity in brown algae. Nature 465:617–621

Courties C, Vaquer A, Troussellier M, Lautier J, Chrétiennot-Dinet MJ, Neveux J, Machado C, Claustre H (1994) Smallest eukaryotic organism. Nature 370:255

Cui L, Wall PK, Leebens-Mack JH, Lindsay BG, Soltis DE, Doyle JJ, Soltis PS, Carlson JE, Arumuganathan K, Barakat A, Albert VA, Ma H, dePamphilis CW (2006) Widespread genome duplications throughout the history of flowering plants. Genome Res 16:738–749

De Smet I, Voss U, Jurgens G, Beeckman T (2009) Receptor-like kinases shape the plant. Nat Cell Biol 11:1166–1173

DeYoung BJ, Bickle KL, Schrage KJ, Muskett P, Patel K, Clark SE (2006) The CLAVATA1-related BAM1, BAM2 and BAM3 receptor kinase-like proteins are required for meristem function in Arabidopsis. Plant J 45:1–16

Dodds PN, Rathjen JP (2010) Plant immunity: towards an integrated view of plant-pathogen interactions. Nat Rev Genet 11:539–548

Endre G, Kereszt A, Kevei Z, Mihacea S, Kalo P, Kiss GB (2002) A receptor kinase gene regulating symbiotic nodule development. Nature 417:962–966

Flannery S, Bowie AG (2010) The interleukin-1 receptor-associated kinases: critical regulators of innate immune signalling. Biochem Pharmacol 80:1981–1991

Force A, Lynch M, Pickett FB, Amores A, Yan YL, Postlethwait J (1999) Preservation of duplicate genes by complementary, degenerative mutations. Genetics 151:1531–1545

Fritz-Laylin LK, Krishnamurthy N, Tor M, Sjolander KV, Jones JD (2005) Phylogenomic analysis of the receptor-like proteins of rice and Arabidopsis. Plant Physiol 138:611–623

Hanada K, Zou C, Lehti-Shiu MD, Shinozaki K, Shiu SH (2008) Importance of lineage-specific expansion of plant tandem duplicates in the adaptive response to environmental stimuli. Plant Physiol 148:993–1003

Handford PA, Baron M, Mayhew M, Willis A, Beesley T, Brownlee GG, Campbell ID (1990) The first EGF-like domain from human factor IX contains a high-affinity calcium binding site. EMBO J 9:475–480

He Z, Wang ZY, Li J, Zhu Q, Lamb C, Ronald P, Chory J (2000) Perception of barssinosteroids by the extracellular domain of the receptor kinase BRI1. Science 288:2360–2363

Hecht PM, Anderson KV (1993) Genetic characterization of tube and pelle, genes required for signaling between Toll and dorsal in the specification of the dorsal-ventral pattern of the Drosophila embryo. Genetics 135:405–417

Hedges SB (2002) The origin and evolution of model organisms. Nat Rev Genet 3:838–849

Hester G, Kaku H, Goldstein IJ, Wright CS (1995) Structure of mannose-specific snowdrop (*Galanthus nivalis*) lectin is representative of a new plant lectin family. Nat Struct Biol 2:472–479

Hinata K, Watanabe M, Yamakawa S, Satta Y, Isogai A (1995) Evolutionary aspects of the S-related genes of the Brassica self-incompatibility system: synonymous and nonsynonymous base substitutions. Genetics 140:1099–1104

Hua Z, Zou C, Shiu SH, Vierstra RD (2011) Phylogenetic comparison of F-Box (FBX) gene superfamily within the plant kingdom reveals divergent evolutionary histories indicative of genomic drift. PLoS One 6:e16219

Hunter T, Lindberg RA, Middlemas DS, Tracy S, van der Geer P (1992) Receptor protein tyrosine kinases and phosphatases. Cold Spring Harb Symp Quant Biol 57:25–41

Janssens S, Beyaert R (2003) Functional diversity and regulation of different interleukin-1 receptor-associated kinase (IRAK) family members. Mol Cell 11:293–302

Jeong S, Trotochaud AE, Clark SE (1999) The Arabidopsis CLAVATA2 gene encodes a receptor-like protein required for the stability of the CLAVATA1 receptor-like kinsae. Plant Cell 11:1925–1934

King N, Westbrook MJ, Young SL, Kuo A, Abedin M, Chapman J, Fairclough S, Hellsten U, Isogai Y, Letunic I, Marr M, Pincus D, Putnam N, Rokas A, Wright KJ, Zuzow R, Dirks W, Good M, Goodstein D, Lemons D, Li W, Lyons JB, Morris A, Nichols S, Richter DJ, Salamov A, Sequencing JG, Bork P, Lim WA, Manning G, Miller WT, McGinnis W, Shapiro H, Tjian

R, Grigoriev IV, Rokhsar D (2008) The genome of the choanoflagellate *Monosiga brevicollis* and the origin of metazoans. Nature 451:783–788

Kobe B, Deisenhofer J (1994) The leucine-rich repeat: a versatile binding motif. Trends Biochem Sci 19:415–421

Kohler S, Delwiche CF, Denny PW, Tilney LG, Webster P, Wilson RJ, Palmer JD, Roos DS (1997) A plastid of probable green algal origin in Apicomplexan parasites. Science 275:1485–1489

Kohorn BD (2001) WAKs; cell wall associated kinases. Curr Opin Cell Biol 13:529–533

Krol E, Mentzel T, Chinchilla D, Boller T, Felix G, Kemmerling B, Postel S, Arents M, Jeworutzki E, Al-Rasheid KA, Becker D, Hedrich R (2010) Perception of the Arabidopsis danger signal peptide 1 involves the pattern recognition receptor AtPEPR1 and its close homologue AtPEPR2. J Biol Chem 285:13471–13479

Krusell L, Madsen LH, Sato S, Aubert G, Genua A, Szczyglowski K, Duc G, Kaneko T, Tabata S, de Bruijn F, Pajuelo E, Sandal N, Stougaard J (2002) Shoot control of root development and nodulation is mediated by a receptor-like kinase. Nature 420:422–426

Lease KA, Walker JC (2006) The Arabidopsis unannotated secreted peptide database, a resource for plant peptidomics. Plant Physiol 142:831–838

Lehti-Shiu MD, Zou C, Hanada K, Shiu SH (2009) Evolutionary history and stress regulation of plant receptor-like kinase/pelle genes. Plant Physiol 150:12–26

Leister D (2004) Tandem and segmental gene duplication and recombination in the evolution of plant disease resistance gene. Trends Genet 20:116–122

Lewis LA, McCourt RM (2004) Green algae and the origin of land plants. Am J Bot 91:1535–1556

Li WH (1983) Evolution of duplicate genes and pseudogenes. In: Nei M, Koehn RK (eds) Evolution of genes and proteins. Sinauer Associates, Sunderland, MA, pp 14–37

Li J (2010) Multi-tasking of somatic embryogenesis receptor-like protein kinases. Curr Opin Plant Biol 13:509–514

Lockton S, Gaut BS (2005) Plant conserved non-coding sequences and paralogue evolution. Trends Genet 21:60–65

Loris R, Hamelryck T, Bouckaert J, Wyns L (1998) Legume lectin structure. Biochim Biophys Acta 1383:9–36

Lynch M, O'Hely M, Walsh B, Force A (2001) The probability of preservation of a newly arisen gene duplicate. Genetics 159:1789–1804

Markmann K, Giczey G, Parniske M (2008) Functional adaptation of a plant receptor-kinase paved the way for the evolution of intracellular root symbioses with bacteria. PLoS Biol 6:e68

Miyakawa T, Sawano Y, Miyazono K, Hatano K, Tanokura M (2007) Crystallization and preliminary X-ray analysis of ginkbilobin-2 from *Ginkgo biloba* seeds: a novel antifungal protein with homology to the extracellular domain of plant cysteine-rich receptor-like kinases. Acta Crystallogr Sect F Struct Biol Cryst Commun 63:737–739

Morillo SA, Tax FE (2006) Functional analysis of receptor-like kinases in monocots and dicots. Curr Opin Plant Biol 9:460–469

Mougel C, Zhulin IB (2001) CHASE: an extracellular sensing domain common to transmembrane receptors from prokaryotes, lower eukaryotes and plants. Trends Biochem Sci 26:582–584

Muller R, Bleckmann A, Simon R (2008) The receptor kinase CORYNE of Arabidopsis transmits the stem cell-limiting signal CLAVATA3 independently of CLAVATA1. Plant Cell 20:934–946

Nei M, Niimura Y, Nozawa M (2008) The evolution of animal chemosensory receptor gene repertoires: roles of chance and necessity. Nat Rev Genet 9:951–963

Nei M, Suzuki Y, Nozawa M (2010) The neutral theory of molecular evolution in the genomic era. Annu Rev Genomics Hum Genet 11:265–289

Nozawa M, Nei M (2007) Evolutionary dynamics of olfactory receptor genes in Drosophila species. Proc Natl Acad Sci USA 104:7122–7127

Ohno S (1970) Evolution by gene duplication. Springer, New York

Osakabe Y, Maruyama K, Seki M, Satou M, Shinozaki K, Yamaguchi-Shinozaki K (2005) Leucine-rich repeat receptor-like kinase1 is a key membrane-bound regulator of abscisic acid early signaling in Arabidopsis. Plant Cell 17:1105–1119

Pan D, Zhang L (2009) Burst of young retrogenes and independent retrogene formation in mammals. PLoS One 4:e5040

Pan CH, Lee EA, Chae YA, Kim SI (1999) Purification of chitinolytic protein from Rehmannia glutinosa showing N-terminal amino acid sequence similarity to thaumatin-like proteins. Biosci Biotechnol Biochem 63:1138–1140

Perrakis A, Tews I, Dauter Z, Oppenheim AB, Chet I, Wilson KS, Vorgias CE (1994) Crystal structure of a bacterial chitinase at 2.3 A resolution. Structure 2:1169–1180

Postel S, Kufner I, Beuter C, Mazzotta S, Schwedt A, Borlotti A, Halter T, Kemmerling B, Nurnberger T (2010) The multifunctional leucine-rich repeat receptor kinase BAK1 is implicated in Arabidopsis development and immunity. Eur J Cell Biol 89:169–174

Qiu Y-L (2008) Phylogeny and evolution of charophytic algae and land plants. J Syst Evol 46:287–306

Reams AB, Neidle EL (2004) Selection for gene clustering by tandem duplication. Annu Rev Microbiol 58:119–142

Rizzon C, Ponger L, Gaut BS (2006) Striking similarities in the genomic distribution of tandemly arrayed genes in Arabidopsis and rice. PLoS Comput Biol 2:e115

Robatzek S, Bittel P, Chinchilla D, Kochner P, Felix G, Shiu SH, Boller T (2007) Molecular identification and characterization of the tomato flagellin receptor LeFLS2, an orthologue of Arabidopsis FLS2 exhibiting characteristically different perception specificities. Plant Mol Biol 64:539–547

Rowe MH, Bergmann DC (2010) Complex signals for simple cells: the expanding ranks of signals and receptors guiding stomatal development. Curr Opin Plant Biol 13:548–555

Santelli E, Schwarzenbache R, McMullan D, Biorac T, Brinen LS, Canaves JM, Cambell J, Dai X, Deacon AM, Elsliger MA, Eshagi S, Floyd R, Godzik A, Grittini C, Grzechnik SK, Jaroszewski L, Karlak C, Klock HE, Koesema E, Kovarik JS, Kreusch A, Kuhn P, Lesley SA, McPhillips TM, Miller MD, Morse A, Moy K, Ouyang J, Page R, Quijano K, Rezezadeh F, Robb A, Sims E, Spraggon G, Stevens RC, van den Bedem H, Velasquez J, Vincent J, von Delft F, Wang X, West B, Wolf G, Xu Q, Hodgson KO, Wooley J, Wilson IA (2004) Crystal structure of a glycerophosphodiester phosphodiesterase (GDPD) from *Thermotoga maritima* (TM1621) at 1.60 A resolution. Proteins 56:167–170

Sasaki G, Katoh K, Hirose N, Suga H, Kuma K, Miyata T, Su ZH (2007) Multiple receptor-like kinase cDNAs from liverwort *Marchantia polymorpha* and two charophycean green algae, *Closterium ehrenbergii* and *Nitella axillaris*: Extensive gene duplications and gene shufflings in the early evolution of streptophytes. Gene 401:135–144

Schnabel E, Journet EP, de Carvalho-Niebel F, Duc G, Frugoli J (2005) The *Medicago truncatula* SUNN gene encodes a CLV1-like leucine-rich repeat receptor kinase that regulates nodule number and root length. Plant Mol Biol 58:809–822

Searle IR, Men AE, Laniya TS, Buzas DM, Iturbe-Ormaetxe I, Carroll BJ, Gresshoff PM (2003) Long-distance signaling in nodulation directed by a CLAVATA1-like receptor kinase. Science 299:109–112

Sharon N, Lis H (2001) The structural basis for carbohydrate recognition by lectins. Adv Exp Med Biol 491:1–16

Shelton CA, Wasserman SA (1993) Pelle encodes a protein kinase required to establish dorsoventral polarity in the Drosophila embryo. Cell 72:515–525

Shimizu T, Nakano T, Takamizawa D, Desaki Y, Ishii-Minami N, Nishizawa Y, Minami E, Okada K, Yamane H, Kaku H, Shibuya N (2010) Two LysM receptor molecules, CEBiP and OsCERK1, cooperatively regulate chitin elicitor signaling in rice. Plant J 64:204–214

Shiu SH, Bleecker AB (2001a) Receptor-like kinases from Arabidopsis form a monophyletic gene family related to animal receptor kinases. Proc Natl Acad Sci USA 98:10763–10768

Shiu SH, Bleecker AB (2001b) Plant receptor-like kinase gene family: diversity, function, and signaling. Sci STKE 113:re22

Shiu SH, Bleecker AB (2003) Expansion of the receptor-like kinase/Pelle gene family and receptor-like proteins in Arabidopsis. Plant Physiol 132:530–543

Shiu SH, Li WH (2004) Origins, lineage-specific expansions, and multiple losses of tyrosine kinases in eukaryotes. Mol Biol Evol 21:828–840

Shiu SH, Karlowski WM, Pan R, Tzeng YH, Mayer KF, Li WH (2004) Comparative analysis of the receptor-like kinase family in Arabidopsis and rice. Plant Cell 16:1220–1234

Shpak ED, McAbee JM, Pillitteri LJ, Torii KU (2005) Stomatal patterning and differentiation by synergistic interactions of receptor kinases. Science 309:290–293

Sivaguru M, Ezaki B, He ZH, Tong H, Osawa H, Baluska F, Volkmann D, Matsumoto H (2003) Aluminum-induced gene expression and protein localization of a cell wall-associated receptor kinase in Arabidopsis. Plant Physiol 132:2256–2266

Soltis DE, Soltis PS, Morgan DR, Swensen SM, Mullin BC, Dowd JM, Martin PG (1995) Chloroplast gene sequence data suggest a single origin of the predisposition for symbiotic nitrogen fixation in angiosperms. Proc Natl Acad Sci USA 92:2647–2651

Stahl Y, Wink RH, Ingram GC, Simon R (2009) A signaling module controlling the stem cell niche in Arabidopsis root meristems. Curr Biol 19:909–914

Stracke S, Kistner C, Yoshida S, Mulder L, Sato S, Kaneko T, Tabata S, Sandal N, Stougaard J, Szczyglowski K, Parniske M (2002) A plant receptor-like kinase required for both bacterial and fungal symbiosis. Nature 417:959–962

Suzaki T, Sato M, Ashikari M, Miyoshi M, Nagato Y, Hirano HY (2004) The gene FLORAL ORGAN NUMBER1 regulates floral meristem size in rice and encodes a leucine-rich repeat receptor kinase orthologous to Arabidopsis CLAVATA1. Development 131:5649–5657

Szyperski T, Fernandez C, Mumenthaler C, Wuthrich K (1998) Structure comparison of human glioma pathogenesis-related protein GliPR and the plant pathogenesis-related protein P14a indicates a functional link between the human immune system and a plant defense system. Proc Natl Acad Sci USA 95:2262–2266

Taguchi-Shiobara F, Yuan Z, Hake S, Jackson D (2001) The fasciated ear2 gene encodes a leucine-rich repeat receptor-like protein that regulates shoot meristem proliferation in maize. Genes Dev 15:2755–2766

Takai R, Isogai A, Takayama S, Che FS (2008) Analysis of flagellin perception mediated by flg22 receptor OsFLS2 in rice. Mol Plant Microbe Interact 21:1635–1642

Tang P, Zhang Y, Sun X, Tian D, Yang S, Ding J (2010) Disease resistance signature of the leucine-rich repeat receptor-like kinase genes in four plant species. Plant Sci 179:399–406

Tor M, Lotze MT, Holton N (2009) Receptor-mediated signalling in plants: molecular patterns and programmes. J Exp Bot 60:3645–3654

Tordai H, Banyai L, Patthy L (1999) The PAN module: the N-terminal domains of plasminogen and hepatocyte growth factor are homologous with the apple domains of the prekallikrein family and with a novel domain found in numerous nematode proteins. FEBS Lett 46:163–167

Torii KU (2000) Receptor kinase activation and signal transduction in plants: an emerging picture. Curr Opin Plant Biol 3:361–367

Towb P, Sun H, Wasserman SA (2009) Tube Is an IRAK-4 homolog in a Toll pathway adapted for development and immunity. J Innate Immun 1:309–321

True JR, Carroll SB (2002) Gene co-option in physiological and morphological evolution. Annu Rev Cell Dev Biol 18:53–80

Tuskan GA, Difazio S, Jansson S, Bohlmann J, Grigoriev I, Hellsten U, Putnam N, Ralph S, Rombauts S, Salamov A, Schein J, Sterck L, Aerts A, Bhalerao RR, Bhalerao RP, Blaudez D, Boerjan W, Brun A, Brunner A, Busov V, Campbell M, Carlson J, Chalot M, Chapma J, Chen GL, Cooper D, Coutinho PM, Couturier J, Covert S, Cronk Q, Cunningham R, Davis J, Degroeve S, Dejardin A, Depamphilis C, Detter J, Dirks B, Dubchak I, Duplessis S, Ehlting J, Elli B, Gendler K, Goodstein D, Gribskov M, Grimwood J, Groover A, Gunter L, Hamberger B, Heinze B, Helariutta Y, Henrissat B, Holligan D, Holt R, Huang W, Islam-Faridi N, Jones S,

Jones-Rhoades M, Jorgensen R, Joshi C, Kangasjarvi J, Karlsson J, Kelleher C, Kirkpatrick R, Kirst M, Kohler A, Kalluri U, Larimer F, Leebens-Mack J, Leple JC, Locascio P, Lou Y, Lucas S, Martin F, Montanini B, Napoli C, Nelson DR, Nelson C, Nieminen K, Nilsson O, Pereda V, Peter G, Philippe R, Pilate G, Poliakov A, Razumovskaya J, Richardson P, Rinaldi C, Ritland K, Rouze P, Ryaboy D, Schmutz J, Schrader J, Segerman B, Shin H, Siddiqui A, Sterky F, Terry A, Tsai CJ, Uberbacher E, Unneberg P, Vahala J, Wall K, Wessler S, Yang G, Yin T, Douglas C, Marra M, Sandberg G, Van de Peer Y, Rokhsar D (2006) The genome of black cottonwood, *Populus trichocarpa* (Torr. & Gray). Science 313:1596–1604

van der Geer P, Hunter T, Lindberg RA (1994) Receptor protein-tyrosine kinases and their signal transduction pathways. Annu Rev Cell Biol 10:251–337

Walker JC (1994) Structure and function of the receptor-like protein kinases of higher plants. Plant Mol Biol 26:1599–1609

Walker JC, Zhang R (1990) Relationship of a putative receptor protein kinase from maize to the S-locus glycoproteins of Brassica. Nature 345:743–746

Wang W, Zheng H, Fan C, Li J, Shi J, Cai Z, Zhang G, Liu D, Zhang J, Vang S, Lu Z, Wong GK, Long M, Wang J (2006) High rate of chimeric gene origination by retroposition in plant genomes. Plant Cell 18:1791–1802

Wrzaczek M, Brosche M, Salojarvi J, Kangasjarvi S, Idanheimo N, Mersmann S, Robatzek S, Karpinski S, Karpinska B, Kangasjarvi J (2010) Transcriptional regulation of the CRK/DUF26 group of receptor-like protein kinases by ozone and plant hormones in Arabidopsis. BMC Plant Biol 10:95

Yamaguchi Y, Pearce G, Ryan CA (2006) The cell surface leucine-rich repeat receptor for AtPep1, an endogenous peptide elicitor in Arabidopsis, is functional in transgenic tobacco cells. Proc Natl Acad Sci USA 103:10104–10109

Yamaguchi Y, Huffaker A, Bryan AC, Tax FE, Ryan CA (2010) PEPR2 is a second receptor for the Pep1 and Pep2 peptides and contributes to defense responses in Arabidopsis. Plant Cell 22:508–522

Yang T, Chaudhuri S, Yang L, Du L, Poovaiah BW (2010) A calcium/calmodulin-regulated member of the receptor-like kinase family confers cold tolerance in plants. J Biol Chem 285:7119–7126

Zhang XC, Wu X, Findley S, Wan J, Libault M, Nguyen HT, Cannon SB, Stacey G (2007) Molecular evolution of lysin motif-type receptor-like kinases in plants. Plant Physiol 144:623–636

Zhang M, Wu YH, Lee MK, Liu YH, Rong Y, Santos TS, Wu C, Xie F, Nelson RL, Zhang HB (2010) Numbers of genes in the NBS and RLK families vary by more than four-fold within a plant species and are regulated by multiple factors. Nucl Acids Res 38:6513–6525

Zhao D (2009) Control of anther cell differentiation: a teamwork of receptor-like kinases. Sex Plant Reprod 22:221–228

Zou C, Lehti-Shiu MD, Thibaud-Nissen F, Prakash T, Buell CR, Shiu SH (2009) Evolutionary and expression signatures of pseudogenes in Arabidopsis and rice. Plant Physiol 151:3–15

Receptor Kinases in Plant Meristem Development

Yvonne Stahl and Rüdiger Simon

Abstract Meristems are groups of cells that promote shoot and root growth, including the growth of new organs, and control vascular development throughout the life of a higher plant. Their continuous proliferation has to be coordinated with the growth requirements of the plant. Signalling systems that facilitate the intercellular communication in meristems have evolved to include the secretion of small signalling peptides, which are perceived by a set of corresponding receptor kinases. Studies on vascular, shoot and root meristems have uncovered surprising similarities and shared functions among these signalling components. One common feature of these meristems is their use of CLE peptides to signal, often *via* redundantly acting RLKs, to regulate the expression of homeodomain transcription factors. These peptide/RLK/homeodomain transcription factor modules control the proliferation and maintenance of stem cells in these meristems.

1 Regulation of Shoot and Floral Meristem Function by RLKs

Meristems are important as the building centres of the plant shoot and root. Primary meristems are initiated during embryogenesis and control growth along the main body axis of the plant. Secondary meristems are generated later and give rise to axillary shoot structures and flowers, or produce lateral roots. Roles for receptor kinases in meristem development were discovered using mutant analysis. Many mutations in meristem functions cause drastic developmental defects, which are obvious and were among the first to be identified and characterized during the early boom of plant developmental genetics and molecular biology in the 1990s. One of the first receptor kinases shown to regulate plant development was CLAVATA1

Y. Stahl • R. Simon (✉)
Institute of Developmental Genetics, Heinrich Heine University, Universitätsstrasse 1, 40225 Düsseldorf, Germany
e-mail: ruediger.simon@hhu.de

F. Tax and B. Kemmerling (eds.), *Receptor-like Kinases in Plants*,
Signaling and Communication in Plants 13,
DOI 10.1007/978-3-642-23044-8_2,

Table 1 Receptors involved in meristem maintenance

Gene	Annotation	Expression domain	Protein family	Putative ligand	Interaction with	Proposed function
CLV1	AT1G75820	OC	LRR-RLK	CLV3/CLEs	CLV1, BAMs	SAM maintenance
CLV2	AT1G65380	SAM	LRR-RLP	CLV3/CLEs	CLV2, CRN	SAM maintenance
CRN	AT5G13290	SAM	MA-RLK	–	CLV2	SAM maintenance
BAM1	AT5G65700	PZ	LRR-RLK	CLEs	unknown	SAM maintenance
BAM2	AT3G49670	PZ	LRR-RLK	CLEs	unknown	SAM maintenance
BAM3	AT4G20270	PZ	LRR-RLK	CLEs?	unknown	SAM maintenance
RPK2	AT3G02130	PZ	LRR-RLK	CLEs?	RPK2	SAM + RM maintenance
PXY	AT5G61480	VAS	LRR-RLK	CLE41/44	unknown	Stem cell maintenance in the vasculature
ACR4	AT3G59420	Distal RM	CR-RLK	CLE40	ACR4	Distal RM maintenance

OC organizing centre, *PZ* peripheral zone, *SAM* shoot apical meristem, *RM* root meristem, *VAS* vasculature, *LRR-RLK* leucine-rich repeat receptor-like kinase, *LRR-RLP* leucine-rich repeat receptor like protein, *MA-RLK* membrane associated receptor-like kinase, *CR-RLK* crinkly repeats receptor-like kinase

(CLV1) from *Arabidopsis thaliana* (Table 1). CLV1 contains an extracellular domain with 21 leucine-rich repeats (LRRs) (Clark et al. 1993, 1997), a single-pass transmembrane domain (TMD) and a cytoplasmic kinase domain with serine/threonine specificity. CLV1 controls the identity and behaviour of stem cells in the vegetative shoot meristem, the inflorescence meristem and the floral meristem. These meristem types share a simple organization (Barton 2009). They contain a small group of stem cells at the tip of the apical dome in the central zone (CZ), which is surrounded by cells of the peripheral zone (PZ) (Fig. 1a). Cell divisions in the CZ will shift daughter cells to a more lateral position into the PZ. At their new position, cells behave differently and start to divide more rapidly. New organ primordia such as leaves or new floral meristems are generated from the PZ. Stem cells within the CZ are controlled by uncharacterized signals from the underlying organizing centre (OC) cells. The expression of the homedomain transcription factor WUSCHEL (WUS) in the OC is pivotal for stem cell maintenance in the CZ (Laux et al. 1996; Mayer et al. 1998; Schoof et al. 2000). CLV1 is expressed in and around the OC, and a *clv1* mutant fails to restrict *WUS* expression in the OC. The resulting expansion of *WUS* expression in *clv1* mutants will cause prolonged maintenance of stem cell identity, thereby causing CZ expansion (Schoof et al. 2000). Such a loss of stem cell control and unrestricted growth of

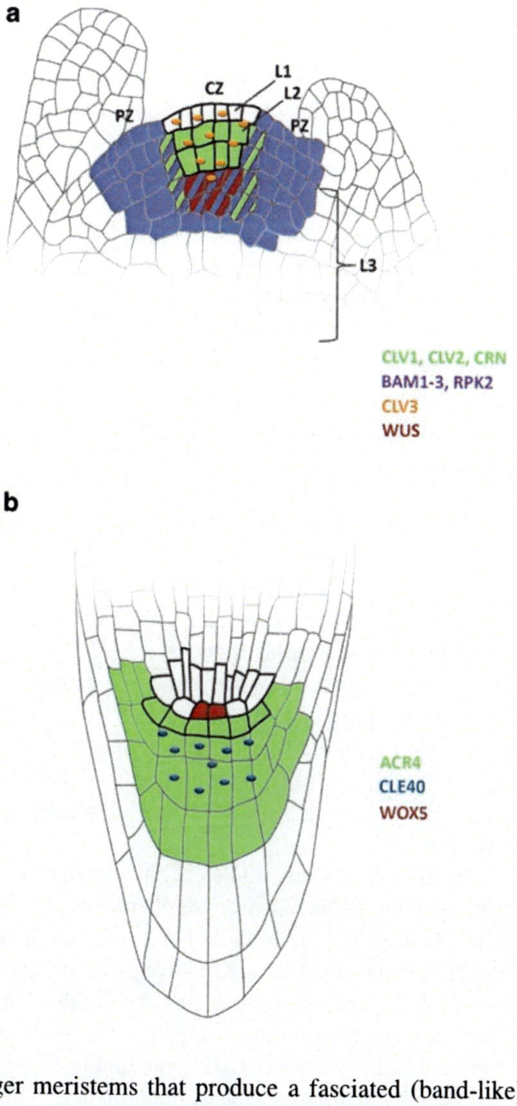

Fig. 1 Expression domains of receptors and ligands involved in shoot and root meristem maintenance. Schematic representation of the shoot apical meristem (**a**) and root meristem (**b**). Stem cells are outlined in *bold*. Expression domains are colour coded. *CZ* central zone, *PZ* peripheral zone

the CZ become evident by larger meristems that produce a fasciated (band-like) stem and flowers with extra floral organs. The production of more carpels, the central organ of the flowers, is very prominent and causes the formation of deformed siliques, reminiscent of clubs, or *clava* in Latin. Mutants in any of the three *CLAVATA* (CLV) genes (Clark et al. 1993, 1995; Kayes and Clark 1998) carry such club-shaped siliques, and this phenotype facilitated the identification of additional mutants affected in stem cell development.

Studies of the three CLV genes led to the model that CLV1 acts as a receptor kinase that is likely to be activated by binding of a small ligand, CLAVATA3 (CLV3), which is secreted from the stem cells of the CZ (Fletcher et al. 1999; Brand et al. 2000) (Table 1). *CLV3* belongs to the large *CLAVATA3/ENDOSPERM*

SURROUNDING REGION (CLE) gene family that encodes small peptides which share a conserved C-terminal amino acid sequence, the CLE motif (Cock and McCormick 2001; Oelkers et al. 2008). Isolation and detailed analysis of CLV3 peptide from transgenic plant tissues revealed that the mature 13 amino acid peptide is proteolytically processed from a larger precusor protein (Kondo et al. 2006; Ohyama et al. 2009). Furthermore, two proline residues carry hydroxyl groups, which can be further modified by the addition of arabinofuranose residues. Such peptide modifications, which were also found for CLE2, could serve to protect the peptides from proteolysis in the extracellular space, or may change their binding affinities for specific receptors.

Further experiments led to a model that a negative feedback system is established which could serve to maintain a constant stem cell population. This model depends on a balance between the expression domains of *CLV3* and *WUS*, respectively. Any increase in stem cell number, for example after a burst of stem cell division activities in the shoot meristem, will cause the production of more CLV3 ligand, which in turn will downregulate *WUS* expression via CLV1 activation. Given that *WUS* serves to maintain stem cells, decreased *WUS* expression will allow cells at the CZ periphery to exit earlier from the stem cell state. The resulting drop in CLV3 levels, following a reduction in stem cell number, should then alleviate *WUS* repression, and allow in turn for more stem cells to be produced. In a similar manner, CLV3-dependent CLV1 signalling would be turned down when too few stem cells are available. WUS levels will then increase again and a normal stem cell number is restored. According to mathematical models, such a simple network based on negative feedback regulation can serve to maintain a stable stem cell population, and therefore guarantee meristem activity, albeit only within a limited range of parameters (Geier et al. 2008; Hohm et al. 2010).

Several *clv1* mutant alleles with missense mutations in the extracellular or kinase domain displayed stronger phenotypes than putative null alleles that disrupt the kinase domain, such as *clv1-6* or *clv1-7* (Dievart et al. 2003). Furthermore, transcriptional cosuppression of strong alleles resulted in partial suppression of the *clavata* phenotype, indicating that the mutant Clv1 proteins encoded by strong alleles can exert a dominant negative function, possibly by interfering with the activity of pathways that signal in parallel to *CLV1*. Screens for other mutants with a *clv* phenotype had previously identified *CLV2*, which encodes a receptor-like protein that contains LRRs, a TMD and a short cytoplasmic C-terminal region (Kayes and Clark 1998; Jeong et al. 1999). Loss-of-function mutants of either *CLV2* or *CLV1* caused an enlargement of the shoot and floral meristems. However, these single mutants were phenotypically less severe than mutants lacking the signalling peptide CLV3. Interestingly, *clv1/clv2* double mutants displayed an enhanced phenotype (Müller et al. 2008). This suggested that both receptors could act in parallel and potentially independent pathways to transmit the CLV3 signal. Both CLV1 and CLV2 proteins were then found to be capable of CLV3 binding (Ogawa et al. 2008; Ohyama et al. 2009; Guo et al. 2010). However, the mode of signal transmission from CLV2 to the nuclear compartment, where *WUS* transcription is downregulated upon CLV3 signalling, remains unclear.

New light on the role of CLV2 in stem cell signalling came from genetic screens. Misexpression of CLV3, CLE40 or CLE19 had previously been found to affect not only stem cell maintenance in the above-ground meristems of *Arabidopsis*, but also root development (Casamitjana-Martinez et al. 2003; Hobe et al. 2003; Fiers et al. 2004, 2005). Seedlings overexpressing CLE peptides, or grown on medium containing synthetic CLE peptides, showed root growth retardation and premature differentiation of root meristem cells, indicating the activity of a CLV-related pathway (Hobe et al. 2003). However, the roots of *clv2* mutants (but not *clv1* mutants) were found to be resistant to treatment with CLE19 and other CLE peptides (Fiers et al. 2005). Two novel mutants (*SUPPRESSOR of LLP1/CLE19* overexpression, *sol1* and *sol2*) were isolated that suppressed the CLE19-dependent growth restriction. *sol1* was found to encode a $Zn2^{+}$-carboxypeptidase that is proposed to process CLE peptide precursors (Casamitjana-Martinez et al. 2003). *sol1* mutant roots were resistant to CLE19 overexpression, but still strongly affected when synthetic CLE peptides (representing the processed version of CLE19) was added externally to the growth medium. Thus, processing of the peptide from a larger precursor molecule was indeed required to generate an actively signalling molecule. The components of such a processing activity were partially purified and characterized from meristem extracts of cauliflower, and shown to be able to process not only CLE peptide precursors, but also related small peptide precursors such as the IDA peptides (Ni and Clark 2006; Stenvik et al. 2008; Ni et al. 2011).

Overexpression of *CLV3* causes shoot meristem arrest (Brand et al. 2000), and suppressor mutants were identified that suppressed the effects of increased CLV3 signalling in the shoot. As expected, not only new alleles of *clv1* and *clv2* were identified, but also mutations in *CORYNE* (*CRN*) (Müller et al. 2008). The predicted CRN protein contains a short extracellular domain, a TMD and a potentially cytoplasmic serine/threonine kinase domain. Kinase activity has not been shown so far, and attempts to complement a *crn* mutant with a presumed non-functional kinase mutant gave conflicting results (Betsuyaku et al. 2011; Nimchuk et al. 2011a). Cloning of *SOL2*, which was isolated based on its insensitivity to *CLE* gene overexpression, revealed that *CRN* and *SOL2* are allelic. This is consistent with the notion that *CRN* and *CLV2* function in a pathway that controls meristem maintenance in both shoot and root tissues (Miwa et al. 2008). *crn/sol2* mutants display enlarged shoot and floral meristems like *clv2* mutants, but are aphenotypic in the root (Müller et al. 2008). Double mutant studies then revealed that *CRN/SOL2* acts with *CLV2* in the same *CLV3* signalling pathway, in parallel to the *CLV1* pathway (Müller et al. 2008). At the molecular level, CRN and CLV2 could interact within membranes via their TMDs and their short juxtamembrane sequences to reconstitute a functional receptor kinase.

Attempts to isolate and identify the components of signalling complexes comprising CLV1 and CLV2 had been reported (Trotochaud et al. 1999). These early experiments suggested the formation of high-molecular-weight CLV complexes,

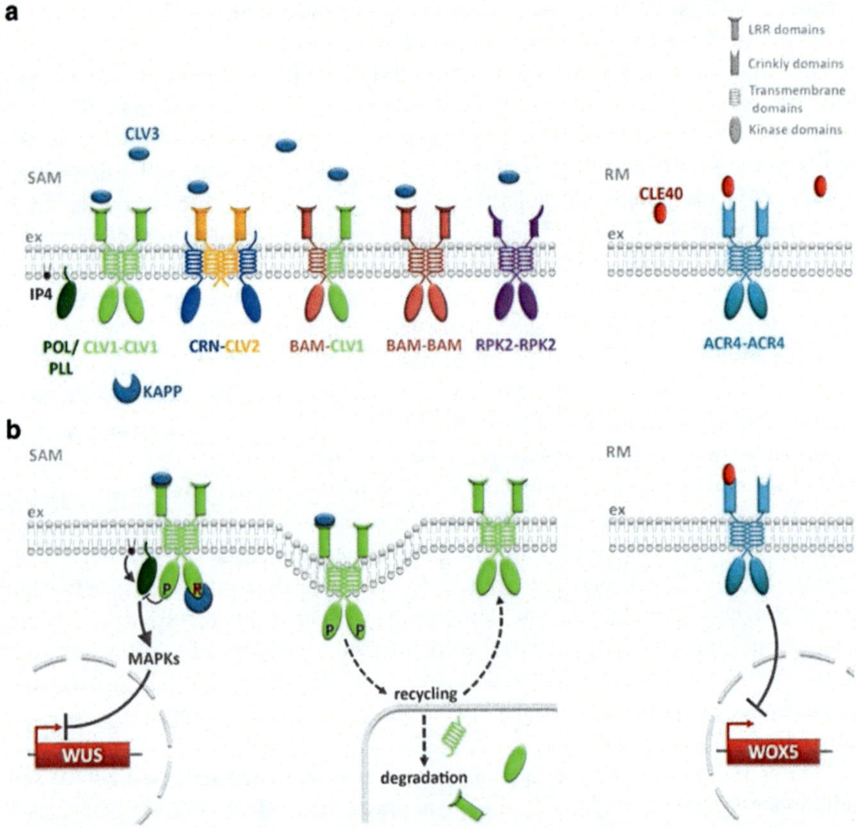

Fig. 2 Receptor complexes involved in meristem maintenance. (**a**) Schematic representations of receptor complexes present in the plasma membrane of the shoot apical meristem (SAM) and the root meristem (RM). The different receptors and their putative ligands are colour coded and shown in their proposed complexes. (**b**) Models of the signalling after receptor/ligand interactions including a proposed mechanism of recycling and degradation of CLV1. *Arrows* indicate positive regulation, *barred lines* indicate negative regulation. *ex* extracellular

however, later observations revealed that LRR containing proteins, such as CLV2 and related RLPs identified as receptors in plant pathogen signalling pathways, showed aberrant behaviour in size fractionation experiments (Rivas et al. 2002; Van Der Hoorn et al. 2003). Interactions between CLV2 and CRN were detected using epitope tagged receptors and co-immunoprecipitation, or the split-luciferase system and transient expression in *Arabidopsis* protoplasts or epidermal cells of *Nicotiana benthamiana* (Zhu et al. 2010) (Fig. 2). However, the significance of weak interactions between CLV1 and CRN remained unclear. Another study analysed the fusions of the receptor proteins CLV1, CLV2 and CRN with the fluorescent reporters GFP and mCherry (fluorescent proteins, FP) in transgenic *Arabidopsis* and by transient expression in *N. benthamiana* (Bleckmann et al. 2010). These experiments showed that receptor proteins tend to form larger and potentially non-functional aggregates in the ER if misexpressed at high levels. Using estradiol-inducible gene expression of receptor fusion proteins, rescue of the corresponding mutants could be shown within a temporally limited window of expression, suggesting that the fusion proteins are fully functional when expressed at low levels. In both *Nicotiana* and *Arabidopsis* cells, CLV1-FPs localized to the plasma membrane, whereas CLV2-FP or CRN-FP remained in the ER when expressed separately. Only when coexpressed, CRN-FP and CLV2-FP were found at the plasma membrane. Interestingly, FRET analysis revealed that the TMDs of CLV2 and CRN are required for their specific interaction, which occurs earlier in the ER. Localization to the plasma membrane requires additional amino acid sequences from the juxtamembrane regions of both proteins. These experiments suggested that at least two independently acting receptor complexes exist: one consisting of CLV1 monomers and a second complex consisting of CLV2 and CRN, which reconstitute a functional receptor via their interaction through the TMDs. Crosstalk between these two complexes can be mediated by an interaction between CLV1 and CRN.

Furthermore, screens for *Arabidopsis* mutants that are insensitive to exogenous CLE peptides uncovered yet another receptor protein contributing to CLV3 signalling. Mutations in the gene *RECEPTOR-LIKE PROTEIN KINASE 2* (*RPK2*), previously identified as *TOADSTOOL2* (*TOAD2*, see Chapter "Experimental Evidence of a Role for RLKs in Innate Immunity") (Nodine et al. 2007), cause an increase of the shoot and floral stem cell domains (Kinoshita et al. 2010). Although phenotypically slightly milder than mutants in *CLV1*, *CLV2*, *CLV3* or *CRN*, *rpk2* mutants are additive with *clv1* and *crn* or *clv2* mutants, suggesting that RPK2 is a component of a third signalling pathway acting in parallel to CLV1 and CLV2/CRN. Protein interaction studies in the *N. benthamiana* transient expression system showed that RPK2 can homomerize at the plasma membrane, but does not interact with the other components of the CLV3 signalling pathways (Betsuyaku et al. 2011). RPK2 contains extracellular Leucine-rich repeats, a TMD and a cytoplasmic serine/threonin kinase domain, but belongs to a different subfamily of LRR-RLKs than CLV1. Whether RPK2 is also able to directly bind CLE peptides has not yet been investigated. RPK2/TOAD2 has multiple roles during plant development, and is required from embryogenesis onwards to specify cell types along the radial axis, generation of the root pole, and in tapetal cell fate specification (Mizuno et al. 2007; Nodine et al. 2007). Interestingly, overexpression of RPK2 in the entire plant caused a reduction in vegetative growth, suggesting that RPK2 function may be rate-limiting in a number of different developmental processes (Kinoshita et al. 2010). Although RPK2 is likely to be involved in CLV3 perception and negative regulation of stem cell fate by repressing *WUS*, increased expression of RPK2 led to a reduction in stem cell number and plant phenotypes resembling those of *wus* mutants. It is possible that RPK2 could either autoactivate when overexpressed, or trigger the activation of the CLV1 or CLV2/CRN pathways independently of CLV3.

1.1 Regulation of Signalling

RPK2 is expressed in most plant tissues analysed, and RNA is found throughout the shoot and floral meristems of *Arabidopsis*. Expression of RPK2-GFP fusion proteins from the RPK2 promoter revealed localization of the receptor at the plasma membrane, but expression levels were clearly reduced in the central zone of inflorescence meristems (Kinoshita et al. 2010). This could suggest that the protein is undergoing cyclic degradation upon perception of the CLV3 signal. A similar scenario was recently proposed for the CLV1 receptor and suggested to be part of a compensatory mechanism that dampens CLV signalling (Nimchuk et al. 2011b) (Fig. 2b). Here, CLV1-GFP fusion proteins were found to be targeted towards the vacuole for degradation in a CLV3-dependent manner. However, overall CLV1 levels were not increased in *clv3* mutants compared with wild-type, suggesting that CLV3-dependent removal of CLV1 from the membrane mainly serves to compensate for the continuous de novo synthesis of CLV1 receptor. Interestingly, *WUS*, the main target of the CLV pathway, directly downregulates *CLV1* expression at the transcriptional level (Busch et al. 2010). Thus, CLV signalling will cause increased *CLV1* expression, which can be compensated by receptor degradation.

1.2 Role of CLV1-Related BAM Receptors in Meristem Signalling

Several LRR-RLKs that are closely related to CLV1 also contribute to meristem maintenance in Arabidopsis. Single mutations in *BARELY ANY MERISTEM1* (*BAM1*), *BAM2* or *BAM3* were aphenotypic, but double mutant combinations of *bam1* with *bam2* carried smaller shoot meristems, while *bam1 bam2 bam3* triple mutants even arrested growth or generated floral meristems lacking organs (DeYoung et al. 2006). Thus, BAM receptors appear to act antagonistically to CLV1 in meristem size control. However, expression of *CLV1* under the control of the *ERECTA* promoter, which is more broadly expressed than *CLV1*, could rescue *bam1 bam2* meristem defects, and expression of *BAM1* or *BAM2* in the shoot meristem was able to partially suppress *clv1* mutant phenotypes, indicating that these receptors are functionally interchangeable. Expression of *CLV1* is mostly confined to the centre of shoot and floral meristems. BAM receptors are expressed preferentially not only at the meristem periphery, but also in other tissues (DeYoung and Clark 2008). For example, BAM1 and BAM2 control cell division and cell type specification during anther development (see Chapter "Experimental Evidence of a Role for RLKs in Innate Immunity"). In meristems, BAM receptors can also affect development of the central meristem domain, because *clv1* mutants showed additive interactions with *bam1* and *bam2*. BAM receptors were reported to bind CLE peptides, although with a lower affinity than CLV1 (Guo et al. 2010) (Fig. 2a). It was proposed that BAM receptors in the outer meristem regions could partially shield the meristem centre from diffusing CLE peptides that originate from the meristem periphery (DeYoung and Clark 2008), and are able to activate CLV signalling similarly to CLV3. The A-type CLE peptides CLE16, 17 and 27 were recently shown to be expressed also in shoot apical meristems, but misexpression of these peptides at high levels in the shoot meristem did not affect meristem maintenance (Jun et al. 2010). It is still possible that other peptides expressed at low levels or divergent in sequence from the CLE family act in parallel with CLV3 and bind the BAM receptors. Alternatively, BAM RLKs could function with the CLV signalling pathway not at the level of ligand sequestration, but further downstream at the regulation of common target genes.

1.3 Signal Transduction to the Nucleus

The precise mechanism of CLV signal transmission from plasma membrane binding sites to transcriptional responses in the nucleus is still not understood. However, genetic screens allowed the isolation of several genes whose products may mediate these signalling pathways. Mutations in the *POLTERGEIST* (*POL*) gene suppress the enlarged meristem phenotypes of *clv1*, *clv2* (Yu et al. 2000) and also *crn* mutants (Müller et al. 2008) (Fig. 2a, b). *POL* acts downstream of the *CLV* genes and encodes a protein phosphatase 2C that, with other members from the plant kingdom, belong to a unique subclass of phosphatases (Yu et al. 2003). POL and the related PLL1 (POLTERGEIST_LIKE-1) protein promote *WUS* expression in shoot and floral meristems (Song and Clark 2005; Song et al. 2006), but are also required for the specification of vascular cells and the root meristem during embryogenesis (Song et al. 2008). As a downstream intermediate, POL was suggested to be repressed by CLV signalling. The finding that POL and PLL1 require N-terminal myristoylation and palmitoylation to be localized to the plasma membrane significantly expanded the understanding of this signalling pathway (Gagne and Clark 2010). POL phosphatase activity is stimulated upon binding of phosphatidylinositol (4) phosphate (PI4P), suggesting that stem cell fate in meristems may be strongly influenced by the lipid composition of the plasma membrane. It is conceivable that preferential activation of CLV receptors by CLV3 at the apical side of a cell could then locally restrict PI4P availability. The resulting polar distribution of PI4P would cause a corresponding polar activation of phosphatases such as POL, which could underly the mechanism that generates asymmetries after stem cell division.

Another CLV signalling component is KAPP, a kinase-associated protein phosphatase that binds to phosphorylated peptides and serves to antagonize diverse RLKs in Arabidopsis (Williams et al. 1997; Stone et al. 1998; Ding et al. 2007). Interaction studies and phosphorylation assays suggest that the MAP kinase pathway is involved in transmitting the CLV3 signal to the nucleus (Betsuyaku et al. 2011). CLV3 was found to trigger CLV1 phosphorylation in an *N. benthamiana* transient expression system and to control MPK6 activity via the CLV receptors also in *Arabidopsis* seedling assays. Interestingly, this study indicates a differential effect of the CLV receptors on MPK6 activity, with RPK2 and CLV2 activating, but

CLV1 repressing MPK6 function (Betsuyaku et al. 2011). However, a more detailed investigation is needed to specifically resolve the roles of MAPKs in stem cell maintenance.

2 ACR4 Regulates Lateral Root Formation and Stem Cell Maintenance

The first indications that RLK-dependent signalling pathways regulate root meristem architecture or maintenance came from CLE peptide misexpression phenotypes in roots (Fig. 1b). As described above, exogenous application of or constitutive misexpression of CLV3, CLE40 or CLE19 caused inhibition of root growth, due to premature differentiation of meristem cells (Casamitjana-Martinez et al. 2003; Hobe et al. 2003; Fiers et al. 2004). These early results suggested that a *CLV*-like signalling pathway also operates in the root. However, knockout mutants in several root-expressed RLKs were aphenotypic, and only *CLV2* and *CRN*, which are both expressed in the vasculature of the root meristem, were found to be required for the growth-limiting phenotypes after high-level misexpression of CLE peptides (Fiers et al. 2005; Müller et al. 2008). Nevertheless, no loss-of-function phenotype was found for these receptors in root development.

Lateral roots are initiated from cells of the pericycle cell layer that are located opposite of the xylem poles. Two asymmetric divisions generate a core group of four small cells, which, after several rounds of periclinal cell divisions, will generate the lateral root primordium. Cell sorting and RNA analysis via microarray resulted in the identification of genes that are specifically expressed during these earliest stages of lateral root formation, among them the membrane-localized RLK *ARABIDOPSIS CRINKLY4* (*ACR4*) (De Smet et al. 2008). *ACR4* is a member of a small gene family, together with four closely related RLKs (*CRR1 – 4*) (Cao et al. 2005). Previously, *ACR4* was found to control the maintenance of the outermost meristem cell layer, the L1, during embryonic development (Tanaka et al. 2002; Gifford et al. 2003; Watanabe et al. 2004). In the root, *ACR4* restricts the number and position of lateral root initiation events (De Smet et al. 2008) acting redundantly with the other family members, but also restricts the proliferation of stem cell layers at the distal position of the root meristem that generate the protective columella cells. A close inspection of *cle40* mutants revealed a similar proliferation defect of the columella stem cell layers as observed in *acr4* mutants (Stahl et al. 2009). *CLE40* is normally expressed in the differentiated columella cells. When *Arabidopsis* roots are grown on media containing synthetic CLE40 peptide, the distal root stem cells also acquired columella cell identity. This *CLE40* overexpression phenotype did not manifest in *acr4* mutants, suggesting that CLE40 could be a potential ligand that signals through ACR4 in columella stem cells.

These parallels to the regulation of the shoot stem cell system through the *CLV* pathway can be even further extended to the targets of the CLE peptide/RLK signalling module. Root stem cells are highly specific and give rise to discrete cell types (Benfey and Scheres 2000). The stem cells surround the quiescent centre (or QC), a group of four cells that inhibits stem cells from differentiating. A candidate gene for stem cell maintenance that is specifically expressed in the QC is *WOX5*, a member of the *WUSCHEL-LIKE HOMEOBOX* (*WOX*) gene family, and WOX5 is very closely related to *WUS* (Sarkar et al. 2007). Indeed, cross-complementation experiments showed that *WUS* and *WOX5* can functionally replace each other when expressed in the corresponding domains. *WOX5* was found to be a target for transcriptional regulation by the proposed *CLE40 – ACR4* module (Stahl et al. 2009) and consistent with this, *wox5* mutants fail to maintain columella stem cells. The CLE40 peptide is most closely related to CLV3 and can replace CLV3 function if expressed from the *CLV3* promoter (Hobe et al. 2003). Thus, the stem cell regulatory units consisting of a CLE peptide (CLV3 or CLE40) and a target transcription factor (WUS or WOX5) are fully conserved between the shoot and the root. However, a different receptor type, one without LRRs, is employed for signal transmission in the root.

ACR4 contains seven CRINKLY repeats with a predicted β-propeller structure and a domain homologous to the Cys-rich repeats of tumor necrosis factor receptor (TNFR) in the extracellular region, followed by a transmembrane domain and a cytoplasmic kinase region. Complementation analysis showed that the TNFR-like domains are not required, and that also a kinase-null version of ACR4 can still rescue *acr4* mutant phenotypes (Gifford et al. 2005). This could indicate that ACR4 interacts with another RLK that provides the kinase activity. On a similar line, CRR1 and CRR2, which were found to be kinase-dead, were shown to be the targets for phosphorylation by ACR4 (Cao et al. 2005). This suggests that RLKs from the ACR4-family may act in larger complexes, and indeed, dimerization via their transmembrane domains has been confirmed for all of them (Stokes and Rao 2008).

Interestingly, ACR4 shows a high degree of turnover, and both the TNFR domain and the cytoplasmic domain appear to be involved in controlling protein stability. Only the functional versions of ACR4 that carry the CRINKLY repeats are rapidly endocytosed, suggesting that the removal of ACR4 protein from the plasma membrane followed by degradation is triggered by ligand binding and serves to dampen signalling (Gifford et al. 2005). *ACR4* expression is also transcriptionally upregulated in the distal stem cell domain upon incubation of roots with excess CLE40 peptide (Stahl and Simon 2009), suggesting that overall ACR4 signalling activity is subject to both positive and negative feedback regulation.

3 Signalling Cell Fate in the Vascular Meristem

The procambium comprises the stem cells of the vascular system. The procambial strands are highly polarized, generating new cell layers along the radial axis that will differentiate into either phloem cells (outside) or into xylem cells (inside). A first hint of a specific role of RLKs in maintaining this organization came from a biochemical tour-de-force, aimed at isolating diffusible factors that control cell differentiation in a tissue culture system (Ito et al. 2006). *Zinnia* cell cultures had long been used to study xylem tracheary element development, with specific emphasis on their ability to link up and form long, continuous vascular strands. A peptide belonging to the CLE family, TDIF (TRACHEARY ELEMENT DIFFERENTIATION INHIBITORY FACTOR), and its *Arabidopsis* homologues CLE41/CLE44 were shown to inhibit xylem cell differentiation. Both CLE41/44 are normally expressed from the phloem. When *Arabidopsis* seedlings were grown in liquid culture containing TDIF peptides, procambium cells proliferated at the expense of xylem differentiation (Hirakawa et al. 2008). A candidate receptor for CLE41 is PXY (PHLOEM INTERCALATED WITH XYLEM), an RLK closely related to CLV1 and the BAM receptors, which is expressed mainly in the procambium (Fisher and Turner 2007). *Pxy* mutants show aberrant arrangement of phloem and xylem cells, indicating that PXY controls oriented cell division of procambium descendants. CLE41 was found to bind to the extracellular domain of PXY (also known as TDR, for TDIF receptor) (Hirakawa et al. 2008) and control PXY/TDR activity in a dosage-dependent manner. A *cle41* mutant formed a thinner stele than wildtype plants (Hirakawa et al. 2010), whereas increased expression of CLE41 from the phloem caused procambium proliferation, albeit with normal vasculature organization (Etchells and Turner 2010). However, *CLE41* misexpression in specific vascular cell types disturbed the normal pattern of vascular development, and also downregulated expression of *PXY* itself. These observations suggest that the CLE41 peptide controls the rate of procambium proliferation and the orientation of cell divisions. This proposed gradient of CLE41 peptide, with phloem cells as a source and signalling competent, PXY-expressing procambial cells as sink would serve to maintain the procambial cells and act to balance phloem and xylem production. One of the targets regulated transcriptionally by the TDIF-PXY/TDR pathway is *WOX4*, another member of the *WOX* gene family (Hirakawa et al. 2010). Furthermore, *WOX4* expression in the procambium and cambium is needed for stem cell maintenance and continuous production of the vasculature. Thus, a CLE/RLK signalling module controls also stem cell behaviour during vascular development, but in contrast to the previous examples, by promoting rather than repressing the expression of a WUS-like homeodomain protein.

4 Evolutionary Considerations and Conclusions

We have discussed here the roles of RLKs in three different meristem systems: the shoot and floral meristems, the primary root meristem and the vascular meristem. In all three systems, RLKs play an important role in controlling the maintenance and proliferation of undifferentiated stem cells, and there are intriguing similarities in the molecular makeup of these signalling pathways. CLE peptides act as ligands that serve to control the communication between differentiated cells and stem cells. In the shoot meristem, three receptor systems have been identified which perceive extracellular CLE peptides via their leucine-rich repeats, and signal through the cytoplasmic kinase domains. Both *CLV1* and *RPK2* have evolved early in the plant lineage, and homologues can be found in bryophytes such as *Physcomitrella patens* (Sawa and Tabata 2011). In *Lotus*, the CLV1 and RPK2 homologues have been identified as HAR1 and KLAVIER, respectively, which interact and control not only shoot meristem development, but also formation of root nodules in a CLE-dependent manner (Okamoto et al. 2009; Miyazawa et al. 2010). The CLV2/CRN composite receptor appears to be a more recent innovation of vascular plants (Miwa et al. 2009). In addition to controlling stem cell fate in the shoot system, CLV2/CRN expression in the root has been exploited by parasitic nematodes as an access point to induce root cell proliferation via production of nematode CLE peptides (Replogle et al. 2011). PXY, belonging to the same subfamily of RLKs as CLV1 and the BAMs, controls vascular development by promoting *WOX4* expression, and homologues from both *PXY/TDR* and *WOX4* were shown to be expressed in the procambium of important crop species such as poplar (Schrader et al. 2004).

For all the signalling systems discussed here, mutant identification provided a starting point for the genetic dissection of pathways. However, we now realize that the regulatory systems are more complex than initially thought. A single signalling molecule, such as CLV3 peptide, can be perceived by at least three different receptor complexes. In addition, further biochemical studies revealed that many more receptors are able to bind the CLV3 ligand and that an array of different CLE peptides can interact with the CLV receptors. An important question arising now is how signalling specificity is maintained. Given the huge number of receptor kinases that are expressed in plants and the tendency of most systems to operate via redundant receptors, which could provide a fail-safe mechanism, it is unlikely that we will be able to unravel the roles of all RLKs through mutant studies. Instead, we will need to study receptor expression, interaction, complex formation and activation *in vivo*, and perform these analyses in individual cells within a tissue context. To do this, novel and more sensitive techniques will have to be developed to allow RLK functional analysis in the developing plant meristems.

References

Barton MK (2009) Twenty years on: the inner workings of the shoot apical meristem, a developmental dynamo. Dev Biol 341:95–113

Benfey PN, Scheres B (2000) Root development. Curr Biol 10:R813–R815

Betsuyaku S, Takahashi F, Kinoshita A, Miwa H, Shinozaki K, Fukuda H, Sawa S (2011) Mitogen-activated protein kinase regulated by the CLAVATA receptors contributes to shoot apical meristem homeostasis. Plant Cell Physiol 52:14–29

Bleckmann A, Weidtkamp-Peters S, Seidel CAM, Simon R (2010) Stem cell signaling in Arabidopsis requires CRN to localize CLV2 to the plasma membrane. Plant Physiol 152:166–176

Brand U, Fletcher JC, Hobe M, Meyerowitz EM, Simon R (2000) Dependence of stem cell fate in Arabidopsis on a feedback loop regulated by CLV3 activity. Science 289:617–619

Busch W, Miotk A, Ariel FD, Zhao Z, Forner J, Daum G, Suzaki T, Schuster C, Schultheiss SJ, Leibfried A, Haubeiss S, Ha N, Chan RL, Lohmann JU (2010) Transcriptional control of a plant stem cell niche. Dev Cell 18:849–861

Cao X, Li K, Suh SG, Guo T, Becraft PW (2005) Molecular analysis of the CRINKLY4 gene family in *Arabidopsis thaliana*. Planta 220:645–657

Casamitjana-Martinez E, Hofhuis HF, Xu J, Liu CM, Heidstra R, Scheres B (2003) Root-specific CLE19 overexpression and the sol1/2 suppressors implicate a CLV-like pathway in the control of Arabidopsis root meristem maintenance. Curr Biol 13:1435–1441

Clark SE, Running MP, Meyerowitz EM (1993) CLAVATA1, a regulator of meristem and flower development in Arabidopsis. Development 119:397–418

Clark SE, Running MP, Meyerowitz EM (1995) CLAVATA3 is a specific regulator of shoot and floral meristem development affecting the same processes as CLAVATA1. Development 121:2057–2067

Clark SE, Williams RW, Meyerowitz EM (1997) The CLAVATA1 gene encodes a putative receptor kinase that controls shoot and floral meristem size in Arabidopsis. Cell 89:575–585

Cock JM, McCormick S (2001) A large family of genes that share homology with CLAVATA3. Plant Physiol 126:939–942

De Smet I, Vassileva V, De Rybel B, Levesque MP, Grunewald W, Van Damme D, Van Noorden G, Naudts M, Van Isterdael G, De Clercq R, Wang JY, Meuli N, Vanneste S, Friml J, Hilson P, Jurgens G, Ingram GC, Inze D, Benfey PN, Beeckman T (2008) Receptor-like kinase ACR4 restricts formative cell divisions in the Arabidopsis root. Science 322:594–597

DeYoung BJ, Clark SE (2008) BAM receptors regulate stem cell specification and organ development through complex interactions with CLAVATA signaling. Genetics 180:895–904

DeYoung BJ, Bickle KL, Schrage KJ, Muskett P, Patel K, Clark SE (2006) The CLAVATA1-related BAM1, BAM2 and BAM3 receptor kinase-like proteins are required for meristem function in Arabidopsis. Plant J 45:1–16

Dievart A, Dalal M, Tax FE, Lacey AD, Huttly A, Li J, Clark SE (2003) CLAVATA1 dominant-negative alleles reveal functional overlap between multiple receptor kinases that regulate meristem and organ development. Plant Cell 15:1198–1211

Ding Z, Wang H, Liang X, Morris ER, Gallazzi F, Pandit S, Skolnick J, Walker JC, Van Doren SR (2007) Phosphoprotein and phosphopeptide interactions with the FHA domain from Arabidopsis kinase-associated protein phosphatase. Biochemistry 46:2684–2696

Etchells JP, Turner SR (2010) The PXY-CLE41 receptor ligand pair defines a multifunctional pathway that controls the rate and orientation of vascular cell division. Development 137:767–774

Fiers M, Hause G, Boutilier K, Casamitjana-Martinez E, Weijers D, Offringa R, van der Geest L, van Lookeren CM, Liu CM (2004) Mis-expression of the CLV3/ESR-like gene CLE19 in Arabidopsis leads to a consumption of root meristem. Gene 327:37–49

Fiers M, Golemiec E, Xu J, van der Geest L, Heidstra R, Stiekema W, Liu CM (2005) The 14-amino acid CLV3, CLE19, and CLE40 peptides trigger consumption of the root meristem in Arabidopsis through a CLAVATA2-dependent pathway. Plant Cell 17:2542–2553

Fisher K, Turner S (2007) PXY, a receptor-like kinase essential for maintaining polarity during plant vascular-tissue development. Curr Biol 17:1061–1066

Fletcher JC, Brand U, Running MP, Simon R, Meyerowitz EM (1999) Signaling of cell fate decisions by CLAVATA3 in Arabidopsis shoot meristems. Science 283:1911–1914

Gagne JM, Clark SE (2010) The Arabidopsis stem cell factor POLTERGEIST is membrane localized and phospholipid stimulated. Plant Cell 22:729–743

Geier F, Lohmann JU, Gerstung M, Maier AT, Timmer J, Fleck C (2008) A quantitative and dynamic model for plant stem cell regulation. PLoS One 3:e3553

Gifford ML, Dean S, Ingram GC (2003) The Arabidopsis ACR4 gene plays a role in cell layer organisation during ovule integument and sepal margin development. Development 130:4249–4258

Gifford ML, Robertson FC, Soares DC, Ingram GC (2005) ARABIDOPSIS CRINKLY4 function, internalization, and turnover are dependent on the extracellular crinkly repeat domain. Plant Cell 17:1154–1166

Guo Y, Han L, Hymes M, Denver R, Clark SE (2010) CLAVATA2 forms a distinct CLE-binding receptor complex regulating Arabidopsis stem cell specification. Plant J 63:889–900

Hirakawa Y, Shinohara H, Kondo Y, Inoue A, Nakanomyo I, Ogawa M, Sawa S, Ohashi-Ito K, Matsubayashi Y, Fukuda H (2008) Non-cell-autonomous control of vascular stem cell fate by a CLE peptide/receptor system. Proc Natl Acad Sci USA 105:15208–15213

Hirakawa Y, Kondo Y, Fukuda H (2010) TDIF peptide signaling regulates vascular stem cell proliferation via the WOX4 homeobox gene in Arabidopsis. Plant Cell 22:2618–2629

Hobe M, Müller R, Grünewald M, Brand U, Simon R (2003) Loss of CLE40, a protein functionally equivalent to the stem cell restricting signal CLV3, enhances root waving in Arabidopsis. Dev Genes Evol 213:371–381

Hohm T, Zitzler E, Simon R (2010) A dynamic model for stem cell homeostasis and patterning in Arabidopsis meristems. PLoS One 5:e9189

Ito Y, Nakanomyo I, Motose H, Iwamoto K, Sawa S, Dohmae N, Fukuda H (2006) Dodeca-CLE peptides as suppressors of plant stem cell differentiation. Science 313:842–845

Jeong S, Trotochaud AE, Clark SE (1999) The Arabidopsis CLAVATA2 gene encodes a receptor-like protein required for the stability of the CLAVATA1 receptor-like kinase. Plant Cell 11:1925–1934

Jun J, Fiume E, Roeder AH, Meng L, Sharma VK, Osmont KS, Baker C, Ha CM, Meyerowitz EM, Feldman LJ, Fletcher JC (2010) Comprehensive analysis of CLE polypeptide signaling gene expression and overexpression activity in Arabidopsis. Plant Physiol 154:1721–1736

Kayes JM, Clark SE (1998) CLAVATA2, a regulator of meristem and organ development in Arabidopsis. Development 125:3843–3851

Kinoshita A, Betsuyaku S, Osakabe Y, Mizuno S, Nagawa S, Stahl Y, Simon R, Yamaguchi-Shinozaki K, Fukuda H, Sawa S (2010) RPK2 is an essential receptor-like kinase that transmits the CLV3 signal in Arabidopsis. Development 137:3911–3920

Kondo T, Sawa S, Kinoshita A, Mizuno S, Kakimoto T, Fukuda H, Sakagami Y (2006) A plant peptide encoded by CLV3 identified by in situ MALDI-TOF MS analysis. Science 313:845–848

Laux T, Mayer KF, Berger J, Jurgens G (1996) The WUSCHEL gene is required for shoot and floral meristem integrity in Arabidopsis. Development 122:87–96

Mayer KF, Schoof H, Haecker A, Lenhard M, Jurgens G, Laux T (1998) Role of WUSCHEL in regulating stem cell fate in the Arabidopsis shoot meristem. Cell 95:805–815

Miwa H, Betsuyaku S, Iwamoto K, Kinoshita A, Fukuda H, Sawa S (2008) The receptor-like kinase SOL2 mediates CLE signaling in Arabidopsis. Plant Cell Physiol 49:1752–1757

Miwa H, Tamaki T, Fukuda H, Sawa S (2009) Evolution of CLE signaling: origins of the CLV1 and SOL2/CRN receptor diversity. Plant Signal Behav 4:477–481

Miyazawa H, Oka-Kira E, Sato N, Takahashi H, Wu GJ, Sato S, Hayashi M, Betsuyaku S, Nakazono M, Tabata S, Harada K, Sawa S, Fukuda H, Kawaguchi M (2010) The receptor-like kinase KLAVIER mediates systemic regulation of nodulation and non-symbiotic shoot development in Lotus japonicus. Development 137:4317–4325

Mizuno S, Osakabe Y, Maruyama K, Ito T, Osakabe K, Sato T, Shinozaki K, Yamaguchi-Shinozaki K (2007) Receptor-like protein kinase 2 (RPK 2) is a novel factor controlling anther development in Arabidopsis thaliana. Plant J 50:751–766

Müller R, Bleckmann A, Simon R (2008) The receptor kinase CORYNE of Arabidopsis transmits the stem cell-limiting signal CLAVATA3 independently of CLAVATA1. Plant Cell 20:934–946

Ni J, Clark SE (2006) Evidence for functional conservation, sufficiency, and proteolytic processing of the CLAVATA3 CLE domain. Plant Physiol 140:726–733
Ni J, Guo Y, Jin H, Hartsell J, Clark SE (2011) Characterization of a CLE processing activity. Plant Mol Biol 75:67–75
Nimchuk ZL, Tarr PT, Meyerowitz E (2011a) An evolutionarily conserved pseudokinase mediates stem cell production in plants. Plant Cell 23:851–854
Nimchuk ZL, Tarr PT, Ohno C, Qu X, Meyerowitz EM (2011b) Plant stem cell signaling involves ligand-dependent trafficking of the CLAVATA1 receptor kinase. Curr Biol 21:345–352
Nodine MD, Yadegari R, Tax FE (2007) RPK1 and TOAD2 are two receptor-like kinases redundantly required for arabidopsis embryonic pattern formation. Dev Cell 12:943–956
Oelkers K, Goffard N, Weiller GF, Gresshoff PM, Mathesius U, Frickey T (2008) Bioinformatic analysis of the CLE signaling peptide family. BMC Plant Biol 8:1
Ogawa M, Shinohara H, Sakagami Y, Matsubayashi Y (2008) Arabidopsis CLV3 peptide directly binds CLV1 ectodomain. Science 319:294
Ohyama K, Shinohara H, Ogawa-Ohnishi M, Matsubayashi Y (2009) A glycopeptide regulating stem cell fate in *Arabidopsis thaliana*. Nat Chem Biol 5:578–580
Okamoto S, Ohnishi E, Sato S, Takahashi H, Nakazono M, Tabata S, Kawaguchi M (2009) Nod factor/nitrate-induced CLE genes that drive HAR1-mediated systemic regulation of nodulation. Plant Cell Physiol 50:67–77
Replogle A, Wang J, Bleckmann A, Hussey RS, Baum TJ, Sawa S, Davis EL, Wang X, Simon R, Mitchum MG (2011) Nematode CLE signaling in Arabidopsis requires CLAVATA2 and CORYNE. Plant J 65:430–440
Rivas S, Romeis T, Jones JD (2002) The Cf-9 disease resistance protein is present in an approximately 420-kilodalton heteromultimeric membrane-associated complex at one molecule per complex. Plant Cell 14:689–702
Sarkar AK, Luijten M, Miyashima S, Lenhard M, Hashimoto T, Nakajima K, Scheres B, Heidstra R, Laux T (2007) Conserved factors regulate signalling in *Arabidopsis thaliana* shoot and root stem cell organizers. Nature 446:811–814
Sawa S, Tabata R (2011) RPK2 functions in diverged CLE signaling. Plant Signal Behav 6:86–88
Schoof H, Lenhard M, Haecker A, Mayer KF, Jurgens G, Laux T (2000) The stem cell population of Arabidopsis shoot meristems in maintained by a regulatory loop between the CLAVATA and WUSCHEL genes. Cell 100:635–644
Schrader J, Nilsson J, Mellerowicz E, Berglund A, Nilsson P, Hertzberg M, Sandberg G (2004) A high-resolution transcript profile across the wood-forming meristem of poplar identifies potential regulators of cambial stem cell identity. Plant Cell 16:2278–2292
Song SK, Clark SE (2005) POL and related phosphatases are dosage-sensitive regulators of meristem and organ development in Arabidopsis. Dev Biol 285:272–284
Song SK, Lee MM, Clark SE (2006) POL and PLL1 phosphatases are CLAVATA1 signaling intermediates required for Arabidopsis shoot and floral stem cells. Development 133:4691–4698
Song SK, Hofhuis H, Lee MM, Clark SE (2008) Key divisions in the early Arabidopsis embryo require POL and PLL1 phosphatases to establish the root stem cell organizer and vascular axis. Dev Cell 15:98–109
Stahl Y, Simon R (2009) Is the Arabidopsis root niche protected by sequestration of the CLE40 signal by its putative receptor ACR4? Plant Signal Behav 4:634–635
Stahl Y, Wink RH, Ingram GC, Simon R (2009) A signaling module controlling the stem cell niche in Arabidopsis root meristems. Curr Biol 19:909–914
Stenvik GE, Tandstad NM, Guo Y, Shi CL, Kristiansen W, Holmgren A, Clark SE, Aalen RB, Butenko MA (2008) The EPIP peptide of INFLORESCENCE DEFICIENT IN ABSCISSION is sufficient to induce abscission in Arabidopsis through the receptor-like kinases HAESA and HAESA-LIKE2. Plant Cell 20:1805–1817
Stokes KD, Rao AG (2008) Dimerization properties of the transmembrane domains of Arabidopsis CRINKLY4 receptor-like kinase and homologs. Arch Biochem Biophys 477:219–226

Stone JM, Trotochaud AE, Walker JC, Clark SE (1998) Control of meristem development by CLAVATA1 receptor kinase and kinase-associated protein phosphatase interactions. Plant Physiol 117:1217–1225

Tanaka H, Watanabe M, Watanabe D, Tanaka T, Machida C, Machida Y (2002) ACR4, a putative receptor kinase gene of Arabidopsis thaliana, that is expressed in the outer cell layers of embryos and plants, is involved in proper embryogenesis. Plant Cell Physiol 43:419–428

Trotochaud AE, Hao T, Wu G, Yang Z, Clark SE (1999) The CLAVATA1 receptor-like kinase requires CLAVATA3 for its assembly into a signaling complex that includes KAPP and a Rho-related protein. Plant Cell 11:393–406

Van Der Hoorn RA, Rivas S, Wulff BB, Jones JD, Joosten MH (2003) Rapid migration in gel filtration of the Cf-4 and Cf-9 resistance proteins is an intrinsic property of Cf proteins and not because of their association with high-molecular-weight proteins. Plant J 35:305–315

Watanabe M, Tanaka H, Watanabe D, Machida C, Machida Y (2004) The ACR4 receptor-like kinase is required for surface formation of epidermis-related tissues in *Arabidopsis thaliana*. Plant J 39:298–308

Williams RW, Wilson JM, Meyerowitz EM (1997) A possible role for kinase-associated protein phosphatase in the Arabidopsis CLAVATA1 signaling pathway. Proc Natl Acad Sci USA 94:10467–10472

Yu LP, Simon EJ, Trotochaud AE, Clark SE (2000) POLTERGEIST functions to regulate meristem development downstream of the CLAVATA loci. Development 127:1661–1670

Yu LP, Miller AK, Clark SE (2003) POLTERGEIST encodes a protein phosphatase 2C that regulates CLAVATA pathways controlling stem cell identity at Arabidopsis shoot and flower meristems. Curr Biol 13:179–188

Zhu Y, Wang Y, Li R, Song X, Wang Q, Huang S, Jin JB, Liu CM, Lin J (2010) Analysis of interactions among the CLAVATA3 receptors reveals a direct interaction between CLAVATA2 and CORYNE in Arabidopsis. Plant J 61:223–233

The Social Network: Receptor Kinases and Cell Fate Determination in Plants

Anthony Bryan, Adriana Racolta, Frans Tax, and Sarah Liljegren

Abstract Cell signaling plays a key role in the determination of cell fate in plants. In many developmental processes, receptor kinases have been identified as necessary for the formation of specific cell types. Many receptor kinase mutants with similar phenotypes have been characterized, and there is abundant experimental evidence for genetic redundancy. In addition, multiple examples of potentially sequential or parallel pathways have been uncovered. In a few instances, mutations in different receptor kinases cause opposite phenotypes. Taken together, these results imply that networks of receptor kinases, rather than single receptors, are necessary for cells to understand their postions with respect to other cells and ultimately to make decisions about their roles. In most cases, it is not yet clear whether receptor kinase networks are characterized by physical interactions between the kinases, by parallel pathways, by receptor kinases controling the localization of other receptor kinases, or by other as yet uncharacterized regulatory mechanisms. Some of the same receptor kinases function in multiple developmental processes and cell types, implying that these networks may be reiterative during development. In this chapter, we will focus on receptor kinase mediated mechanisms that establish and maintain cell fates in two broad contexts: (1) the determination of radial identity, particularly epidermal cell fate and the radial layers of the anther and (2) activation of cell separation within organ abscission zones, an enactment of cell fate that also spans radial layers.

A. Bryan • A. Racolta • F. Tax
Department of Molecular and Cellular Biology, University of Arizona, Tucson, AZ 85721, USA

S. Liljegren (✉)
Department of Biology, University of North Carolina, Chapel Hill, NC 27514, USA
e-mail: liljegren@olemiss.edu

F. Tax and B. Kemmerling (eds.), *Receptor-like Kinases in Plants*,
Signaling and Communication in Plants 13,
DOI 10.1007/978-3-642-23044-8_3, © Springer-Verlag Berlin Heidelberg 2012

1 Introduction to Cell Fate

Pattern formation in both plants and animals requires precise coordination of cell proliferation and differentiation, leading to complex arrangements of cells in space and time. In animal systems, the identity of specific cells is determined through a variety of cellular processes including cell migration, asymmetric cell division, signaling between cells and proliferation. In plants, cells typically do not move, which means there is more emphasis on asymmetric cell division and intercellular signaling for specifying cell fates. The final identity of a cell can be dictated by cues from its lineage, cues from its neighbors or by integrating cues from both sources. Experimental evidence from *Arabidopsis* suggests that positional signals rather than lineage play a more important role in plant cell fate determination (van den Berg et al. 1995; Saulsberry et al. 2002). The identity of an individual plant cell is provided by its position in the apical/basal axis, the radial axis and the adaxial/abaxial axis, as well as its specific functions within the group of cells that surround it. However, the identity of these positional cues and the signaling pathways governing these processes are poorly understood.

In the *Arabidopsis thaliana* genome, 610 genes have been annotated as members of the Receptor-Like Kinase (RLK) family including at least 125 Receptor-Like Cytoplasmic Kinases (RLCKs) (Shiu and Bleecker 2001; Shiu et al. 2004). The sheer number of receptor kinases in *Arabidopsis* and other plants such as rice suggests that signaling through receptor proteins plays a fundamental role in plant growth and development. Over the past 15 years, functional analysis of RLKs in model plants has revealed numerous examples of roles in cell fate specification. In this chapter, we will focus on RLK-mediated mechanisms that establish and maintain radial identity, that determine epidermal cell fates and that act in specifying cell fates across layers with different radial identities.

2 Origin of Tissue Fate Begins During Embryogenesis

Cells first begin receiving cues about their positions relative to other cells during embryogenesis. The apical-basal axis becomes evident after the first cell division of the single-celled zygote. The zygote divides asymmetrically to give rise to a smaller apical cell and a larger basal cell (Mansfield and Briarty 1991). Auxin specifies the apical pro-embryo cell, thereby establishing the apical-basal axis (Friml et al. 2003). The apical cell continues to divide and gives rise to an eight-cell embryo proper (octant) with an apical and central domain. The basal cell produces a file of cells called the suspensor, an extra-embryonic structure that provides a physical link to the surrounding maternal tissue. Auxin is transported from the maternal tissue through the suspensor and accumulates in the apical embryo to trigger specification of the apical domain, and later reverses directional flow to specify the basal pole of the root (Friml et al. 2003).

The radial axis, which will provide the basis for much of the adult body form, is initially established after the transition from the octant (8-cell) stage to the dermatogen (16-cell) stage of embryo development, with tangential cell divisions delineating the outer cell layer (protoderm) from the inner cell layers (Laux et al. 2004). The protoderm layer is maintained through periclinal cell divisions along the embryo surface. The two inner cell layers are generated through further tangential cell divisions during the early globular stage, producing the provasculature and ground tissue initials. During the globular to heart stage transition, the apical domain of the embryo transforms from a radial structure to a bilaterally symmetric heart-shaped structure. The nearly invariant cell lineage of *Arabidopsis* has made it a very useful system for studying embryogenesis and many recent analyses have indicated that intercellular signaling by members of the RLK family plays a central role in directing critical stages of embryo development (Mansfield and Briarty 1991; reviewed by Nodine et al. 2011).

2.1 Markers for Epidermal Patterning in Embryos

Cell-type specific markers are very useful for interpreting the phenotypes of mutants that affect cell fate. Expression of *Arabidopsis thaliana MERISTEM LAYER 1* (*AtML1*), a HD-GL2 class homeobox gene of the HD-ZIP family, has been used to track protodermal (cells that will become epidermal) cell identity (Nodine et al. 2007; Takada et al. 2007). *AtML1* expression can intially be detected in both cells of the 2-cell stage embryo via *in situ* hybridization or by analysis of reporter gene fusions to the *AtML1* promoter region. By the dermatogen stage and thereafter, *AtML1* is restricted to cells within the outer cell layer of the embryo (Takada and Jürgens 2007). Although the mechanism for how *AtML1* becomes restricted to the outer cell layer is not yet known, maintenance of protodermal cell identity is now known, at least in part, to be regulated by RLK signaling.

The AtML1 and closely related PROTODERMAL DETERMINING FACTOR 2 (PDF2) transcription factors are both required for epidermal cell fate determination (Abe et al. 2003). In *atml1 pdf2* mutants, the apical domain of the embryo develops abnormally and an outer to inner cell fate transformation occurs in the resulting seedlings, with the epidermal cells of the cotyledons taking on the identity of the internal mesophyll cells. AtML1 and PDF2 both bind to the "L1 box", an 8-base pair cis-regulatory motif found in the promoters of epidermal-specific genes (Abe et al. 2001). *AtML1* and *PDF2* also have L1 box binding sites within their own promoters, suggesting that once the protoderm is specified in the dermatogen stage, its identity is maintained, at least in part, via a positive feedback loop.

3 Maintaining the Protoderm Cell Identity: RLKs Pave the Way

There is currently very little understanding of the mechanisms that establish the identity of the protoderm during the dermatogen stage of embryo development, or the mechanisms that maintain *AtML1* expression in the epidermal layer and lead to its disappearance in the inner cell layers. However, recent evidence indicates that signaling through the RECEPTOR PROTEIN KINASE 1 (RPK1) and RPK2/TOADSTOOL2 (TOAD2) RLKs is required for maintaining protoderm cell identity during the globular stage, and for signaling to the inner radial layers (Nodine et al. 2007). In the *rpk1 toad2* double mutant and in approximately half of *rpk1 toad2*/+ mutants, early embryonic lethality results in a phenotype of brown shrunken seeds. Mutant embryos show extra tangential cell divisions in the protoderm layer, growth cessation at the globular stage and display a mushroom shape dubbed the "Toadstool" phenotype. "Toadstool" embryos show a loss of *AtML1* expression at the early globular stage, although *AtML1* is initially expressed and correctly restricted to the protoderm during the dermatogen stage (Nodine et al. 2007). These data demonstrate that RLK signaling by the redundant RLKs RPK1 and TOAD2 is necessary to maintain protoderm cell identity. It is possible that RLKs are also required for specifying the protoderm when it is first formed at the dermatogen stage. However, the specific RLKs that might be candidates for signaling between the protoderm and the inner layer are not known.

3.1 Maintaining the Epidermis Requires a Network of RLKs

The epidermal cell layer needs to be properly specified for normal plant development. Recent reports suggest a common mechanism for the maintenance of the epidermis in adult tissues and the maintenance of the embryonic protoderm. CRINKLY LEAF-4 (CR4) was the first RLK reported to function in epidermal development; maize *cr4* mutants have defects in the adult leaf epidermis and also in the epidermal-like aleurone layer of the seed endosperm (Becraft et al. 1996). Five *CR4*-related genes are encoded in the *Arabidopsis* genome with the closest homolog designated *Arabidopsis CR4* (*ACR4*) (Tanaka et al. 2002). *acr4* mutants were shown to have similar epidermal defects in adult leaves and ovule integuments as seen in the maize *cr4* mutant, including crinkly leaf surfaces, fusions between organs, abnormal outer layer morphology of ovules and altered cuticle development (Becraft et al. 2006; Watanabe et al. 2004). Since *ACR4* is also expressed during embryo development, primarily in the protoderm, it is possible that ACR4 acts during additional stages of embryo development to maintain epidermal identity. While the arrested growth of *acr4* mutant embryos is likely due to their integument defects, embryos expressing *ACR* anti-sense RNA showed severe abnormal embryonic morphological defects, possibly due to knocking down of the expression of other closely related ACR4-like genes (Tanaka et al. 2002). These data suggest that *ACR4* may act redundantly with other *ACR*-related genes in the maintenance of protoderm cell fate. *ACR4* may act early in embyonic development to maintain the protoderm either in the *RPK1/TOAD2* pathway or in an alternate pathway that is also necessary to maintain this tissue layer.

Genetic analysis has revealed that the ABNORMAL LEAF SHAPE 2 (ALE2) RLK acts with ACR4 during embryo development (Tanaka et al. 2007). *ale2* mutant plants show similar adult defects as those seen in the maize *cr4* and Arabidopsis *acr4* mutants including crinkly leaves, defective cuticle and abnormal ovule development (Becraft et al. 1996; Tanaka et al. 2002, 2007; Watanabe et al. 2004). Genetic and biochemical analyses of ACR4 and ALE2 have shown that they physically interact and function in the same pathway to maintain protoderm identity during the early stages of embryo development. Double *acr4 ale2* mutants resemble *ale2* mutants, indicating along with biochemical interaction that these RLKs likely act in the same pathway. However, genetic analysis of *ale2* or *acr4* mutants with mutations in ALE1, a secreted protease, shows earlier defects in embryo development that are not apparent in *acr4* or *ale2* single mutants. Double mutants of *ale1* with *acr4* or *ale2* are defective in protoderm development, as seen by morphological defects including embryo lethality at heart stage embryos as well as by loss of protodermal markers, including expression of *AtML1*, *PDF1* and *PDF2* as early as the heart stage. Since ALE1 and either ALE2 or ACR4 are required for embryo growth by the early heart stage transition, maintenance of protodermal identity at early globular and heart transition stages embryos may be essential for coordinating growth of the entire embryo.

As described above, RPK1, TOAD2, ALE2 and ACR4 are required for the maintenance of epidermal identity during early and intermediate stages of embryogenesis, and ALE2 and ACR4 are also required for maintaining epidermal identity during adult stages of *Arabidopsis* development. There is evidence that two additional closely related and redundantly-acting LRR RLKs, GASSHO1 (GSO1) and GASSHO2 (GSO2) are required for epidermal identity by the late heart stage of embryogenesis and during early seedling development. The *gso1 gso2* double mutants display morphological defects during embryogenesis, including adherence of the embryo to the inner seed integument and abnormal cell divisions in early seedling stages, resulting in seedling lethality soon after germination (Tsuwamoto et al. 2008). Although the identity of the epidermal cell layer has not been analyzed for these mutants, it is apparent that their epidermal development is impaired. *gso1 gso2* mutant seedlings show abnormal permeability to the dye Toluidine Blue, likely due to defects in the formation of the cuticle, a waxy, protective covering produced by plant epidermal cells. The lethality of *gso1 gso2* seedlings is probably due to the dehydration caused by defects in epidermal cell differentiation.

These studies indicate that a large number of receptor kinases (RPK1, TOAD2, ALE2, ACR4, GSO1 and GSO2) are required for the maintenance of epidermal cell fate during embryonic and adult development. There may well be many more RLKs that play important roles (Nodine et al. 2011). Whether the maintenance of epidermal cell fate involves the activation of a series of independent ligand/receptor pathways, or sequential physical interactions between these receptor kinases remains to be uncovered.

3.2 Signaling Between Layers and Domains in Embryonic Development

From the early stages of embryogenesis, cell layers must be able to communicate with each other to coordinate plant growth. If coordinated growth relies on numerous conversations between cell layers, it is predicted that developmental disruption of one cell layer could have non-cell autonomous consequences on the development of its neighboring cell layers. As discussed in Chapter "Receptor Kinases in Plant Meristem Development", signaling through RLK-mediated pathways from the outer L1 layer is required to maintain the inner stem cell population of the shoot apical meristem (SAM). A number of experimental observations have demonstrated that disruption of the protodermal layer affects the development of inner cell types during many stages of development. Chemically induced disruption of the protoderm in globular embryos resulted in improper division of the hypophysial cell and its progeny (Baroux et al. 2001; Weijers et al. 2003). *rpk1 toad2* double mutant embryos, in addition to their defects in epidermal identity, also show a lack of patterning of inner cell types (Nodine et al. 2007). The expanded expression of markers specific to the inner cell layers (*SHORTROOT* (*SHR*) for provascular cell identity, *SCARECROW* (*SCR*) for ground tissue initials) in *rpk1 toad2* embryos suggest that the fates of the inner cell layers are affected. These results suggest that protodermal identity, mediated by RPK1/TOAD2 signaling, inhibits inner cell identitites in the protoderm and in the ground tissues. One model suggests that signals from the protoderm, the outer radial layer, are perceived by receptors in the inner layers (Nodine et al. 2007).

3.3 BR Perception in the Epidermal Layer Is Required to Signal Growth in Underlying Cell Layers

Analysis of the regulation of growth in adult plants suggest that signals from the epidermis control the growth of inner layers. The plant steroid hormone Brassinosteroid (BR) is known to promote growth through stimulation of cell expansion and cell division, and also plays a role in vascular cell fates (Caño-Delgado et al. 2004). Loss of production or perception of BR results in reduced growth and a dwarf phenotype (Li and Chory 1997; Nakaya et al. 2002; Choe et al. 1999). BR binds the RLK BRASSINOSTEROID INSENSITIVE 1 (BRI1) (Kinoshita et al. 2005). When *BRI1* expression is restricted to the epidermal layer by driving expression using the *AtML1* promoter, the dwarf phenotype of *bri1* mutants was rescued (Savaldi-Goldstein et al. 2007). However, the vascular bundle organization defect of *bri1* mutants was not rescued, suggesting BR perception is still required for specification of inner cell types. In contrast, when BRI1 or the production of steroids was driven by inner layer promoters, there was significantly less rescue of the dwarf phenotype. These data led to a model in which BRI1-mediated perception and signaling in the epidermis signal to inner cell types in order to coordinate the growth of the entire plant.

Brassinosteroids also regulate meristem activity and growth of root apices (Hacham et al. 2011; González-García et al. 2011). Treatment of plants with brassinolide, the most active BR, also affected the identity of the quiescent center (QC) in the root apical meristem (González-García et al. 2011). In these experiments, several markers for QC identity showed an expanded expression domain, while other markers for QC identity showed reduced expression in the presence of brassinolide or the enhanced BR signaling mutant *bes1-D*. The effects of BR on meristem size and activity can be attributed to the perception of BR solely in the epidermal layer. By expressing BRI1 in specific radial layers using the *GL2*, *SCR* and *SHR* promoters, BR signaling in the outer layer (GL2 promoter) was shown to have the most significant effect on regulating meristem size and QC identity (Hacham et al. 2011). These data, combined with known effects of BR perception in the SAM, suggest perception of BR in the outer layer regulates plant growth through potential feedback mechanisms to the meristem cells, providing a mechanism to coordinately regulate the overall size of the plant.

4 Position-Dependent RLK Signaling Pathways Regulate Root Epidermal Cell Differentiation

After the identity of the epidermal cell layer is established and successfully maintained, cells within the epidermis differentiate into specialized cell types. While the aerial epidermis is patterned into trichomes, guard cells and pavement cells, root epidermal cells differentiate into hair (root hair) and non-hair cells. Root hair cell patterning is one of the best studied systems for understanding plant cell differentiation, and the transcriptional networks that control this process have been well characterized (reviewed in Ishida et al. 2008).

In *Arabidopsis* roots, epidermal cell differentiation is regulated by a position-dependent mechanism in which epidermal cells that overlay two cortex cells become hair cells (H-position), and epidermal cells that overlay a single cortex cell become non-hair cells (N-position) (Dolan et al. 1993; Galway et al. 1994). The first inroads into the mechanisms that produce H- and N-cells came from the analysis of mutants with fewer or additional root hairs. Mutants with extra root hairs were isolated in the *GLABRA2* (*GL2*) gene, encoding a homeodomain-leucine zipper transcription factor, and *WEREWOLF* (*WER*), encoding an R2R3 Myb transcription factor (Di Cristina et al. 1996; Masucci et al. 1996; Lee and

Schiefelbein 1999). *GL2*, which is specifically expressed in non-hair cells, and *WER*, whose transcripts preferentially accumulate in non-hair cells, both act to prevent hair cell formation in the non-hair cell position (reviewed in Schiefelbein et al. 2009). CAPRICE (CPC), an R3 Myb protein lacking a transcriptional activation domain, is expressed in non-hair cells, and CPC is thought to move to hair cells where it negatively regulates WER (Wada et al. 1997; Lee and Schiefelbein 2002; Kurata et al. 2005). Both positive and negative feedback loops based on interactions of these and other transcription factors further refine the alternating patterning of H- and N-cells laterally within the epidermal layer. Lateral movement of factors, including CPC, has been demonstrated and contributes to the feedback mechanisms in the lateral inhibiton of cell fates (Wada et al. 2002; Bernhardt et al. 2005). Although epidermal root hair cell differentiation is clearly based on the orientation of epidermal cells with respect to their neighboring cortex cell layer, how this positional information was perceived until recently remained a mystery.

A series of recent studies have determined that the positional-dependent regulation of root hair cell differentiation is influenced, at least in part, by SCRAMBLED (SCM), an LRR RLK. *scm* was identified in a screen for mutants that disrupted the expression pattern of *GL2* in epidermal cells; mutations in *SCM* caused ectopic expression of *GL2* in hair cells and a decrease in the frequency of N-cells expressing *GL2* (Fig. 1) (Kwak et al. 2005). The expression patterns of several other root hair/non-root hair cell identity genes including *CPC*, *WER* and *ENHANCER of GLABRA 3* (*EGL3*) were also affected in the *scm* mutant (Kwak et al. 2005). It is worth noting that although WER GFP and GUS markers show

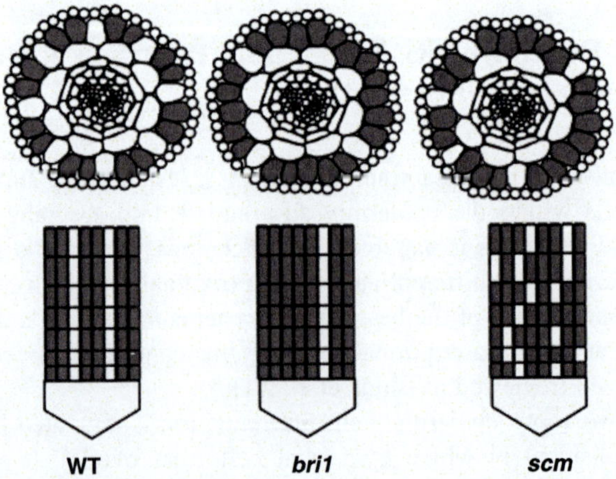

Fig. 1 Alteration of epidermal cell patterning in *bri1* and *scm* roots. Diagram depicting the cell fates of non-hair cells in the roots of *bri1* and *scm* mutants compared to wild-type. *Grey shading* represents the non-root hair cell fate of epidermal cells with the cross-section view shown above and longitudinal surface representation shown below. *bri1* mutants exhibit a transformation of hair cells to a non-hair cell fate as shown by epidermal cells that overlay two cortical cells displaying a non-hair cell fate. *scm* mutants are named for a "scrambled" epidermal epidermal cell fate with both non-hair cells and hair cells mis-expressing their molecular cell fate

specific accumulation in non-hair cells, in situ hybridization of the *WER* transcript shows some accumulation in hair cells (Lee and Schiefelbein 1999). The conclusion from these studies is that SCM likely regulates epidermal cell fate through its negative regulation of *WER* in cells in the hair cell position.

Based on these and other studies, there are two clear alternative models for the action of SCM. The first model is that SCM is required in the epidermal cell layer to perceive a signal necessary from underlying cortical cells to determine non-root hair cell fate. Support for this model comes from in situ hybridization and GUS reporter experiments that show that *SCM* is expressed in the epidermal layer and that a 4 kilobase (kb) *SCM* promoter region drives expression of GUS in all radial layers of the root (Kwak et al. 2005). Furthermore, an SCM-GFP fusion protein under the control of the same promoter region is sufficient to rescue *scm* mutant defects (Kwak and Schiefelbein 2008). Expression of an SCM-GFP fusion protein driven by the promoter of *GL3*, a gene expressed in H-cells, rescues the disrupted *GL2* expression pattern in *scm* roots, whereas altered *GL2* expression is still observed when SCM-GFP is driven by inner cell-specific promoters such as *SCR* and *SHR* (Kwak and Schiefelbein 2008). In this model, SCM activation would result from inside outside signals emanating from inner layers.

An alternative model is that SCM, also known as STRUBBELIG (SUB), functions in inner cell layers and is responsible for generating an inner to outer signal. This is the model supported by independent studies of SUB (Chevalier et al. 2005; Yadav et al. 2008). SUB also has roles in stem, floral and ovule development; for example, the outer integument fails to enclose the ovule leading to infertility. A 3.5 kb promoter region of *SUB* and 0.2 kb of its 3′-untranslated region was found to drive expression of GUS fused to a cyclin destruction box (CDB-GUS transcripts will be degraded after mitosis) in all cell layers of ovules and roots, although the expression in the root epidermal layer was somewhat variable (Yadav et al. 2008). However, a SUB-enhanced GFP fusion protein driven by the same regulatory regions was only found in the inner cells of the ovule and in the root vasculature. These results suggest that SCM/SUB is regulated at the level of translation. To test the layer-specific regulation of SUB, its expression was driven using layer-specific promoters. The results were somewhat inconsistent with the previously described model: SUB was determined to primarily function in inner cell layers to coordinate growth of the outer integument (Yadav et al. 2008). However, surprisingly, expression of *SUB* using the epidermal *AtML1* promoter rescued defects in ovules, flowers and stems, although root epidermal patterning was not tested. With the exception of this last result, these experiments lead to the conclusion that SCM/SUB can act non-cell autonomously, and that activation of SCM in inner cells leads to signals that regulate outer fates.

Both of these models emphasize an important role for signaling from inner cell layers to outer layers. However, establishing whether SCM/SUB is at the initiating or receiving end of this signaling mechanism needs to be determined. Both groups tested different *SCM/SUB* regulatory sequences. The transcriptional fusion using the CDB was intended to remove problems associated with excess stability of GUS, but these sequences may have had other effects on SCM/SUB transcript stability. In addition, the genetic backgrounds in which these experiments were carried out differed significantly. The SUB experiments were conducted in the Landsberg *erecta* accession, which contains a weak mutation in the *ER* receptor kinase gene (see below). This receptor functions in the epidermis and interacts with other signaling pathways. Additional experiments that rely on the same constructs in the same genetic backgrounds are necessary to determine the precise role of SCM/SUB in signaling between layers. It is also possible that both models are correct: that SCM/SUB functions both in the epidermis and in signaling from inside layers to outwards. Identifying other components of the SCM/SUB pathway and perhaps parallel pathways will ultimately be useful in understanding how epidermal patterning is established and maintained.

4.1 BRs and BRI1 Signaling Regulate Cell Fate Decisions in Epidermal Root Cell Differentiation

We previously described the role of BRs in regulating plant growth through their function in the protodermal or epidermal layer. BRs also have a role in cell fate specification of root epidermal cells. Global transcription analysis of whole seedlings has revealed that *WER* gene expression is induced upon treatment with BRs (Nemhauser et al. 2004). The BR signaling pathway is one of the best characterized signaling pathways in *Arabidopsis*; BR is perceived by the BRI1, BRL1 and BRL3 LRR-RLKs (see Chapter "Receptor Trafficking in Plants" by Beck and Robatzek). By analyzing the expression of the root hair cell specific gene *CPC* and *GL2*, the effects of BR on root non-hair cell specification were shown to require BRI1 (Fig. 1) (Kuppusamy et al. 2009). This report showed that BRs and the BRI1 receptor act to suppress non-hair cell (N) formation in the hair cell (H) position. *bri1* mutant seedlings carrying a transgene of the *GL2* promoter fused to the β-glucuronidase reporter gene (*GL2::GUS)* show ectopic expression of *GUS* in hair cells. Furthermore, *CPC* expression was reduced in a *bri1* background consistent with the role of CPC in inhibiting *GL2* expression. These results provide strong evidence for BRs and BRI1 in regulating the cell fate of epidermal cells. Further experiments will be necessary to determine if BRL1 and BRL3 act redundantly with BRI1 in this process.

5 Postembryonic Radial Patterning Through Cell Fate Determination

De novo patterning of developing plant organs requires precise temporal coordination of cell proliferation and differentiation to allow a complex assemblage of cell fates. Cell–cell communication has a major role in cell fate specification, and receptor-mediated signaling has recently been implicated as the mechanism controlling spatiotemporal cell-type determination in numerous organs. Here we review the importance of RLKs in cell signaling by highlighting recent discoveries of their involvement in *Arabidopsis* anther and abscission zone cell fate specification.

5.1 *Stamen Specification During Flower Development*

As the male reproductive organs of all angiosperm flowers, stamens have two morphologically and functionally distinct parts: the anther, which contains the microsporangia (or pollen sacs), the site of male gametophyte development, and the filament, a supportive stalk-like structure (Sanders et al. 1999). In eudicots, stamens are the third whorl of floral organs produced by the floral meristem. Our fundamental understanding of stamen specification comes from seminal studies of homeotic mutants conducted more than 20 years ago that led to the ABC model of floral organ identity (reviewed in Coen and Meyerowitz 1991). Stamen specification was found to require the combined function of both B (APETALA3 and PISTILLATA) and C (AGAMOUS) class MADS-domain transcription factors. Much of the research that followed this discovery has focused on characterizing the contributions of the *SEPALLATA* genes to B and C class function, identifying the factors acting downstream of the homeotic transcription factors, and discovering the signaling mechanisms involved in patterning specific cell types within the anther (reviewed in Scott et al. 2004; Ma 2005; Kaufmann et al. 2009; Ge et al. 2010; Ma and Sundaresan 2010).

5.2 *Morphological Features of Anther Development*

Based on specific morphological, cellular and molecular events, anther development has been divided into 14 stages (Sanders et al. 1999). During the early phase (stages 1–10) of anther development, cell division and differentiation lead to the formation of five concentric layers of cells, the epidermis, endothecium, middle layer, tapetum and microsporocytes (see Fig. 2). These cell lineages can be traced back to the L1, L2 and L3 "histological layers" of the floral meristem. The outermost layer of the anther, the epidermis, forms from the L1, the four inner cell layers, including the somatic (endothecium, middle layer and tapetum) and sporogenic (microsporocytes) cells, originate from the L2 and the connective and vascular tissues are derived from the L3 (Sanders et al. 1999). Interestingly, the four L2-derived layers are formed from single cells after a finite number of cell divisions and differentiation of the cell lineages generated without the involvement of a canonical population of meristematic cells (see Chapter "Receptor kinases in plant meristem development"). During the late phase of anther development (stages 10–14), growth of the specified cell layers continues, the middle layer and tapetum

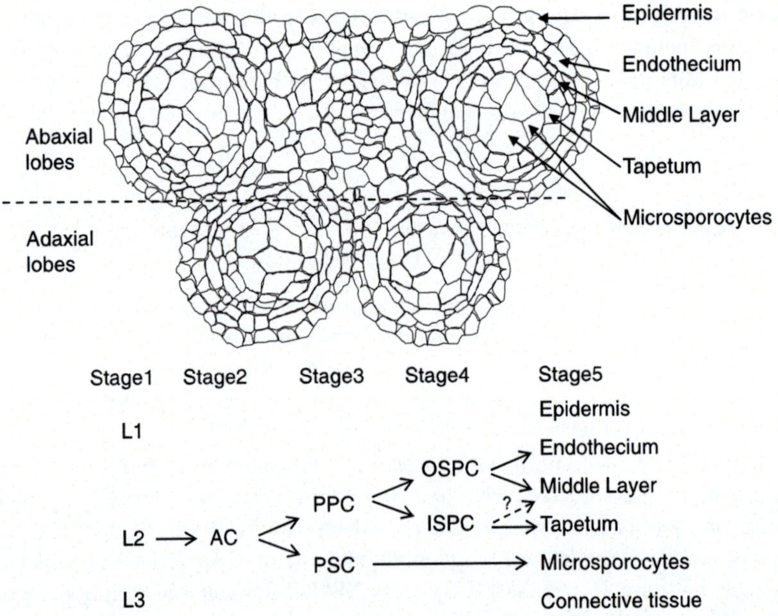

Fig. 2 Structure and cell layer differentiation in Arabidopsis anthers. The cell lineages that produce the five concentric cell layers of the anther visible at stage 5 (**a**) can be traced back to the histological layers of the floral meristem (**b**). RLKs regulate various steps in patterning the anther: BARELY ANY MERISTEM1 (BAM1) and BAM2 affect formation of inner somatic cell layers; EXCESS MICROSPOROCYTES1/EXTRA SPOROGENOUS CELLS (EMS1/EXS) and SOMATIC EMBRYOGENESIS RECEPTOR-LIKE KINASE1 (SERK1) and SERK2 regulate sporogenous cells proliferation and tapetum formation; ERECTA family members (ER and ERL1,2) control cell proliferation at multiple stages and TOADSTOOL2 (TOAD2) mediates the specification of middle layer and degeneration of tapetum. *L2* Histological L2 layer, *AC* archesporial cell, *PPC* primary parietal cell, *PSC* primary sporogenous cell, *OSPC* outer secondary parietal cells, *ISPC* inner secondary parietal cells. Figure adapted from Zhao (2009)

are degraded, and certain cell types (septum and stomium) within the endothecium and epidermis undergo programmed cell death to facilitate pollen release.

Arabidopsis anthers have two pairs of adaxial and abaxial microsporangia, creating a four-lobed structure. The microsporangium produced within each lobe originate from an archesporial cell (AC) in the L2 layer (Fig. 2). Each of the AC cells divides periclinally to form a primary parietal cell (PPC) subjacent to the L1, and a primary sporogenous cell (PSC), which then undergoes proliferation and differentiation to generate the microsporocytes in the center of the anther locules. The PPC continues to divide forming the outer and inner secondary parietal cells (OSPC, ISPC); two of the inner somatic cell layers become the endothecium and middle layer originate from the OSPC, while the third somatic cell layer, the tapetum, is derived from the ISPC. Whether the ISPC also contributes to the middle layer is still an open debate (Sanders et al. 1999; Sorensen et al. 2002; Yang et al. 1999; Zhao et al. 2002; Ma 2005).

The mechanism of successive periclinal divisions of cells derived from an AC cell cannot itself account for the radial symmetry of each microsporangium (Nanda and Gupta 1978; Goldberg et al. 1993). An interesting hypothesis has been put forth (Scott et al. 2004) proposing that the PSC has an essential role in establishing a radial signaling "field" that affects the adjacent PPC cell and its cell lineages that surround the PSC-derived microsporocytes. Although the hypothetical signal(s) generated by the PSC and its descendants is unknown, a good candidate is a small signaling peptide produced by the microsporocytes, TAPETUM DETERMINANT1 (TPD1). TPD1 is the putative ligand for the EXCESS MICROSPOROCYTES1/ EXTRA SPOROGENOUS CELLS (EMS1/EXS) LRR-RLK (Yang et al. 2003).

5.3 *Roles of RLKs in Patterning the Anther*

Because pollen needs to be produced and released at the right time in order for successful reproduction to occur, anther development is a critical stage in a plant's life cycle. The identification of six LRR-RLKs affecting anther cell differentiation supports the concept that cell-cell signaling has a major role in patterning the anther (Table 1) (Canales et al. 2002; Zhao et al. 2002; Albrecht et al. 2005; Colcombet et al. 2005; Hord et al. 2006; Mizuno et al. 2007).

Two LRR RLKs, BARELY ANY MERISTEM1 (BAM1) and BAM2, closely related to CLAVATA1 (CLV1), affect one of the earliest events in anther cell fate specification: the formation of somatic cell layers. While the *bam1* and *bam2* single mutants do not have readily observable phenotypes, *bam1 bam2* double mutants have abnormally reduced meristem size, leaf defects, male sterility and reduced female fertility (DeYoung et al. 2006). The anthers of *bam1 bam2* mutants lack the inner somatic cell layers (tapetum, middle layer, endothecium) and produce only microsporocytes that fail to complete meiosis and develop into pollen grains (Hord et al. 2006). Analysis of *bam1 bam2* mutant phenotypes including the expression of molecular markers has revealed that BAM1/2 are required not only to specify the somatic cell layers but also to repress proliferation of sporogenic cells (Hord et al. 2006). The *BAM1/2* genes are regulated by SPOROCYTELESS/NOZZLE (SPL/ NZZ), an AG-regulated transcription factor that acts very early during anther development (Wijeratne et al. 2007). BAM1/2 and SPL function in a feedback loop: BAM1/2 activity restricts the expression of *SPL* to the innermost cells of the anther locules (the sporogenous cells), and SPL positively regulates the expression of *BAM1/2* (Hord et al. 2006). This signaling may create a regulatory feedback loop similar to that produced by CLV1-WUS signaling in shoot meristems (Hord et al. 2006). However, evidence that SPL acts directly in the BAM1/2 pathway is still lacking, as genetic interaction and mutant analyses have not yet been published. In other contexts, genetic analysis has been used to elucidate the interactions of BAM receptors with CLV pathway components during stem cell and floral organ specification (DeYoung and Clark 2008).

Table 1 RLK mutants that affect anther patterning

Mutant	Expression pattern of corresponding genes		Anther mutant phenotypes			
	Before meiosis (stages 1–6)	After meiosis (stages 6–10)	Microspores	Tapetum (T); Middle cell layer (MCL)	Pollen	Anther dehiscence
bam1 bam2	Sporogenic cells (stages 2–4)	Sporogenic cells and tapetum (stages 6–7)	Not formed, few cells enter meiosis	T-absent, (endothecium also absent)	Not formed	NR
ems1 OR *tpd1*	Sporogenic and somatic cells	Sporogenic cells and tapetum	Excess number	T-absent (pre-tapetal cells present)	Not formed, meiocytes degenerate	NR
serk1 serk2	Sporogenic and somatic cells	Tapetum and middle layer	Excess number	T-absent (pre-tapetal cells present); MCL- highly vacuolated	Not formed, meiocytes degenerate	NR
toad2	Not detected	Tapetum	NR	T-hypertrophic, non-degenerating after meiosis; MCL- absent	Trinucleated, viable, aggregated, non-germinating	Indehiscent
er erl1 erl2	All anther primordia cells	NR	Fewer, in 1–2 locules per anther	T-hypertrophic, disorganized, cells attached	Fewer,viable, adherent to anther wall	Indehiscent

NR-not reported

The first LRR-RLK found to play an important role in microsporangial patterning was EMS1 (Canales et al. 2002; Zhao et al. 2002). *ems1* anthers do not produce tapetal cells and instead produce an increased number of microsporocytes that undergo meiosis but do not develop into mature pollen. The *EMS1* transcript is detected broadly during early development in the inflorescence and flower meristem but is restricted to the microsporangia and tapetal cells at the time of meiosis. Interestingly, mutations in *TPD1*, which encodes a small secreted protein produced by the microsporocytes, lead to a nearly identical phenotype as those of *ems1* mutants. TPD1 has been found to interact with the EMS1 RLK in vitro and in vivo (Yang et al. 2005; Jia et al. 2008) and probably serves as a directional signal and a molecular "checkpoint" confirming the presence of properly developed microsporocytes (Feng and Dickinson 2010).

Although the exact mechanism of tapetal cell fate specification is still under debate, it has been proposed that EMS1 controls the number of cells that become AC initials and therefore the total number of microsporocytes produced (Canales et al. 2002). Another proposed mechanism is that in the absence of *EMS1* transcript, cells resulting from the division of ISPC take on the fate of microsporocytes instead of differentiating into tapetum, which implies a role for EMS1 in tapetal cell fate maintenance (Zhao et al. 2002). More recently, it was shown that when *ems1* mutant plants are engineered to ectopically express *EMS1* after the start of meiosis and the formation of the anther wall layers (which is later in development than it would normally be expressed), a small number of cells at the periphery of the anther locules start proliferating and form a functional tapetal layer; consequently, a reduced number of microsporocytes is observed (Feng and Dickinson 2010). This rescue of tapetum function implies that the peripheral cells are specified early as "pre-tapetal" cells but fail to proliferate and become fully committed to a tapetal fate in the absence of EMS1. The "tug of war" between tapetal and microsporangial cells over the locular space in the anther is also affected by EMS1, as in its absence the microsporangial cells occupy the space of tapetal cells and push their unproliferating precursors to the periphery of the locules. These results suggests that EMS1 mediates cell–cell communication required for tapetal cell differentiation from their pre-tapetal precursors, and restrict microsporocyte proliferation.

Mutations in two closely related LRR RLKs, *SOMATIC EMBRYOGENESIS RECEPTOR-LIKE KINASE1* (*SERK1*) and *SERK2*, also result in production of extra sporogenous cells and absence of the tapetal cell layer (Colcombet et al. 2005; Albrecht et al. 2005). *serk1 serk2* mutants are male sterile, with fast-degrading meiocytes and a highly vacuolated middle cell layer. SERK1/2 proteins are primarily found in the plasma membrane of the tapetum and middle cell layer at the end of meiosis II but are also present in most microspores and the parietal layers at earlier stages (Colcombet et al. 2005; Albrecht et al. 2005). *SERK1/2* transcripts are detected in the surrounding tapetal layer rather than in the meiocytes at the stage of their degradation in the *serk1 serk2* mutant. This led to the hypothesis that SERK1/2 signaling between the tapetum and sporocytes is required for proper cell fate specification of both layers. As with the defects in *ems1* and *tpd1* anthers, the failure to specify tapetal cells in the *serk1 serk2* double mutant results in a block in pollen development.

Mutations in another LRR RLK *TOADSTOOL2* (*TOAD2*), also known as *RECEPTOR PROTEIN KINASE2* (*RPK2*), result in a male sterile phenotype with anthers displaying only three parietal layers, the epidermis, endothecium and tapetum. In addition to a missing middle layer and defects in pollen maturation and germination, the tapetal layer of *toad2* mutant anthers does not degenerate after meiosis. *TOAD2* transcripts accumulate in the tapetal layer for a short developmental window, between stages 7 and 10 (Mizuno et al. 2007). Unlike the *serk1 serk2*, *ems1*, and *tpd* mutants described above, the *toad2* mutant is able to produce trinucleated pollen, albeit at a delayed rate and in an asynchronous manner. The effect of *toad2* mutations on the number of microsporocytes produced has not been reported; so it is not known whether increased proliferation of this cell type occurs. Another phenotypic difference from the *ems1* and *serk1 serk2* mutants is that tapetal cells appear to be present in the *toad2* mutant, and show abnormal enlargement (hypertrophy) and are not degraded after pollen maturation as in wild type anthers. Since the identity of the enlarged cells has not yet been confirmed with molecular markers, it can be concluded that a third cell layer containing highly vacuolated cells is present. Assuming that these cells retain a tapetal cell fate, TOAD2 might be required to receive a signal to restrict the growth of tapetal cells and promote their degradation upon pollen maturation. Consistent with the hypothesis that a functional tapetum is required for normal pollen development, the pollen in these mutants displays abnormal aggregation, delayed maturation and release, and germination defects.

Although the LRR RLKs of the ERECTA (ER) family, ER, ERECTA-LIKE1 (ERL1) and ERL2, are best known for their functions in specifying stomatal cell fates, the *er erl1 erl2* triple mutant was also found to be sterile due to defects in anther development and pollen release (Shpak et al. 2004; Hord et al. 2008). The anthers of the *er erl1 erl2* mutant show a range of patterning defects, from a complete absence of somatic and sporogenic tissue to the presence of enlarged, disorganized cells and delays in middle layer and tapetum development (Hord et al. 2008). Although the *er erl1 erl2* mutants are able to form viable pollen at a reduced rate, anther indehiscence blocks its release. This broad range of developmental defects suggests that the ER family genes may regulate the links between anther cell proliferation and differentiation.

In summary, a set of RLKs has been identified that enable communication of the complex information required for patterning the anther. Some receptor kinases, such as ER, ERL1 and ERL2, affect cell proliferation in multiple cell layers, while others, such as EMS1, SERK1/2 and TOAD2 affect proliferation within the tapetum and microsporocytes. Formation and maintenance of specific cell types are regulated by receptor kinase signaling. BAM1 and BAM2 are required for the somatic layers, TOAD2 controls the middle layer and tapetum, SERK1/2 and EMS1 are required to specify the tapetum and microsporocytes and ER, ERL1 and ERL2 function primarily in the tapetum, middle layer and microsporocytes (Fig. 2). Table 1, it remains to be determined whether the receptors that specify particular radial cell layers interact, and what regulatory mechanisms allow receptors to function in specifying multiple anther cell types.

6 Coordinated Cell Fate Decisions During Organ Abscission and RLK Signaling

Sophisticated mechanisms must be in place to allow cell fate decisions to occur in a synchronized fashion across multiple cell layers. In plants, coordinated decision making is exemplified by the process of cell separation within distinct zones of cells at organ boundaries. In *Arabidopsis* flowers, separation or abscission zones are composed of a few cell layers spanning the bases of the sepals, petals and stamens that allow for their shedding when these organs are no longer required to protect and pollinate the central pistil, respectively. Patterning of floral organ boundaries is a key prerequisite for abscission zone differentiation (McKim et al. 2008). Converging developmental and hormonal pathways are thought to tightly control the initiation of cell separation in abscission zones, leading to the release of cell-wall modifying and hydrolytic enzymes that loosen the cell walls, dissolve the pectin-rich middle lamella between adjacent cells and deposit substances to protect the newly exposed cell wall surfaces (reviewed in Roberts et al. 2000; Leslie et al. 2007; Binder and Patterson 2009; Swain et al. 2011).

Activation of organ abscission in the outer floral organs has been demonstrated to be dependent on signaling pathways controlled by the functionally redundant HAESA (HAE) and HAESA-LIKE2 (HSL2) LRR RLKs (Jinn et al. 2000; Cho et al. 2008). HAE and HSL2 act upstream of a Mitogen-Activated Protein Kinase (MAPK) cascade including MAPK Kinase 4 (MKK4) and MKK5, and MAPK 6 (MPK6) and MPK3 (Cho et al. 2008). One of the ligands for HAE/HSL2 is proposed to be a small, secreted peptide called INFLORESCENCE DEFICIENT IN ABSCISSION (IDA) (Butenko et al. 2003; Cho et al. 2008; Stenvik et al. 2008). In the *hae hsl2* double mutant and the *ida* single mutant, differentiation of the organ abscission zones appears to occur normally but cell separation is blocked. Ectopic expression of IDA causes visibly expanded collars of abscission zone tissue to form at the base of *35S::IDA* flowers and can also result in premature organ separation (Stenvik et al. 2006). An elegant set of genetic experiments strongly supports a linear sequence of the IDA, HAE/HSL2, MKK4/MKK5 and MPK3/MPK6 activities. Organ abscission is blocked and enlarged abscission zones are not found in *35S::IDA* flowers with mutations in *HAE* and *HSL2* (Cho et al. 2008; Stenvik et al. 2008), indicating that IDA signaling is dependent on downstream HAE/HSL2 activity. Flowers with mutations in either *IDA* or *HAE* and *HSL2* show reduced activity of MPK6, suggesting that it acts downstream (Cho et al. 2008). Furthermore, the presence of constitutively active MKK4 or MKK5 mutants driven by a steroid-inducible promoter is sufficient to rescue abscission in *ida* or *hae hsl2* mutant flowers (Ren et al. 2002; Cho et al. 2008). The mechanism of how HAE/HSL2 activates this downstream MAPK cascade and whether intermediate MAPK kinase kinase(s) are involved remains to be determined.

Recently another pair of LRR-RLKs, EVERSHED (EVR) and SERK1, was found to modulate the extent and timing of organ separation, potentially by regulating the activity or localization of HAE and HSL2 (Leslie et al. 2010;

Lewis et al. 2010). Mutations in the *EVR* and *SERK1* genes were identified through a suppressor screen for mutants that restored organ abscission in *nevershed (nev)* mutant flowers. As in the *hae hsl2* and *ida* mutants, floral organ abscission is blocked in the *nev* mutant. Mutations in *NEV* also appear to block the terminal decision of abscission zone cells to undergo cell separation, rather than the earlier decision of cells at the bases of the outer floral organs to differentiate as abscission zone cells (Liljegren et al. 2009). NEV encodes an ADP-ribosylation factor-GTPase-activating protein (ARF GAP) whose activity is likely required to promote trafficking of key components essential for cell separation.

In addition to restoring organ abscission, mutations in the *EVR* or *SERK1* LRR RLK genes were found to partially rescue the membrane trafficking defects of *nev* mutant flowers (Leslie et al. 2010; Lewis et al. 2010). At the time of organ shedding, multi-lamellar circularized structures are present in the sepal abscission zone regions of *nev* mutant flowers instead of the flat stacks of Golgi cisternae and the transiently associated vesicular structures of the *trans*-Golgi network (TGN) observed in wild-type *Arabidopsis* cells (Liljegren et al. 2009). The unusual structures found in *nev* cells closely resemble those of concanamycin-induced chimeric Golgi-TGN fusions (Viotti et al. 2010). Clusters of extracellular vesicles, which are not frequently observed in wild-type cells, are also found between the plasma membrane and cell wall of *nev* cells (Liljegren et al. 2009). In *nev evr* and *nev serk1* abscission zone regions, the Golgi and TGN were found to have a normal appearance (Leslie et al. 2010; Lewis et al. 2010). The partial rescue of *nev*-associated trafficking defects by mutations in either *EVR* or *SERK1* suggests that disruption of traffic flow through the TGN/early endosome is likely responsible for blocking organ separation in *nev* flowers. Candidate molecules affected by trafficking defects in *nev* flowers include the HAE and HSL2 LRR RLKs known to activate organ abscission (Fig. 3).

EVR was found to act as a temporal inhibitor of abscission, as floral organs are shed about a day earlier in *nev evr* plants as compared to wild type (Leslie et al. 2010). Interestingly, the abscission zones regions of *nev evr* and *nev serk1* flowers are significantly enlarged like those of *35S::IDA* flowers (Stenvik et al. 2006; Leslie et al. 2010; Lewis et al. 2010), suggesting that EVR and SERK1 spatially restrict the domains of cells in the abscission zone regions that undergo cell separation as well as the extent of abscission zone cell expansion. These results also suggest that mis-regulated HAE/HSL2 RLK signaling in *nev evr* and *nev serk1* flowers is capable of activating cell separation and directing ectopic cell expansion in the absence of NEV activity. The current model proposed is that EVR and SERK1 may inhibit organ abscission by interacting with HAE/HSL2 at the plasma membrane and promoting internalization of receptor complexes prior to ligand binding (Fig. 3). Whether EVR, SERK1 and other receptor-like kinases redundantly or cooperatively regulate organ abscission remains an unanswered question.

Although the genetic analysis of EVR and SERK1 conducted to date is consistent with the hypothesis that EVR and SERK1 may interact with and inhibit HAE/HSL2 activity, the relationship between NEV, IDA and HAE/HSL2 remains unresolved (Leslie et al. 2010; Lewis et al. 2010). One of the complexities in this

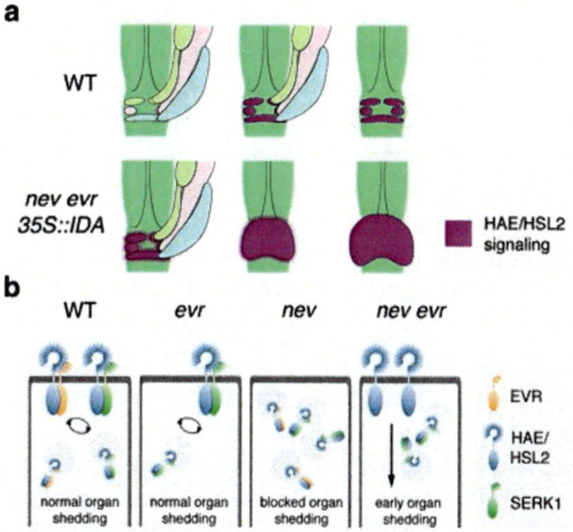

Fig. 3 Regulation of organ separation by RLK signaling and membrane traffic. (**a**) Ectopic signaling by the HAESA (HAE) and HAESA-like2 (HSL2) LRR-RLKs may be responsible for the rescue of organ abscission, premature organ shedding and enlarged abscission zone regions of *nevershed evershed (nev evr)* flowers. Flowers constitutively expressing INFLORESCENCE DEFICIENT IN ABSCISSION (IDA), the predicted peptide ligand of HAE/HSL2, also have enlarged abscission zones and early organ separation compared to wild-type (wt) flowers. Figure adapted from Lewis et al. (2010). (**b**) The EVR and SOMATIC EMBYROGENESIS RECEPTOR-LIKE KINASE1 (SERK1) LRR-RLKs may inhibit organ shedding by promoting internalization of HAE and HSL2 from the cell surface. In wt flowers, EVR and SERK1 may control the pools of available HAE/HSL2 receptors at the plasma membrane. Organ abscission is not affected in either *evr* or *serk1* single mutants, raising the possibility that EVR and SERK1 act redundantly with each other or with unknown RLKs to control HAE/HSL2 localization. In *nev* flowers, recycling of HAE and HSL2 receptor complexes to the cell surface may be blocked, preventing organ separation. Disruption of EVR function in *nev* flowers may reduce HAE/HSL2 internalization, and thereby lead to ectopic signaling of HAE/HSL2 and rescue of organ separation. Figure adapted from Leslie et al. (2010)

analysis has been the discovery that although organ abscission is blocked in the *ida serk1*, *hae hsl2 serk1* and *nev ida* mutants, floral organs are shed in *nev ida serk1* flowers (Lewis et al. 2010). These results suggest that NEV and SERK1 function downstream of IDA, and either act upstream of HAE/HSL2, downstream of HAE/ HSL2, or in a parallel pathway to modulate organ separation. A critical test of these possibilities will be to generate the *nev serk1 hae hsl2* and *nev evr hae hsl2* quadruple mutants. One would predict that if the rescue of organ abscission and the ectopic expansion of abscission zone cells in *nev evr* and *nev serk1* flowers is due to mis-regulated HAE/HSL2 activity, organ abscission should be blocked when this activity is removed in the quadruple mutants.

An intriguing idea derived from analyzing *35S::IDA*, *nev evr* and *nev serk1* flowers is that ectopic, prolonged signaling of the HAE/HSL2 LRR RLKs may be sufficient to induce cell separation in the gynophore, the stem-like structure between the pistil and receptacle which includes the abscission zones and nectary glands.

Although the accepted theory in the field has been that abscission zone differentiation was a prerequisite for acquiring competence to undergo cell separation in *Arabidopsis* flowers, an increased area of the gynophore is occupied by cells with the appearance of abscission zone cells in *35S::IDA*, *nev evr* and *nev serk1* flowers after organ detachment (Bleecker and Patterson 1997; Stenvik et al. 2006; Leslie et al. 2007, 2010; Lewis et al. 2010). Although the nectaries can still be detected after abscission, as fruit development progresses they appear to be either enveloped by expanding and proliferating abscission zone cells or subsumed by continued waves of cell separation signaling. If analysis of *35S::IDA*, *nev evr* and *nev serk1* flower sections supports the idea that cell separation can indeed occur throughout the epidermal cell layers of the gynophore, it will also be important to test whether hydrolytic enzymes associated with cell separation are ectopically active.

In wild-type flowers, modulation of HAE/HSL2 RLK activity by positive and negative regulators may directly control both the timing of cell separation and the spatial domains of cells that undergo separation as their terminal cell fate (Fig. 3). Notably, two of the proposed regulators of HAE and HSL2, EVR and SERK1, are also RLKs. Recently, a third candidate, CAST AWAY (CST), has emerged from the same genetic screen used to identify EVR and SERK1 (Lewis et al. 2010). Disruption of *CST*, which encodes an RLCK associated with the plasma membrane through N-terminal myristoylation and palmitoylation, was found to restore organ abscission and rescue the subcellular defects of *nev* mutant flowers (Burr et al. 2011). Like EVR and SERK1, CST is thought to regulate the spatial extent of cell separation by modulating HAE/HSL2 activity; *nev cst* flowers also exhibit enlarged abscission zone regions. Physical interactions between a subset of these receptor kinases were tested in *Arabidopsis* protoplasts using the split-YFP assay (Burr et al. 2011). CST was found to uniformly interact with EVR and to interact with HAE in distinct subdomains of the plasma membrane. These studies suggest a sequential model in which CST may facilitate interactions between the EVR and HAE/HSL2 RLKs at the cell surface prior to HAE/HSL2 internalization (Burr et al. 2011). Determining how the activity and localization of HAE/HSL2 can potentially be cooperatively regulated by a set of receptor kinases to restrict cell separation will be an important step in understanding how populations of plant receptor kinases can be controlled prior to ligand activation.

7 Conclusions

Networks of receptor kinases play important roles in many key cell fate decisions in plants. These receptor kinases can share overlapping functions with each other, physically interact, regulate different stages of a developmental process and act as positive or negative regulators of the same process. Given the large number of as yet uncharacterized receptor kinases, we predict that complex, sophisticated networks participate in the signaling events required for cell fate specification.

References

Abe M, Takahashi T, Komeda Y (2001) Identification of a cis-regulatory element for L1 layer-specific gene expression, which is targeted by an L1-specific homeodomain protein. Plant J 26:487–494

Abe M, Katsumata H, Komeda Y, Takahashi T (2003) Regulation of shoot epidermal cell differentiation by a pair of homeodomain proteins in *Arabidopsis*. Development 130:635–643

Albrecht C, Russinova E, Hecht V, Baaijens E, de Vries S (2005) The *Arabidopsis thaliana* SOMATIC EMBRYOGENESIS RECEPTOR-LIKE KINASES1 and 2 control male sporogenesis. Plant Cell 17:3337–3349

Baroux C, Blanvillain R, Moore IR, Gallois P (2001) Transactivation of *BARNASE* under the *AtLTP1* promoter affects the basal pole of the embryo and shoot development of the adult plant in *Arabidopsis*. Plant J 28:503–515

Becraft PW, Stinard PS, McCarty DR (1996) CRINKLY4: a TNFR-like receptor kinase involved in maize epidermal differentiation. Science 273:1406–1409

Bernhardt C, Zhao M, Gonzalez A, Lloyd A, Schiefelbein J (2005) The bHLH genes GL3 and EGL3 participate in an intercellular regulatory circuit that controls cell patterning in the *Arabidopsis* root epidermis. Development 132:291–298

Binder BM, Patterson SE (2009) Ethylene-dependent and -independent regulation of abscission. Stewart Postharv Rev 5:1–10

Bleecker AB, Patterson SE (1997) Last exit: senescence, abscission, and meristem arrest in Arabidopsis. Plant Cell 9:1169–1179

Burr CB, Leslie ME, Orlowski SK, Chen I, Wright CE, Daniels MJ, Liljegren SJ (2011) CAST AWAY, a membrane-associated receptor kinase, inhibits organ abscission in Arabidopsis. Plant Physiol 156:1837–1850

Butenko MA, Patterson SE, Grini PE, Stenvik GE, Amundsen SS, Mandal A, Aalen RB (2003) *INFLORESCENCE DEFICIENT IN ABSCISSION* controls floral organ abscission in Arabidopsis and identifies a novel family of putative ligands in plants. Plant Cell 15:2296–2307

Canales C, Bhatt AM, Scott R, Dickinson H (2002) EXS, a putative LRR receptor kinase, regulates male germline cell number and tapetal identity and promotes seed development in Arabidopsis. Curr Biol 12:1718–1727

Caño-Delgado A, Yin Y, Yu C, Vafeados D, Mora-Garcia S, Cheng J-C, Nam KH, Li J, Chory J (2004) BRL1 and BRL3 are novel brassinosteroid receptors that function in vascular differentiation in Arabidopsis. Development 131:5341–5351

Chevalier D, Batoux M, Fulton L, Pfister K, Yadav RK, Schellenberg M, Schneitz K (2005) STRUBBELIG defines a receptor kinase-mediated signaling pathway regulating organ development in Arabidopsis. Proc Natl Acad Sci USA 102:9074–9079

Cho SK, Larue CT, Chevalier D, Wang H, Jinn T-L, Zhang S, Walker JC (2008) Regulation of floral organ abscission in *Arabidopsis thaliana*. Proc Natl Acad Sci USA 105:15629–15634

Choe S, Noguchi T, Fujioka S, Takatsuto S, Tissier CP, Gregory BD, Ross AS, Tanaka A, Yoshida S, Tax FE, Feldmann KA (1999) The Arabidopsis *dwf7/ste1* mutant is defective in the Δ^7 sterol C-5 desaturation step leading to brassinosteroid biosynthesis. Plant Cell 11:207–222

Clouse SD, Sasse JM (1998) Brassinosteroids: essential regulators of plant growth and development. Annu Rev Plant Physiol Plant Mol Biol 49:427–451

Coen ES, Meyerowitz EM (1991) The war of the whorls: genetic interactions controlling flower development. Nature 353:31–37

Colcombet J, Boisson-Dernier A, Ros-Palau R, Vera CE, Schroeder JI (2005) *Arabidopsis* SOMATIC EMBRYOGENESIS RECEPTOR KINASES 1 and 2 are essential for tapetum development and microspore maturation. Plant Cell 17:3350–3361

DeYoung BJ, Bickle KL, Schrage KJ, Muskett P, Patel K, Clark SE (2006) The CLAVATA1-related BAM1, BAM2 and BAM3 receptor kinase-like proteins are required for meristem function in Arabidopsis. Plant J 45:1–16

DeYoung BJ, Clark SE (2008) BAM receptors regulate stem cell specification and organ development through complex interactions with CLAVATA signaling. Genetics 180:895–904

Di Cristina M, Sessa G, Dolan L, Linstead P, Baima S, Ruberti I, Morelli G (1996) The Arabidopsis Athb-10 (GLABRA2) is an HD-Zip protein required for regulation of root hair development. Plant J 10:393–402

Dolan L, Janmaat K, Willemsin V, Linstead P, Poethig S, Roberts K, Sheres B (1993) Cellular organisation of the *Arabidopsis thaliana* root. Development 119:71–84

Feng X, Dickinson HG (2010) Tapetal cell fate, lineage and proliferation in the Arabidopsis anther. Development 137:2409–2416

Friml J, Vieten A, Sauer M, Weijers D, Schwarz H, Hamann T, Offringa R, Jürgens G (2003) Efflux-dependent auxin gradients establish the apical-basal axis of Arabidopsis. Nature 426:147–153

Galway ME, Masucci JD, Lloyd AM, Walbot V, Davis RW, Schiefelbein JW (1994) The TTG gene is required to specify epidermal cell fate and cell patterning in the Arabidopsis root. Dev Biol 166:740–754

Ge X, Chang F, Ma H (2010) Signaling and transcriptional control of reproductive development in Arabidopsis. Curr Biol 20:R988–R997

Goldberg RB, Beals TP, Sanders PM (1993) Anther development: basic principles and practical applications. Plant Cell 5:1217–1229

González-García M-P, Vilarrasa-Blasi J, Zhiponova M, Divol F, Mora-García S, Russinova E, Caño-Delgado AI (2011) Brassinosteroids control meristem size by promoting cell cycle progression in Arabidopsis roots. Development 138:849–859

Hacham Y, Holland N, Butterfield C, Ubeda-Tomas S, Bennett MJ, Chory J, Savaldi-Goldstein S (2011) Brassinosteroid perception in the epidermis controls root meristem size. Development 138:839–848

Hord CLH, Chen C, DeYoung BJ, Clark SE, Ma H (2006) The BAM1/BAM2 receptor-like kinases are important regulators of Arabidopsis early anther development. Plant Cell 18:1667–1680

Hord CLH, Yu-Jin Sun Y-J, Pillitteri LJ, Torii KU, Wang H, Zhang S, Ma H (2008) Regulation of Arabidopsis early anther development by the mitogen-activated protein kinases, MPK3 and MPK6, and the ERECTA and related receptor-like kinases. Mol Plant 1:645–658

Ishida T, Kurata T, Okada K, Wada T (2008) A genetic regulatory network in the development of trichomes and root hairs. Annu Rev Plant Biol 59:365–386

Jia G, Liu X, Owen HA, Zhao D (2008) Signaling of cell fate determination by the TPD1 small protein and EMS1 receptor kinase. Proc Natl Acad Sci USA 105:2220–2225

Jinn TL, Stone JM, Walker JC (2000) *HAESA*, an Arabidopsis leucine-rich repeat receptor kinase, controls floral organ abscission. Genes Dev 14:108–117

Kaufmann K, Muiño JM, Jauregui R, Airoldi CA, Smaczniak C, Krajewski P, Angenent GC (2009) Target genes of the MADS transcription factor SEPALLATA3: integration of developmental and hormonal pathways in the Arabidopsis flower. PLoS Biol 7:e1000090

Kuppusamy KT, Chen AY, Nemhauser JL (2009) Steroids are required for epidermal cell fate establishment in Arabidopsis roots. Proc Natl Acad Sci USA 106:8073–8076

Kinoshita T, Caño-Delgado A, Seto H, Hiranuma S, Fujioka S, Yoshida S, Chory J (2005) Binding of brassinosteroids to the extracellular domain of plant receptor kinase BRI1. Nature 433:167–171

Kurata T, Ishida T, Kawabata-Awai C, Noguchi M, Hattori S, Sano R, Nagasaka R, Tominaga R, Koshino-Kimura Y, Kato T et al (2005) Cell-to-cell movement of the CAPRICE protein in Arabidopsis root epidermal cell differentiation. Development 132:5387–5398

Kwak S-H, Schiefelbein J (2008) A feedback mechanism controlling SCRAMBLED receptor accumulation and cell-type pattern in Arabidopsis. Curr Biol 18:1949–1954

Kwak S-H, Shen R, Schiefelbein J (2005) Positional signaling mediated by a receptor-like kinase in Arabidopsis. Science 307:1111–1113

Laux T, Würschum T, Breuninger H (2004) Genetic regulation of embryonic pattern formation. Plant Cell 16:S190–S202

Lee MM, Schiefelbein J (1999) WEREWOLF, a MYB-related protein in Arabidopsis, is a position-dependent regulator of epidermal cell patterning. Cell 99:473–483

Lee MM, Schiefelbein J (2002) Cell pattern in the Arabidopsis root epidermis determined by lateral inhibition with feedback. Plant Cell 14:611–618

Leslie ME, Lewis ML, Liljegren SJ (2007) Organ abscission. In: Roberts J, Gonzalez-Carranza Z (eds) Plant cell separation and adhesion. Blackwell, Oxford, UK, pp 106–136

Leslie ME, Lewis MW, Youn J-Y, Daniels MJ, Liljegren SJ (2010) The EVERSHED receptor-like kinase modulates floral organ shedding in Arabidopsis. Development 137:467–476

Lewis MW, Leslie ME, Fulcher EH, Darnielle L, Healy PN, Youn J-Y, Liljegren SJ (2010) The SERK1 receptor-like kinase regulates organ separation in Arabidopsis flowers. Plant J 62:817–828

Li J, Chory J (1997) A putative leucine-rich repeat receptor kinase gene involved in brassinosteroid signal transduction. Cell 90:929–938

Liljegren SJ, Leslie ME, Darnielle L, Lewis MW, Taylor SM, Luo R, Geldner N, Chory J, Randazzo PA, Yanofsky MF, Ecker JR (2009) Regulation of membrane trafficking and organ separation by the NEVERSHED ARF-GAP protein. Development 136:1909–1918

Ma H (2005) Molecular genetic analyses of microsporogenesis and microgametogenesis in flowering plants. Annu Rev Plant Biol 56:393–434

Ma H, Sundaresan V (2010) Development of flowering plant gametophytes. Curr Top Dev Biol 91:379–412

Mansfield SG, Briarty LG (1991) Early embryogenesis in *Arabidopsis thaliana*. II. The developing embryo. Can J Bot 69:461–476

Masucci JD, Rerie WG, Foreman DR, Zhang M, Galway ME, Marks MD, Schiefelbein JW (1996) The homeobox gene *GLABRA 2* is required for position-dependent cell differentiation in the root epidermis of *Arabidopsis thaliana*. Development 122:1253–1260

McKim SM, Stenvik G-S, Butenko MA, Kristiansen W, Cho SK, Hepworth SR, Aalen RB, Haughn GW (2008) The *BLADE-ON-PETIOLE* genes are essential for abscission zone formation in Arabidopsis. Development 135:1537–1546

Mizuno S, Osakabe Y, Maruyama K, Ito T, Osakabe K, Sato T, Shinozaki K, Yamaguchi-Shinozaki K (2007) Receptor-like protein kinase 2 (RPK 2) is a novel factor controlling anther development in *Arabidopsis thaliana*. Plant J 50:751–766

Nakaya M, Tsukaya H, Murakami N, Kato M (2002) Brassinosteroids control the proliferation of leaf cells of *Arabidopsis thaliana*. Plant Cell Physiol 43:239–244

Nanda K, Gupta SC (1978) Studies in the Bignoniaceae. II. Ontogeny of the dimorphic anther tapetum in *Tecoma*. Am J Bot 65:400–405

Nemhauser JL, Mockler TC, Chory J (2004) Interdependency of brassinosteroid and auxin signaling in Arabidopsis. PLoS Biol 2:1460–1471

Nodine MD, Yadegari R, Tax FE (2007) *RPK1* and *TOAD2* are two receptor-like kinases redundantly required for Arabidopsis embryonic pattern formation. Dev Cell 12:943–956

Nodine MD, Bryan AC, Racolta A, Jerosky KV, Tax FE (2011) A few standing for many: embryo receptor-like kinases. Trends Plant Sci 16:211–217

Ren D, Yang H, Zhang S (2002) Cell death mediated by MAPK is associated with hydrogen peroxide production in Arabidopsis. J Biol Chem 277:559–565

Roberts JA, Whitelaw CA, Gonzalez-Carranza ZH, McManus MT (2000) Cell separation processes in plants–models, mechanisms and manipulation. Ann Bot 86:223–235

Sanders PM, Bui AQ, Weterings K, McIntire KN, Hsu Y-C, Lee PY, Truong MT, Beals TP, Goldberg RB (1999) Anther developmental defects in *Arabidopsis thaliana* male-sterile mutants. Sex Plant Reprod 11:297–322

Saulsberry A, Martin PR, O'Brien T, Sieburth LE, Pickett FB (2002) The induced sector Arabidopsis apical embryonic fate map. Development 129:3403–3410

Savaldi-Goldstein S, Peto C, Chory J (2007) The epidermis both drives and restricts plant shoot growth. Nature 446:199–202

Schiefelbein J, Kwak S-H, Wieckowski Y, Barron C, Bruex A (2009) The gene regulatory network for root epidermal cell-type pattern formation in Arabidopsis. J Exp Bot 60:1515–1521

Scott RJ, Spielman M, Dickinson HG (2004) Stamen structure and function. Plant Cell 16: S46–S60

Shiu S-H, Bleecker AB (2001) Receptor-like kinases from Arabidopsis form a monophyletic gene family related to animal receptor kinases. Proc Natl Acad Sci USA 98:10763–10768

Shiu SH, Karlowski WM, Pan R, Tzeng YH, Mayer KF, Li WH (2004) Comparative analysis of the receptor-like kinase family in Arabidopsis and rice. Plant Cell 16:1220–1234

Shpak ED, Berthiaume CT, Hill EJ, Torii KU (2004) Synergistic interaction of three ERECTA-family receptor-like kinases controls Arabidopsis organ growth and flower development by promoting cell proliferation. Development 131:1491–1501

Sorensen A, Guerineau F, Canales-Holzeis C, Dickinson HG, Scott RJ (2002) A novel extinction screen in *Arabidopsis thaliana* identifies mutant plants defective in early microsporangial development. Plant J 29:581–594

Stenvik G-E, Butenko MA, Urbanowicz BR, Rose JK, Aalen RB (2006) Overexpression of *INFLORESCENCE DEFICIENT IN ABSCISSION* activates cell separation in vestigial abscission zones in Arabidopsis. Plant Cell 18:1467–1476

Stenvik G-E, Tandstad NM, Guo Y, Shi C-L, Kristiansen W, Holmgren A, Clark SE, Aalen RB, Butenko MA (2008) The EPIP peptide of INFLORESCENCE DEFICIENT IN ABSCISSION is sufficient to induce abscission in Arabidopsis through the receptor-like kinases HAESA and HAESA-LIKE2. Plant Cell 20:1805–1817

Swain S, Kay P, Ogawa M (2011) Preventing unwanted breakups: using polygalacturonases to regulate cell separation. Plant Signal Behav 6:93–97

Takada S, Jürgens G (2007) Transcriptional regulation of epidermal cell fate in the Arabidopsis embryo. Development 134:1141–1150

Tanaka H, Watanabe M, Watanabe D, Tanaka T, Machida C, Machida Y (2002) ACR4, a putative receptor kinase gene of *Arabidopsis thaliana*, that is expressed in the outer cell layers of embryos and plants, is involved in proper embryogenesis. Plant Cell Physiol 43:419–428

Tanaka H, Watanabe M, Sasabe M, Hilroe T, Tanaka T, Tsukaya H, Ikezaki M, Machida C, Machida Y (2007) Novel receptor-like kinase ALE2 controls shoot development by specifying epidermis in Arabidopsis. Development 134:1643–1652

Tsuwamoto R, Fukuoka H, Takahata Y (2008) GASSHO1 and GASSHO2 encoding a putative leucine-rich repeat transmembrane-type receptor kinase are essential for the normal development of the epidermal surface in Arabidopsis embryos. Plant J 54:30–42

van den Berg C, Willemsen V, Hage W, Weisbeek P, Scheres B (1995) Cell fate in the Arabidopsis root meristem determined by directional signalling. Nature 378:62–65

Viotti C, Bubeck J, Stierhof Y-D, Krebs M, Langhans M, van den Berg W, van Dongen W, Richter S, Geldner N, Takano J, Jürgens G, de Vries SC, Robinson DG, Schumacher K (2010) Endocytic and secretory traffic in Arabidopsis merge in the trans-Golgi network/early endosome, an independent and highly dynamic organelle. Plant Cell 22:1344–1357

Wada T, Tachibana T, Shimura Y, Okada K (1997) Epidermal cell differentiation in *Arabidopsis* determined by a Myb homolog, CPC. Science 277:1113–1116

Wada T, Kurata T, Tominaga R, Koshino-Kimura Y, Tachibana T, Goto K, Marks MD, Shimura Y, Okada K (2002) Role of a positive regulator of root hair development, *CAPRICE*, in Arabidopsis root epidermal cell differentiation. Development 129:5409–5419

Watanabe M, Tanaka H, Watanabe D, Machida C, Machida Y (2004) The ACR4 receptor-like kinase is required for surface formation of epidermis-related tissues in *Arabidopsis thaliana*. Plant J 39:298–308

Weijers D, van Hamburg J-P, van Rijn E, Hooykaas PJJ, Offringa R (2003) Diptheria toxin-mediated cell ablation reveals interregional communication during Arabidopsis seed development. Plant Physiol 133:1882–1892

Wijeratne AJ, Zhang W, Sun Y, Liu W, Albert R, Zheng Z, Oppenheimer DG, Zhao D, Ma H (2007) Differential gene expression in Arabidopsis wild-type and mutant anthers: insights into anther cell differentiation and regulatory networks. Plant J 52:14–29

Yadav RK, Fulton L, Batoux M, Schneitz K (2008) The Arabidopsis receptor-like kinase STRUBBELIG mediates inter-cell-layer signaling during floral development. Dev Biol 323:261–270

Yang W-C, Ye D, Xu J, Sundaresan V (1999) The *SPOROCYTELESS* gene of Arabidopsis is required for initiation of sporogenesis and encodes a novel nuclear protein. Genes Dev 13:2108–2117

Yang SL, Xie LF, Mao HZ, Puah CS, Yang WC, Jiang L, Sundaresan V, Ye D (2003) *TAPETUM DETERMINANT1* is required for cell specialization in the Arabidopsis anther. Plant Cell 15:2792–2804

Yang SL, Jiang L, Puah CS, Xie LF, Zhang XQ, Chen LQ, Yang WC, Ye D (2005) Overexpression of *TAPETUM DETERMINANT1* alters the cell fates in the Arabidopsis carpel and tapetum via genetic interaction with *EXCESS MICROSPOROCYTES/EXTRA SPOROGENOUS CELLS*. Plant Physiol 139:186–191

Zhao D-Z, Wang G-F, Speal B, Ma H (2002) The *EXCESS MICROSPOROCYTES1* gene encodes a putative leucine-rich repeat recepter protein kinase that controls somatic and reproductive cell fates in the Arabidopsis anther. Genes Dev 16:2021–2031

Zhao D (2009) Control of anther cell differentiation: a teamwork of receptor-like kinases. Sex Plant Reprod 22:221–228

Experimental Evidence of a Role for RLKs in Innate Immunity

Thomas Boller

Abstract Four lines of experimental evidence point to a major role of RLKs in plant innate immunity. First, several RLKs are located at the plasma membrane and perceive specific "microbe-associated molecular patterns" (MAMPs), such as bacterial flagellin, bacterial EF-Tu or fungal chitin. The high affinity and specificity of these RLKs for their respective ligands, and the absence of endogenous ligands in plants, strongly indicate that these RLKs serve as "pattern recognition receptors" (PRRs) to signal the presence of microbes. Second, mutational loss of individual PRRs can lead to a reduced resistance against pathogens. Third, biotechnological transfer of a PRR from one given plant species to another may lead to increased resistance against pathogens. Fourth, and most importantly, successful pathogens produce effectors that inhibit the PRRs themselves or prevent the signal transduction pathways activated upon stimulation of the PRRs.

1 Introduction

For many decades, plant disease resistance to microbial pathogens was thought to be essentially based on resistance genes, which served as recognition systems for the so-called "avirulence gene products" of pathogens (Keen 1990). Indeed, paradoxically as it may sound, plant pathogenic bacteria were experimentally proved to possess avirulence genes which could be transferred to an aggressive virulent bacterium to render it avirulent on a host possessing the corresponding resistance gene (Staskawicz et al. 1984).

Although it remained unexplainable at this stage why pathogens should possess avirulence genes, it appeared logical that the plant's resistance genes should encode

T. Boller (✉)
Zürich-Basel Plant Science Center, Botanical Institute, University Basel, Hebelstrasse 1, CH-4056 Basel, Switzerland
e-mail: thomas.boller@unibas.ch

F. Tax and B. Kemmerling (eds.), *Receptor-like Kinases in Plants*, Signaling and Communication in Plants 13,
DOI 10.1007/978-3-642-23044-8_4,

receptors for the products of avirulence genes. When the first plant resistance genes were cloned, their similarity to receptors was noticed. However, only one of them, Xa21, conferring resistance of rice to the bacterial pathogen *Xanthomonas oryzae*, encoded an RLK (Song et al. 1995). Two more (Cf9 and Cf2), conferring resistance of tomato to the fungal pathogen *Cladosporium fulvum*, were receptor-like proteins (RLP) without intracellular kinase domains (Jones et al. 1994; Dixon et al. 1996). By contrast, the bulk of resistance genes, cloned in plenty after the initial breakthroughs, including some of the famous flax resistance genes directed against the rust fungus *Melampsora lini* (Ellis et al. 2000), turned out to be intracellular proteins of the NB-LRR class (colloquially called "nibblers"). All of these "nibblers" had nothing to do with cell surface receptors in the style of RLKs or RLPs but appeared to be localized in the cytoplasm, and to recognize a pathogen attack from within the plant cell (Dangl and Jones 2001).

At about this time, it became apparent that plants have surface receptors of the RLK class which do not detect avirulence factors of specific pathogens, but microbe-associated molecular patterns (MAMPs) in general, such as the flagellin receptor FLS2 of *Arabidopsis thaliana*, which perceives a highly conserved domain of bacterial flagellin (Gómez-Gómez and Boller 2000; Asai et al. 2002). Shortly thereafter, a mammalian innate immunity receptor, TLR5, was also found to perceive bacterial flagellin (Hayashi et al. 2001). In the plant community, this led to a renaissance of the concept that plants have perception systems on the cell surface to detect approaching microbes through "elicitors" or "MAMPs" (Mackey and McFall 2006), and that this is a first line of defense against unwanted intruders (reviewed in Boller and Felix 2009; Zipfel and Robatzek 2010). Experimental evidence of a function for RLKs perceiving MAMPs was provided when it was shown that Arabidopsis mutants lacking the FLS2 receptor were more susceptible to disease caused by *Pseudomonas syringae* DC3000 (Zipfel et al. 2004).

Soon after, it was shown that the constitutive expression of a single bacterial "avirulence gene", *AvrPto*, in transgenic *A. thaliana* plants suppressed "normal" defense responses and rendered the transgenic plant more susceptible to attack by the bacterial pathogen (Hauck et al. 2003). Thus, one of the roles of "avirulence genes", which was counterintuitive for such a long time, now turned out to consist of the suppression of the "normal" defense responses: Avirulence gene products such as AvrPto acted as "effectors" to suppress the response to elicitors.

On the basis of these and other findings, it was proposed, in three simultaneous seminal reviews (Abramovitch et al. 2006; Chisholm et al. 2006; Jones and Dangl 2006), that there are two lines of defense in plants against pathogens, namely PAMP-triggered immunity (PTI) and effector-triggered immunity (ETI). Note that in this review, PTI is spelled out as "pattern-triggered immunity", as the pattern triggering the response is clearly not a "pathogen-associated molecular pattern" (PAMP) but rather a MAMP, see Mackey and McFall 2006; Boller and He 2009).

According to the PTI/ETI theory, the first line of defense is the perception of microbes through a surveillance system consisting of PRRs at the plasma membrane, targeted to detect MAMPs. This first line of defense is sufficient to ward off most of the unwanted microbes. However, successful pathogens have evolved effectors, which they "inject" into the plant cells to suppress recognition of the MAMPs or to manipulate the ensuing signal transduction by other means. This second line of defense is based on the resistance genes described earlier, the "nibblers" which render the plant cell aware of these effectors.

Here, I highlight experimental evidence that RLKs of the PRR type indeed form the plants' first line of defense that this first line of defense is effective to ward off microbes of all kinds, and that successful pathogens need to breach this first line of defense through their effectors, the products of the time-honored avirulence genes (Abramovitch et al. 2006; Chisholm et al. 2006; Jones and Dangl 2006). It is these effectors that are recognized by the plants' second line of defense, the "nibblers" encoded by the resistance genes (Collier and Moffett 2009).

2 MAMP Perception by PRRs Shows a High Degree of Selectivity and Specificity

In early work on plant–pathogen interactions, it was found that plants have highly sensitive chemoperception systems for characteristic microbial molecules, which acted as "elicitors" to induce the synthesis of phytoalexins and other defense responses (reviews: Darvill and Albersheim 1984; Boller 1995). Already in this early work, it was apparent that these chemoperception systems were highly specific and selective, and that they were targeted to microbe-associated molecular patterns distinct from the plant's own molecules. For example, the first "elicitor" to be chemically characterized and synthesized, a molecule derived from cell walls of the pathogen *Phytophthora megasperma* inducing defense responses in soybean, corresponded to a heptaglucan with a β-1-6 backbone typical of oomycete cell walls (Darvill and Albersheim 1984). Similarly, the perception system of tomato cells for chitin, a typical constituent of cell walls of fungi and of the exosceleton of insects, responded to chitin oligomers of four or more *N*-acetylglucosamine units at subnanomolar levels but did not respond to *N*-acetylglucosamine dimers, which may occur in the plant itself (Felix et al. 1993, reviewed recently by Hamel and Beaudoin 2011).

This specificity and selectivity became more apparent when the first peptide "elicitors" were investigated (see also Albert et al. 2010; Müller and Felix 2011). The peptide "pep13", derived from an extracellular protein of the oomycete *P. megasperma*, was active at nanomolar levels in parsley (Nürnberger et al. 1994), the peptide "flg22", derived from bacterial flagellin, induced defense responses at nanomolar levels in various plants (Felix et al. 1999), the peptide "cps13" derived from cold-shock protein of bacteria (Felix and Boller 2003) was an elicitor at nanomolar levels in tobacco, and the peptide "elf18", derived from the bacterial elongation factor EF-Tu, induced defense responses at nanomolar levels in Arabidopsis (Kunze et al. 2004). In all these cases, the plant perception system targeted a highly conserved domain of the microbial proteins in question that was completely absent or evolutionarily diverged in the plant itself.

All these findings strongly indicate that perception of characteristic "non-self" MAMPs play an important role in the plant's capability to recognize and monitor microbial ingression (Boller 1995; Segonzac and Zipfel 2011). They also give an indication of how precisely and selectively evolution shapes these perception systems to target the most conserved, invariant domains of the microbial proteins in question (Boller and Felix 2009; Segonzac and Zipfel 2011). However, they do not provide any experimental proof that these perception systems play a role in actual plant pathogen interactions. In the case of flagellin and EF-Tu perception, molecular identification of the PRRs involved was crucial for delivering such an experimental proof, as discussed in the next two subchapters. Similarly, the identification, cloning and manipulation of PRRs for other MAMPs will help in our understanding of "pattern recognition", the first line of defense of plants, as indicated in a recent review (Lehti-Shiu et al. 2009).

3 Mutation of Plant PRRs Can Lead to Enhanced Disease Susceptibility

Genetics provide a cornerstone for the experimental proof of a biological function of a protein in general. A loss-of-function mutation should disable the biological function in question. In the case of plant resistance genes, this was easy and evident; loss of resistance to a pathogen carrying an avirulence gene was almost invariably as a result of a mutation of the corresponding resistance gene (Dangl and Jones 2001). This was also true for *Xa21*, one of the first plant "resistance genes" to be cloned in the mid-nineties (Song et al. 1995). Ironically, *Xa21* is now no longer seen as a classic "resistance gene" but more as a pattern recognition receptor for a MAMP (namely, for a bacterial quorum-sensing factor of *Xanthomonas* bacteria, Park et al. 2010), and thus qualifies as an element of PTI, the first line of plant defense (Ronald and Beutler 2010).

By contrast, the role of receptors for MAMPs remained enigmatic. When we found, also in the mid-nineties, that plants perceive a highly conserved part of bacterial flagellin, the flg22 domain, our discovery met with considerable resistance. Reviewers considered our results irrelevant as long as we did not demonstrate that flagellin perception was important in defense against bacterial pathogens. Nevertheless, we were finally able to publish our findings (Felix et al. 1999). Subsequently, we cloned *FLS2*, the *Arabidopsis* gene responsible for flg22 perception, which turned out to encode a typical RLK. Trying to publish our findings was again difficult because we did not present evidence for a role of FLS2 in defense against bacterial pathogens, but we finally succeeded in the year 2000

(Gómez-Gómez and Boller 2000). It may be worth noticing that a year later, scientists of the medical field cloned the animal receptor for flagellin, encoded by *TLR5*; this work was published without any evidence for a functional role of the *TLR5* gene (Hayashi et al. 2001).

Obviously, having cloned FLS2, it became possible for us to test its role in plant–pathogen interactions using a genetic approach. Surprisingly, *Arabidopsis fls2* mutants did not show any phenotype toward infection by *P. syringae* DC3000 when tested by the commonly used infiltration technique (Zipfel et al. 2004). However, they showed a strongly enhanced susceptibility when subjected to spray inoculation (Zipfel et al. 2004). This work showed, for the first time, that the loss of a single PRR can lead to enhanced disease susceptibility, or in other words, that PRRs have a role in disease resistance. Interestingly, the reason why *Arabidopsis* mutants lacking FLS2 are more susceptible to the pathogen *P. syringae* DC3000 appears to have to do with stomatal function. Wild type *Arabidopsis* plants close their stomata rapidly when treated with flg22 (Melotto et al. 2006, 2008).

Cloning of the *Arabidopsis* gene *EFR*, encoding the receptor for EF-Tu, an RLK related to FLS2, enabled a similar approach to demonstrating the biological role of this receptor (Zipfel et al. 2006). The experimental results of this work showed that *efr* mutant plants were more susceptible to *Agrobacterium tumefaciens*, because they displayed a much higher transformation rate when inoculated with *A. tumefaciens* carrying a β-glucuronidase (GUS) construct, a finding of considerable potential in biotechnological applications (Zipfel et al. 2006).

It should be mentioned that the enhanced susceptibility of *fls2* mutant plants toward spray infection with *P. syringae* DC3000 was verified not only in the work of several other scientists (e.g., Melotto et al. 2006), but also in various different experimental setups in our laboratory, both in practical courses and in dedicated experimental approaches. Although the experimental finding of an enhanced susceptibility of *fls2* plants upon spray infection with *P. syringae* DC3000 was extremely robust, even in the relatively uncontrolled settings of practical courses, both the *fls2* plants and the *efr* plants did not show a significant increase in susceptibility toward different bacteria in several other assays (unpublished results). It is a problem of scientific progress that "negative results" are usually not published; thus, the importance of the RLKs may be overrated. On the other hand, redundancy in the "first line of defense" may partly explain the lack of effect of elimination of a single receptor. Another explanation may be that some well-adapted pathogens "camouflage" their MAMPs. Specifically, well-adapted pathogens of *Arabidopsis* and other members of the family of the Brassicaceae may avoid detection by FLS2 through mutations in the highly conserved flg22 epitope (Pfund et al. 2004; Sun et al. 2006; Boller and Felix 2009). In the field of animal innate immunity, it has proved very difficult, too, to demonstrate an enhanced susceptibility of an organism lacking a single innate immunity receptor such as TLR5 (review: Akira and Takeda 2004), in part because bacterial pathogens "camouflage" their flagellin as well (Andersen-Nissen et al. 2005).

4 Transgenic Expression of PRRs May Lead to Enhanced Disease Resistance

The *Arabidopsis* cultivar Ws-0 has been considered particularly sensitive to bacterial disease. Interestingly, this natural cultivar is blind to flagellin (Gómez-Gómez et al. 1999), because of a mutation in the FLS2 gene, as it turned out subsequently (Zipfel et al. 2004). Therefore, an attempt was made to complement this defect by genetic transformation. Indeed, the Ws-0 ecotype, transformed with a construct of FLS2 with a myc-tag, expressed under the native FLS2 promoter, showed clearly enhanced resistance to *P. syringae* DC3000 (Zipfel et al. 2004).

Flagellin perception through FLS2 and its orthologs is a property that is apparently shared by most seed plants (Boller and Felix 2009). Therefore, expression of *At*FLS2 in other plants is not expected to contribute to resistance against bacteria. Unexpectedly, it even proved to be difficult to express FLS2 orthologs from other plants, e.g., from tomato (*Solanum lycopersicum*), to demonstrate functional complementation in *Arabidopsis fls2* mutants (Robatzek et al. 2007).

By contrast, perception of EF-Tu appears to be restricted to the Brassicaceae (Kunze et al. 2004). In an initial work on the receptor for EF-Tu, it was therefore possible to introduce the RLK identified as being essential for EF-Tu perception, named EFR, into *Nicotiana benthamiana* by transient expression, and to show that this rendered *N. benthamiana* sensitive to the EF-Tu derived peptide elf18 (Zipfel et al. 2006). These results also demonstrated that while the signal perception capability of EFR is specific to *Arabidopsis*, the signal transduction brought into motion by ligand–receptor interaction is evolutionarily conserved. This is in agreement with the finding that both FLS2 and EFR signal transductions depend on BAK1 in *Arabidopsis* (Chinchilla et al. 2007; Schulze et al. 2010) as well as in *Nicotiana* (Heese et al. 2007).

In subsequent work, *N. benthamiana* was stably transformed with the Arabidopsis EFR gene and tested for disease susceptibility (Lacombe et al. 2010). The transformants became more resistant to tumor formation by *A. tumefaciens*; this result was somewhat expected because the *efr* mutant in *Arabidopsis* proved to be more susceptible to transformation by *A. tumefaciens* (Zipfel et al. 2006). What was unexpected, however, was the fact that the transformed *N. benthamiana* plants were much more resistant to two *P. syringae* pathovars, *P. syringae* pv. *syringae* B728a, and *P. syringae* pv. *tabaci* 11528, both of which cause severe damage on *N. benthamiana* (Lacombe et al. 2010). Furthermore, transgenic expression of EFR rendered the tomato cultivar "Moneymaker", which was crippled by the bacterial pathogen *Ralstonia solanacearum*, almost completely resistant to this pathogen, and also conferred enhanced resistance to the bacterium *Xanthomonas perforans* (Lacombe et al. 2010).

These experiments not only demonstrate very clearly that RLKs dedicated to the perception of MAMPs can have an important role in disease resistance. They also indicate that transgenic expression of highly specialized RLKs such as EFR, which appears to exist only in the Brassicaceae, may be particularly promising. Pathogens that are successful on a host plant of the *Solanaceae* family must have countered all the defenses of this host plant, including the recognition systems provided by the host plant's RLKs. An entirely new perception capability, such as the perception capability for EF-Tu provided by the transgenic EFR from *Arabidopsis*, may provide a durable resistance against such pathogens.

5 Plant RLKs Are Targeted by Pathogen Effectors

Perhaps the most compelling evidence for a biological role of RLKs in plant defense is the fact that successful pathogens use effectors to target and blindfold them. Typical bacterial plant pathogens inject around 30–40 proteins into the plant cell by way of their type three secretion system (Espinosa and Alfano 2004). It came as a big surprise when it became clear that more than half of these injected proteins functioned in the suppression of PTI (Boller and He 2009). Expression of these bacterial proteins in plants demonstrate beyond doubt that they very strongly suppress some or all of the typical defense responses induced by MAMPs, and thereby inactivate PTI.

Two well-studied cases are the effectors AvrPto and AvrPtoB (which had obtained their names in times of the prevalence of the "avirulence gene" model). These two proteins are encoded in the genome of *P. syringae* DC3000, and although they have almost the same name and almost the same function, they have no similarity at all in their primary sequence, indicating that they have arisen by convergent evolution (Kim et al. 2002; Lin and Martin 2005). Clearly, these two bacterial effectors block signal transduction through RLKs, but there is currently a controversy how this happens (Xiang et al. 2008, 2011; Shan et al. 2008; Boller 2008). Some researchers provided evidence for an interaction of AvrPto with FLS2 (Xiang et al. 2008, 2011), while others showed data indicating an interaction of AvrPto and AvrPtoB with BAK1, a co-receptor for FLS2 and EFR (Shan et al. 2008). Clearly, both AvrPto and AvrPtoB have multiple functions, and the puzzle is not yet solved. However, all available evidence shows that at least AvrPto inhibits RLK-mediated signaling induced by flg22 and elf18 as well as by chitin, but chitin signaling, in contrast to flg22 and elf18 signaling, does not require BAK1 (Heese et al. 2007). Therefore, it appears that AvrPto, in addition to its effects on FLS2 and BAK1, may target a common downstream element of the MAMP-induced signaling cascades (Boller 2008).

Another interesting case is the effector HopAI1 of *P. syringae* DC3000. This effector suppresses MAPK activation, one of the important signaling events downstream of RLK activation, and it has been shown to act as a phosphothreonine lyase that irreversibly dephosphorylates MAPK (Zhang et al. 2007)

In addition to the proteins injected through the type III secretion system, bacteria also dispose of micromolecules that inhibit PTI signaling. For example, as mentioned above, FLS2 activation causes a closure of stomata (Melotto et al. 2006; Zeng and He 2010). Virulent strains of *P. syringae* produce coronatine, a structural mimic of jasmonate–isoleucine conjugate, which counteracts flg22-induced stomatal closure (Melotto et al. 2006, 2008).

6 Conclusions

Plants possess a number of RLKs and RLPs that function as PRRs to perceive MAMPs. These receptors form a first line of defense as they perceive the presence of microbes in their environment through elicitors from the outside of the plant cell (Boller and He 2009; Albert et al. 2010). Typically, they display an extremely high specificity and selectivity for their presumed targets, perceiving "non-self" molecules at nanomolar or even subnanomolar concentrations, while they do not recognize the plants' "self" molecules even if they are chemically or structurally closely related. This high specificity and selectivity are in itself evidence for a biological function of these PRRs in non-self detection.

Experimental evidence for the importance of these PRRs by mutational analysis in actual plant–pathogen interactions is still relatively limited. It appears that in many real plant–pathogen interactions, the lack or presence of a single PRR does not influence the outcome significantly (Boller and Felix 2009; Albert et al. 2010). However, this should probably have been expected because of the redundancy of the system. In innate immunity in the animal field, too, there is rarely any change in the outcome of a pathogen attack when a single toll-like-receptor is eliminated by mutation (Akira and Takeda 2004).

Clean and clear results have recently been obtained by the opposite approach, expressing a PRR specific to the Brassicaceae, EFR, in heterologous systems (Lacombe et al. 2010). The Arabidopsis EFR receptor, when transgenically expressed in *N. benthamiana* or tomato, i.e., in plants that do not recognize bacterial EF, strongly contributes to an enhanced disease resistance. This transgenic approach, which uses plant genes in plants, might be considered as a turning point also in the current political debate about the usefulness of transgenic approaches in plant breeding. The deployment of plant PRRs, shaped by millions of years of evolution, should be inherently more "biological" and "natural" in order to protect crop plants than the spraying of fungicides or bactericides.

In terms of fundamental science, probably the strongest evidence for a role of PRRs in pathogen resistance is the finding that successful pathogens target these PRRs and their signal transduction chains. They do so by injection of proteins, the so-called *avirulence gene products* or, more narrowly defined, the *effectors*, directly into the cytoplasm of the plant cell (Cui et al. 2009). However, these effectors pose a risk to the pathogen as well. The plant has evolved *resistance proteins* which can recognize effectors and mount an efficient defense against pathogenic bacteria (Abramovitch et al. 2006; Chisholm et al. 2006; Jones and Dangl 2006).

In addition to this direct attack on the plant's RLKs and their signaling cascades, the pathogens have another option too. They may camouflage the molecules that can be detected by the plant RLKs effectors, using a sort of "stealth" strategy.

Indeed, work in the animal as well as in the plant field has indicated that some successful pathogens modify their flagellins to avoid detection by the corresponding PRRs (Pfund et al. 2004; Andersen-Nissen et al. 2005; Sun et al. 2006; Boller and Felix 2009).

Overall, current knowledge supports a bipartite immune system of plants, consisting of a first line of active defense elicited by RLKs, which serve as PRRs to perceive the presence of MAMPs, followed by a second line of defense that more specifically recognizes effectors that are produced by well-adapted pathogens to overcome the first line of defense.

References

Abramovitch RB, Anderson JC, Martin GB (2006) Bacterial elicitation and evasion of plant innate immunity. Nature Rev Mol Cell Biol 7:601–611

Akira S, Takeda K (2004) Toll-like receptor signalling. Nat Rev Immunol 4:499–511

Albert M, Jehle AK, Lipschis M, Mueller K, Zeng Y, Felix G (2010) Regulation of cell behaviour by plant receptor kinases: pattern recognition receptors as prototypical models. Eur J Cell Biol 89:200–207

Andersen-Nissen E, Smith KD, Strobe KL, Barrett SLR, Cookson BT, Logan SM, Aderem A (2005) Evasion of Toll-like receptor 5 by flagellated bacteria. Proc Natl Acad Sci USA 102:9247–9252

Asai T, Tena G, Plotnikova J, Willmann MR, Chiu WL, Gómez-Gómez L, Boller T, Ausubel FM, Sheen J (2002) MAP kinase signalling cascade in *Arabidopsis* innate immunity. Nature 415:977–983

Boller T (1995) Chemoperception of microbial signals in plant cells. Ann Rev Plant Physiol Plant Mol Biol 46:189–214

Boller T (2008) Stabbing in the BAK – an original target for avirulence genes of plant pathogens. Cell Host Microbe 4:5–7

Boller T, Felix G (2009) A renaissance of elicitors: perception of microbe-associated molecular patterns and danger signals by pattern-recognition receptors. Annu Rev Plant Biol 60:379–406

Boller T, He SY (2009) Innate immunity in plants: an arms race between pattern recognition receptors in plants and effectors in microbial pathogens. Science 324:742–744

Chinchilla D, Zipfel C, Robatzek S, Kemmerling B, Nürnberger T, Jones JDG, Felix G, Boller T (2007) A flagellin-induced complex of the receptor FLS2 and BAK1 initiates plant defence. Nature 448:497–500

Chisholm ST, Coaker G, Day B, Staskawicz BJ (2006) Host-microbe interactions: shaping the evolution of the plant immune response. Cell 124:803–814

Collier SM, Moffett P (2009) NB-LRRs work a "bait and switch" on pathogens. Trends Plant Sci 14:521–529

Cui HT, Xiang TT, Zhou JM (2009) Plant immunity: a lesson from pathogenic bacterial effector proteins. Cell Microbiol 11:1453–1461

Dangl JL, Jones JDG (2001) Plant pathogens and integrated defence responses to infection. Nature 411:826–833

Darvill AG, Albersheim P (1984) Phytoalexins and their elicitors – a defense against microbial infection in plants. Ann Rev Plant Physiol Plant Mol Biol 35:243–275

Dixon MS, Jones DA, Keddie JS, Thomas CM, Harrison K, Jones JDG (1996) The tomato Cf-2 disease resistance locus comprises two functional genes encoding leucine-rich repeat proteins. Cell 84:451–459

Nürnberger T, Nennstiel D, Jabs T, Sacks WR, Hahlbrock K, Scheel D (1994) High affinity binding of a binding of a fungal oligopeptide elicitor to parsley plasma membranes triggers multiple defense responses. Cell 78:449–460

Park CJ, Han SW, Chen XW, Ronald PC (2010) Elucidation of Xa21-mediated innate immunity. Cell Microbiol 12:1017–1025

Pfund C, Tans-Kersten J, Dunning FM, Alonso JM, Ecker JR, Allen C, Bent AF (2004) Flagellin is not a major defense elicitor in *Ralstonia solanacearum* cells or extracts applied to *Arabidopsis thaliana*. Mol Plant-Microbe Interact 17:696–706

Robatzek S, Bittel P, Chinchilla D, Köchner P, Felix GS, Boller HST (2007) Molecular identification and characterization of the tomato flagellin receptor LeFLS2, an orthologue of Arabidopsis FLS2 exhibiting characteristically different perception specificities. Plant Mol Biol 64:539–547

Ronald PC, Beutler B (2010) Plant and animal sensors of conserved microbial signatures. Science 330:1061–1064

Schulze B, Mentzel T, Jehle AK, Mueller K, Beeler S, Boller T, Felix G, Chinchilla D (2010) Rapid heteromerization and phosphorylation of ligand-activated plant transmembrane receptors and their associated kinase BAK1. J Biol Chem 285:9444–9451

Segonzac C, Zipfel C (2011) Activation of plant pattern-recognition receptors by bacteria. Curr Opin Microbiol 14:54–61

Shan L, He P, Li J, Heese A, Peck SC, Nürnberger T, Martin GB, Sheen J (2008) Bacterial effectors target BAK1 to disrupt MAMP receptor signaling complexes and impede plant innate immunity. Cell Host Microbe 4:17–27

Song WY, Wang GL, Chen LL et al (1995) A receptor kinase-like protein encoded by the rice disease resistance gene, *Xa21*. Science 270:1804–1806

Staskawicz BJ, Dahlbeck D, Keen NT (1984) Cloned avirulence gene of *Pseudomonas syringae* pv. *glycinea* determines race-specific incompatibilty on *Glycine max* (L) Merr. Proc Natl Acad Sci USA 81:6024–6028

Sun WX, Dunning FM, Pfund C, Weingarten R, Bent AF (2006) Within-species flagellin polymorphism in *Xanthomonas campestris* pv. *campestris* and its impact on elicitation of *Arabidopsis FLAGELLIN SENSING2*-dependent defenses. Plant Cell 18:764–779

Xiang T, Zong N, Zou Y et al (2008) *Pseudomonas syringae* effector AvrPto blocks innate immunity by targeting receptor kinases. Curr Biol 18:74–80

Xiang TT, Zong N, Zhang J, Chen JF, Chen MS, Zhou JM (2011) BAK1 Is not a target of the *Pseudomonas syringae* effector AvrPto. Mol Plant-Microbe Interact 24:100–107

Zeng WQ, He SY (2010) A prominent role of the flagellin receptor FLAGELLIN-SENSING2 in mediating stomatal response to *Pseudomonas syringae* pv *tomato* DC3000 in Arabidopsis. Plant Physiol 153:1188–1198

Zhang J, Shao F, Cui H et al (2007) A *Pseudomonas syringae* effector inactivates MAPKs to suppress PAMP-induced immunity in plants. Cell Host Microbe 1:175–185

Zipfel C, Robatzek S (2010) Pathogen-associated molecular pattern-triggered immunity: Veni, vidi . . . ? Plant Physiol 154:551–554

Zipfel C, Robatzek S, Navarro L, Oakeley EJ, Jones JDG, Felix G, Boller T (2004) Bacterial disease resistance in *Arabidopsis* through flagellin perception. Nature 428:764–767

Zipfel C, Kunze G, Chinchilla D, Caniard A, Jones JDG, Boller T, Felix G (2006) Perception of the bacterial PAMP EF-Tu by the receptor EFR restricts *Agrobacterium*-mediated transformation. Cell 125:749–760

Cell-Death Control by Receptor Kinases in *Arabidopsis thaliana*

Jia Li, Junbo Du, Kai He, and Xiaoping Gou

Abstract *Arabidopsis* contains at least 223 leucine-rich repeat receptor-like protein kinases (LRR-RLKs). The somatic embryogenesis receptor kinases (SERKs) form a small subfamily of the LRR-RLK family. The BRI1-associated receptor kinase (BAK1 or SERK3) was identified as a co-receptor of the brassinosteroid (BR) receptor BRI1. Loss-of-function genetic analysis indicates that BAK1 and its closest paralog BKK1 are independently involved in BRI1-mediated BR signaling and cell-death control signaling pathways. More recently, BAK1 has been identified as a co-receptor interacting with a number of distinct ligand-binding RLKs to regulate multiple signaling pathways. This chapter will mainly discuss how BAK1 and BKK1 were identified as key regulators in controlling cell death, and the possible mechanisms involved in this process.

1 Introduction

Receptor-like kinases (RLKs) are transmembrane protein kinases critically involved in a variety of distinct biological processes. A typical RLK contains an extracellular receptor domain perceiving small chemical signals, a single-pass hydrophobic transmembrane domain attaching the protein on the plasma membrane, and a cytoplasmic kinase domain delivering extracellular signals to downstream regulatory components primarily through protein phosphorylation. Based on

J. Li (✉) • X. Gou
School of Life Sciences, Lanzhou University, Lanzhou 730000, People's Republic of China
e-mail: lijia@lzu.edu.cn

J. Du
School of Life Sciences, Sichuan University, Chengdu 610064, People's Republic of China

K. He
Department of Plant and Microbial Biology, University of California at Berkeley, Berkeley, CA 94720, USA

F. Tax and B. Kemmerling (eds.), *Receptor-like Kinases in Plants*,
Signaling and Communication in Plants 13,
DOI 10.1007/978-3-642-23044-8_5,

two-dimensional thin layer electrophoresis/thin layer chromatography (2D-TLE/TLC) and phylogenetic analyses, it was proposed that RLKs are serine/threonine protein kinases (Horn and Walker 1994). However, recent studies by using more sensitive phosphoprotein analysis tools such as proteomics and anti-phosphotyrosine antibodies in immunoblotting assays indicate that some of the RLKs such as the brassinosteroid (BR) receptor BRI1 and co-receptor BAK1 also possess tyrosine kinase activity. Tyrosine phosphorylation is responsible for at least a subset of the signaling events in BR signaling (Clouse 2011; Oh et al. 2009, 2010). The structural characteristics suggest that RLKs play crucial roles in cell-to-cell and cell-to-environmental communications.

In *Arabidopsis*, there are at least 610 RLKs, representing about 2.2% of total proteins encoded by the entire *Arabidopsis* genome, including more than 400 typical transmembrane RLKs and nearly 200 receptor-like cytoplasmic kinases (RLCKs), whose extracellular and/or transmembrane domains have been lost during evolution (*Arabidopsis* Genome Initiative 2000; Shiu and Bleecker 2001). Based on extracellular domain structures, the RLK superfamily can be further classified into 13 families, and leucine-rich repeat RLKs (LRR-RLKs) belong to the largest family containing about 223 members (Gou et al. 2010). Currently, the LRR-RLKs are the best-characterized family among all RLKs. Within the last 2 decades, the biological functions of more than 30 LRR-RLKs have been gradually revealed (Gou et al. 2010). These RLKs are involved in almost all known physiological processes controlling plant growth and development, as well as various defense responses (Albrecht et al. 2005; Chinchilla et al. 2007; Clark et al. 1997; Clay and Nelson 2002; Colcombet et al. 2005; Fisher and Turner 2007; Gao et al. 2009; Godiard et al. 2003; Gomez-Gomez and Boller 2000; He et al. 2007; Heese et al. 2007; Jinn et al. 2000; Kemmerling et al. 2007; Li and Chory 1997; Li et al. 2002; Nam and Li 2002; Nodine et al. 2007; Osakabe et al. 2005; Torii et al. 1996; Tsuwamoto et al. 2008; Zhao et al. 2002).

The *Arabidopsis* somatic embryogenesis receptor kinases (SERKs) belong to subfamily II of the LRR-RLK family, containing only five members, named SERK1 to SERK5 (Hecht et al. 2001). *SERK* was first isolated in *Daucus carota* (carrot). The expression of *DcSERK* was found to be associated with the transition of identities from somatic to embryonic cells in carrot cell culture (Schmidt et al. 1997). All five predicted SERK proteins contain a cleavable N-terminal signal peptide, leucine zippers, five tandemly repeated LRRs, a single-pass transmembrane domain, and a cytoplasmic kinase domain possessing dual serine/threonine and tyrosine kinase activities (Oh et al. 2009, 2010). Among the five Arabidopsis SERKs, at least SERK1 is thought to be involved in embryogenesis based on gene expression pattern analyses and gain-of-function genetic studies (Hecht et al. 2001). However, loss-of-function genetic evidence to confirm that SERKs are indeed essential to embryogenesis in plants is still lacking. The best-characterized SERK protein in *Arabidopsis* is SERK3, also named as BAK1 for *BRI1*-associated receptor kinase (Li et al. 2002; Nam and Li 2002). The role of BAK1 in regulating BR signaling was independently identified by an activation tagging genetic screening for suppressors of a weak *BRI1* mutant allele, *bri1-5*, and by a yeast two-hybrid analysis for BRI1 kinase domain interacting proteins (Li et al. 2002; Nam and Li 2002). Genetic and biochemical studies also revealed that at least SERK1 and SERK4 (also named BKK1 for BAK1-like kinase) play redundant roles with BAK1 in mediating the BR signaling pathway (He et al. 2007; Karlova et al. 2006). A detailed molecular mechanism for BAK1 in mediating BR signaling pathway was recently described (Wang et al. 2008). It was proposed that BAK1and BRI1 follow a reciprocal and sequential phosphorylation mechanism during the early events of BR perception and signal transduction (Wang et al. 2008).

Loss-of-function genetic analysis indicates that BAK1 and BKK1 can regulate BR-dependent plant growth and BR-independent cell-death pathways. Several laboratories found that BAK1 also plays roles in regulating plant defense against various pathogens (Boller and Felix 2009). Considering that several other chapters cover the functions of BAK1 in regulating plant defense responses, this chapter will mainly focus on BAK1 and BKK1 in mediating the control of spontaneous cell-death and early senescence.

2 The *bak1 bkk1* Double Null Mutant Shows a Spontaneous Cell-Death Phenotype

Although BAK1 was identified as a co-receptor of BRI1 in mediating BR signaling, *bak1* null mutants only exhibit a subtle *bri1*-like phenotype, suggesting either the paralogs of BAK1 play functionally redundant roles with BAK1, or that BAK1 and its paralogs are less important than BRI1 in regulating BR signaling. To resolve this puzzle, *SERK4* and *SERK5* were first overexpressed in *bri1-5* to test whether they can suppress *bri1-5* phenotypes, mainly because SERK4 and SERK5 are the two closest paralogs of BAK1 in SERK subfamily. The results of these experiments indicate that *SERK4* but not *SERK5*, from accession Col-0, can suppress *bri1-5* phenotypes when overexpressed (He et al. 2007). SERK4 was therefore renamed BKK1 for BAK1-LIKE1. Sequence analysis of SERK5 from Col-0 revealed that it contains a natural point mutation, which alters one of the amino acids at position 401 within the critical "RD" kinase motif "RD" to "LD". This substitution was not identified in other accessions such as Wassilewskiji 2 (WS2). Site-directed mutagenesis analysis changing "LD" back to "RD" in Col-0 SERK5 can partially rescue the function of SERK5 in BR signaling, as the mutated *SERK5* (L401R) can partially suppress the defective phenotypes of *bri1-5* when overexpressed (He et al. 2007).

To further test the roles of BAK1 and BKK1 in BR signaling, multiple T-DNA insertion lines for both *BAK1* and *BKK1* were isolated from the *Arabidopsis* Biological Resource Center (ABRC) (http://abrc.osu.edu/). All *bak1* null alleles showed a weak *bri1* phenotype. All *bkk1* null alleles, on the contrary, did not show any phenotypic defects. The surprise was that *bak1-4 bkk1-1* double null mutant lines are virtually lethal (Fig. 1, He et al. 2007), which is unexpected

Fig. 1 Phenotypes of *bak1-4 bkk1-1* double mutants. Seeds collected from self-pollinated *bak1-4* (+/−) *bkk1-1* plants were planted on 1/2 MS medium and incubated under 16 h light/ 8 h dark condition with a constant temperature of 22°C. The plants were photographed 3 weeks after germination. Homozygous double mutants showed a seedling lethality phenotype with a segregation ration of approximately 1:3

because all the BR signaling or BR defective mutants show a prolonged life span, opposite of the *bak1-4 bkk1-1* mutant phenotype. For the first 4–5 days after germination, the *bak1-4 bkk1-1* seedlings appeared perfectly healthy. One week after germination, however, the growth of the *bak1 bkk1* seedlings was completely arrested before they finally died. This seedling lethality occurs through cell death and is accompanied by the accumulation of reactive oxygen species (ROS). Overexpression of either *BAK1* or *BKK1* driven by 35S promoter completely complemented the lethal phenotype of the *bak1-4 bkk1-1* double null mutant (He et al. 2007). Most of the complemented plants show prolonged life span and delayed leaf senescence phenotypes compared to wild-type plants. Close examination of the developing embryos produced from plants with genotypes that were homozygous for one gene and heterozygous for the other gene suggested that the defects occurred postembryonically (He et al. 2007). Trypan blue staining was used to test whether cell death is involved in the seedling lethality phenotype (Shirasu et al. 1999). The results clearly indicate that spontaneous cell death is triggered in the double null mutant but not in wild-type, or in various single mutants. Furthermore, using the 3, 3′-diaminobenzidine (DAB) staining technique (Thordal-Christensen et al. 1997), it was demonstrated that ROS (such as H_2O_2) accumulation in the double mutant is independent of the attenuated BR signaling.

Among the isolated *bak1* T-DNA mutants, *bak1-3* is likely a leaky mutant, although it shows a subtle *bri1*-like phenotype indistinguishable to the null mutant *bak1-4*. RT-PCR analysis indicated that reduced amount of full-length *BAK1* cDNA was detected from *bak1-3* but not from *bak1-4*. Sequence analysis indicated that *bak1-3* mutant still can transcribe a small amount of wild type such as *BAK1* cDNA (our unpublished results), which were not detectable in Northern blots (Chinchilla et al. 2007). Consistently, *bak1-3 bkk1-1* also showed an early leaf senescence phenotype and a strongly enhanced symptom development after *Alternaria* infection. These results confirmed that *bak1-3* is a strong but slightly leaky mutant, whereas *bak1-4* is a null mutant.

To further investigate the molecular mechanisms causing cell death in the double mutant, an Affymetrix microarray analysis was conducted to compare the gene expression profiles of wild-type and *bak1-4 bkk1-1* seedlings. Eight-day-old light-grown seedlings were used for the analysis. Genes upregulated in the double null mutant are mainly those related to defense responses and senescence. Representative genes include *PR1*, *PR2*, *PR5*, *ACS2*, and *ACS6* (He et al. 2007). Because the cell death is light dependent and is always accompanied by the accumulation of excessive amount of ROS, we specifically analyzed all known genes associated with ROS scavenging. Nine out of 39 known major *Arabidopsis* ROS-scavenging genes were significantly downregulated (by 2.3-fold to 5.5-fold). Only one of the 39 genes was upregulated. The downregulated genes encode five different classes of enzymes, including one glutathione peroxidase [(GPX1 (At2g25080)], two ascorbate peroxidases [APX4 (At4g09010) and tAPX (At1g77490)], one catalase [(CAT2 (At4g35090)], one superoxide dismutase [CSD3 (At5g18100)], and four peroxiredoxins [(2-cys PrxR A (At 3 g11630), 2-cys PrxR B (At5g06290), PrxR Q (At3g26060), type 2 PrxR E (At3g52960)]. The products of almost all of these genes are localized in the chloroplasts (only CSD3 is localized in peroxisome), which is 42% of the all known 19 major chloroplast-localized ROS-scavenging enzymes (Mittler et al. 2004). Semiquantitative RT-PCR was used to further confirm the microarray data. The RT-PCR results are consistent with those from the microarray analyses.

3 BAK1 and BKK1 Involved in Cell-Death Control Is SA Dependent

Many known cell-death mutants, such as *lsd1*, show salicylic acid (SA)-dependent cell death. To test whether cell-death seen in *bak1-4 bkk1-1* is also SA dependent, the bacterial gene *NahG* was overexpressed in the double mutant plants. *NahG* encodes a salicylate hydroxylase, which converts a bioactive SA molecule to an inactive catechol molecule (Delaney et al. 1994). Our results indicated that *NahG* can partially suppress the defective phenotypes of the double null mutant (He et al. 2007). When *NahG* was overexpressed in *bak1-3 bkk1-1*, it can completely suppress the early senescence phenotype (data not published). These data suggested that the cell death in the double mutant is at least partially SA dependent (He et al. 2007).

4 BAK1 and BKK1 Involved in Cell-Death Control Is Light Dependent

To investigate whether light contributes to the cell death seen in *bak1-4 bkk1-1*, cotyledons from 8-day-old Col-0 and *bak1-4 bkk1-1* seedlings grown in the dark and long-day lighting conditions were stained with Trypan blue. All 20 *bak1-4 bkk1-1* seedlings grown in darkness did not show any cell-death symptoms in their cotyledons. Conversely, all 20 *bak1-4 bkk1-1* seedlings grown under long-day lighting condition showed severe cell-death symptoms. Even 3 weeks after germination, no cell death was identified in *bak1-4 bkk1-1*cotyledons grown in darkness, whereas the light-grown double mutant seedlings were already completely dead. These results suggest that light-grown plants may produce factors, which can trigger cell death, and the double mutant has lost its capability to protect the plants from cell death.

5 Cell-Death/Early Senescence in Weak Double Mutant *bak1-3 bkk1-1* Is Environment Dependent

The leaf cell death symptom of weak double mutant *bak1-3 bkk1-1* is highly dependent on the environment the plants are exposed to. For example, seedlings grown on ½ MS medium look healthy, similar to wild-type ones. When grown in potting soil and in the greenhouse, however, the weak double mutant seedlings show an accelerated early senescence phenotype, similar to lesion mimic mutants. These results suggest that the cell death (or senescence) can be triggered by unknown environmental factors. In addition, the cell death symptoms of the weak double mutant can be partially suppressed by high temperature. The plants growing in a 28°C growth chamber are much healthier than the ones growing in a 22°C growth chamber (data not published). The ability to produce seeds from the weak double mutant will greatly facilitate the screening of extragenic suppressors to identify other genes that regulate cell death.

6 Spreading Cell Death Can Be Induced by Pathogens in *bak1* Single Mutant Background

bak1 single null mutant shows a subtle *bri1*-like phenotype and reduced response to exogenously applied BR. No cell death, however, can be detected on single mutant seedlings under normal growth conditions. Upon pathogen infection, *bak1*, but not wild-type or *bri1* mutants, showed a spreading necrosis phenotype due to uncontrolled cell death in *bak1* mutant background (Kemmerling et al. 2007). This observation indicates that *bak1* is more vulnerable to pathogen attacks compared to wild-type plants. This result is consistent with the spontaneous cell death phenotypes seen in *bak1 bkk1* double mutant background (He et al. 2007). Both data suggest that BAK1 and BKK1 play key roles in plant cell-death control.

7 BAK1 and Its Paralogs Interact with BIR1 and Control Cell Death

To isolate genes crucial for innate immunity, Gao et al. attempted to identify genes whose expression can be upregulated upon the treatment of the bacterial pathogen *Pseudomonas syringae pv. maculicola* ES4326 (Gao et al. 2009). One of the genes identified is an *RLK*, loss-of-function mutant in that gene showed a constitutive defense response, cell death, and seedling lethality phenotypes, similar to the phenotypes exhibited by *bak1-4 bkk1-1* double null mutant. Co-immunoprecipitation analysis identified SERK proteins as interactors, including SERK1, SERK2, BAK1, and BKK1. The gene was subsequently named BIR1 for BAK1-interacting receptor-like kinase (Gao et al. 2009). In addition, a calcium-dependent phospholipid binding protein, BON1, was found to interact with both BAK1 and BIR1 (Wang et al. 2011). BON1 can be phosphorylated by BAK1 in vitro. The phenotypic resemblance of *bir1*, *bak1-4 bkk1-1*, and *bon1* and the reciprocal physical interactions among BIR1, BAK1 and BON1 suggest that these proteins may regulate the same signaling pathway to control cell death. Elucidation of detailed molecular mechanisms, however, relies on the identification of authentic ligand (a "survival signal") of the BIR1-BAK1 receptor complex in the future.

8 Kinase Activity of BAK1 Determines Its Specificity

It appears that BAK1 and its paralogs independently participate in several different signaling pathways. One of the most intriguing questions would be how the specificity has been determined. Recent identification of a novel BAK1 mutant allele, *bak1-5*, provides significant insights into our understanding how BAK1 is able to differentiate distinct signaling pathways (Schwessinger et al. 2011). *bak1-5* bears a single amino acid substitution, C408Y, in subdomain VIa of the BAK1 cytoplasmic kinase domain, preceding the activation loop. Both BR and cell-death control signaling pathways have not been altered in *bak1-5*. But BAK1-associated innate immunity responses, such as FLS2- and EFR-associated signaling pathways, have been severely blocked. Biochemical analyses indicated that the attenuation of these innate immunity signaling pathways was not caused by the weakened association between FLS2 and bak1-5, or between EFR and bak1-5. It is more likely resulted from the altered phosphorylation activity of bak1-5 relative to the wild-type BAK1. Detailed molecular mechanisms will rely on the determination of differential phosphorylation sites of FLS2 or EFR in wild-type and *bak1-5* plants. Given the fact that the phosphorylation levels of FLS2 and EFR are extremely low, analyzing phosphorylation sites on these two RLKs will be a tough task.

9 A Model for the Roles of BAK1 and BKK1 in Cell-Death Control

Recent studies indicated that BAK1 and its paralogs are involved in multiple signaling pathways to control plant growth and development as well as various defense responses (Li 2010). Accumulating evidence indicates that BAK1 and BKK1 regulate a cell-death control signaling pathway distinct from either the BRI1-mediated or the FLS2-mediated pathways. But it is likely that the cell-death pathway and multiple defense-related pathways are converged at a certain signaling point, leading to the expression of same or largely overlapped set of defense-related genes. Based on the roles of BAK1 in BRI1-BAK1 and FLS2-BAK1 signaling pathways, we propose that BAK1 and BKK1 may interact with a ligand-binding receptor, which can sense a "surviving signal" yet to be isolated (Fig. 2). The "surviving signal" could be a polypeptide or could be a small proteinaceous molecule. The identification of BIR1 suggests that it could be the ligand-binding receptor, and that BIR 1 controls cell death with its co-receptor BAK1 or BKK1. This notion is supported by the fact that BIR1 belongs to the LRR-RLK subfamily X, which also contains some well-studied ligand-binding receptors such as BRI1 and EMS1/EXS (Canales et al. 2002; Li and Chory 1997; Zhao et al. 2002). Downstream components of BIR1-BAK1-mediated signaling could include MEKK1-MKK1/MKK2-MPK4 (Gao et al. 2009; Li 2010). *mekk1* mutants also show a cell death phenotype similar to that of *bir1* and *bak1-4 bkk1-1* double mutants, suggesting that BAK1/BKK1 and MEKK1 may indeed be involved in the same signaling pathway (Ichimura et al. 2006; Suarez-Rodriguez et al. 2007). MEKK1, MKK1/MKK2, and MPK4 are also the known components in FLS2-BAK1 regulated signaling pathway (Ichimura et al. 2006; Suarez-Rodriguez et al. 2007). Therefore, MEKK1 could be the convergence point between the two signaling pathways (Fig. 2). In wild-type plants, the extracellular domain of BIR1 (or an unknown ligand binding RLK) may perceive a "survival signal", initiating the formation of a functional receptor complex. The signal then can be delivered to downstream components such as MAPKs likely via protein phosphorylation, leading to the suppression of SA accumulation and maintaining ROS at certain levels. In the *bak1-4 bkk1-1* double mutant background, on the contrary, SA biosynthesis cannot be inhibited, leading to the constitutive expression of defense genes and downregulation of antioxidant genes, which finally results in the accumulation of ROS in chloroplasts and cell death (Fig. 2).

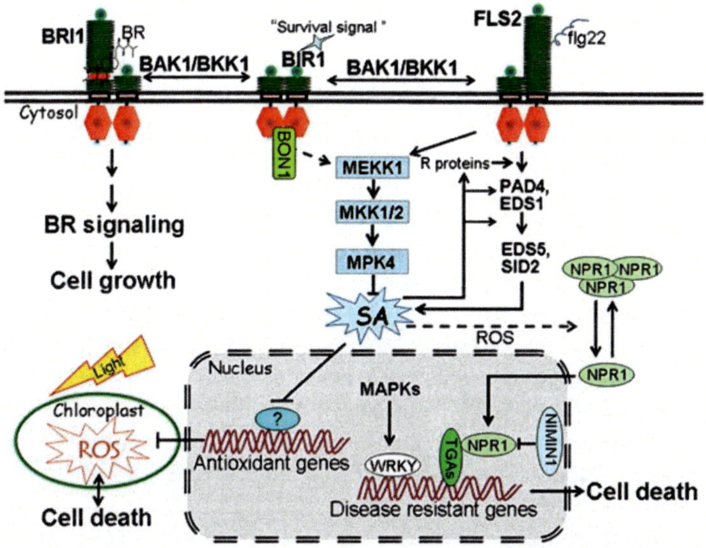

Fig. 2 A current model of BAK1 in regulating BR, cell-death control, and FLS2-mediated innate immunity signaling pathways. BAK1 and BKK1 play redundant roles in the early events of BR signal transduction. BRs are perceived by the cell surface receptor BRI1 via its direct interaction with the extracellular domain of BRI1. The interaction triggers the autophosphorylation of BRI1, which then recruits and transphosphorylates BAK1. The activated BAK1 in turn phosphorylates BRI1 on residues at the juxtamembrane and C-terminal regions, initiating the BR signaling cascade and cell growth (Wang et al. 2008; Kim and Wang 2010; Clouse 2011). In addition, BAK1 and BKK1 also mediate cell-death control via its interaction with BIR1. Since BIR1 is a member of the LRR-RLK X subfamily, it likely plays a role as a ligand-binding RLK. The ligand for BIR1 could be a "surviving signal" that is not been elucidated. A cell-death-related protein, BON1, can interact with both BIR1 and BAK1, suggesting these three proteins control the same cell-death signaling pathway (Wang et al. 2011). The downstream components of this pathway may include MAPK members, MEKK1, MKK1/MKK2, and MPK4 (Gao et al. 2009; Li 2010); the *mekk1* mutant shows a cell death phenotype similar to that of *bak1 bkk1*, *bir1*, and *bon1*. The activated MPK4 negatively regulates the accumulation of SA (Ichimura et al. 2006; Gao et al. 2009). In the *bak1 bkk1* double mutant, the accumulated SA leads to the downregulation of antioxidant genes, which then results in the accumulation of ROS in chloroplasts and cell death. Moreover, BAK1 also participates in innate immunity responses through its association with pathogen-associated molecular patterns (PAMPs) recognition receptors such as FLS2 and EFR (Gomez-Gomez and Boller 2002; Zipfel et al. 2006). FLS2 and EFR recognize flg22 and elf18 peptide signals, respectively. Activated FLS2/BAK1 receptor complex results in the accumulation of SA through the PAD4-EDS1-, EDS5- and SID2-mediate biosynthetic pathways (Shah 2003; Chinchilla et al. 2007; Dunning et al. 2007). The accumulation of SA then regulates the production of R proteins, which regulate PAD4, EDS1, EDS5, and SID2 through a positive feedback loop (Shah 2003; Dunning et al. 2007). FLS2-BAK1 signaling also activates the MEKK1-MKK1/MKK2-MPK4 cascade (Ichimura et al. 2006; Suarez-Rodriguez et al. 2007). The activated MAP kinases regulate their downstream WRKY transcription factors, promoting the defense gene expression (Pandey and Somssich 2009; Rushton et al. 2010). SA can promote the expression of disease-resistant genes in an NPR1-dependent signaling pathway, which is negatively regulated by NIMIN1, producing immune responses (Mou et al. 2003; Spoel et al. 2009). This process alters the cellular redox state through increased SA, which promotes deoligomerization of cytosolic NPR1. The NPR1 monomers subsequently moved into the nucleus where they interact with TGA transcription factors to upregulate the *PR* gene expression, immune response and cell-death (Mou et al. 2003; Spoel et al. 2009)

10 Future Perspectives

Identification of the RLKs BAK1 and BKK1 as key regulators of cell-death control is exciting. However, several key questions need to be answered in the near future before the true roles of BAK1 and BKK1 in the regulation of cell death can be fully understood. These questions include, (1) is BIR1 really the ligand-binding receptor that can sense a "survival signal"? (2) What is a "survival signal"? The "survival signal" could be a small molecule, which can interact with the extracellular domain of the ligand-binding receptor. Alternatively, ROS may serve as "survival signals," which can be sensed by RLK via redox reactions on paired cysteine residues commonly observed at the flanking regions of LRR repeats. The functions of these paired cysteine residues have not been elucidated. (3) What are the downstream components in the BAK1-mediated cell-death control signaling pathway? (4) How is the specificity of BAK1 and BKK1 in regulating cell-death control and innate immunity responses determined? Identification and functional elucidation of *bak1-5* clearly indicate that BAK1-invloved cell-death control is distinctive from the FLS2- and EFR-associated innate immunity pathways (Schwessinger et al. 2011), although these pathways may likely share some common downstream regulatory components. The model presented in this chapter is mainly based on current available data. Elucidation of the functions of BAK1 and BKK1 in cell-death control relies on the future identification of additional signaling proteins in this pathway. The weak double mutant *bak1-3 bkk1-1* will provide a useful tool for future genetic screening for extragenic suppressors and the study of genes in the BAK1 BKK1 signaling pathways. The authors' group has conducted a large-scale activation tagging screen in *bak1-3 bkk1-1* background. A number of authentic genetic suppressors have been isolated, suggesting that activation tagging is indeed a feasible approach to identify novel signaling components in controlling BAK1-mediated cell-death pathway.

Acknowledgements The authors' research group is currently supported by National Basic Research Program of China Grant 2011CB915401 (to J.L.), National Natural Science Foundation of China Grants 90917019 (to J.L.) and 31070283 (to X.G.).

References

Albrecht C, Russinova E, Hecht V, Baaijens E, de Vries S (2005) The *Arabidopsis thaliana* SOMATIC EMBRYOGENESIS RECEPTOR-LIKE KINASES1 and 2 control male sporogenesis. Plant Cell 17:3337–3349

Arabidopsis Genome Initiative (2000) Analysis of the genome sequence of the flowering plant *Arabidopsis thaliana*. Nature 408:796–815

Boller T, Felix G (2009) A renaissance of elicitors: perception of microbe-associated molecular patterns and danger signals by pattern-recognition receptors. Annu Rev Plant Biol 60:379–406

Canales C, Bhatt AM, Scott R, Dickinson H (2002) EXS, a putative LRR receptor kinase, regulates male germline cell number and tapetal identity and promotes seed development in Arabidopsis. Curr Biol 12:1718–1727

Chinchilla D, Zipfel C, Robatzek S, Kemmerling B, Nurnberger T, Jones JD, Felix G, Boller T (2007) A flagellin-induced complex of the receptor FLS2 and BAK1 initiates plant defence. Nature 448:497–500

Clark SE, Williams RW, Meyerowitz EM (1997) The *CLAVATA1* gene encodes a putative receptor kinase that controls shoot and floral meristem size in Arabidopsis. Cell 89:575–585

Clay NK, Nelson T (2002) VH1, a provascular cell-specific receptor kinase that influences leaf cell patterns in Arabidopsis. Plant Cell 14:2707–2722

Clouse SD (2011) Brassinosteroid signal transduction: from receptor kinase activation to transcriptional networks regulating plant development. Plant Cell 23:1219–1230

Colcombet J, Boisson-Dernier A, Ros-Palau R, Vera CE, Schroeder JI (2005) Arabidopsis SOMATIC EMBRYOGENESIS RECEPTOR KINASES1 and 2 are essential for tapetum development and microspore maturation. Plant Cell 17:3350–3361

Delaney TP, Uknes S, Vernooij B, Friedrich L, Weymann K, Negrotto D, Gaffney T, Gut-Rella M, Kessmann H, Ward E, Ryals J (1994) A central role of salicylic Acid in plant disease resistance. Science 266:1247–1250

Dunning FM, Sun W, Jansen KL, Helft L, Bent AF (2007) Identification and mutational analysis of Arabidopsis FLS2 leucine-rich repeat domain residues that contribute to flagellin perception. Plant Cell 19:3297–3313

Fisher K, Turner S (2007) PXY, a receptor-like kinase essential for maintaining polarity during plant vascular-tissue development. Curr Biol 17:1061–1066

Gao M, Wang X, Wang D, Xu F, Ding X, Zhang Z, Bi D, Cheng YT, Chen S, Li X, Zhang Y (2009) Regulation of cell death and innate immunity by two receptor-like kinases in Arabidopsis. Cell Host Microbe 6:34–44

Godiard L, Sauviac L, Torii KU, Grenon O, Mangin B, Grimsley NH, Marco Y (2003) ERECTA, an LRR receptor-like kinase protein controlling development pleiotropically affects resistance to bacterial wilt. Plant J 36:353–365

Gomez-Gomez L, Boller T (2000) FLS2: an LRR receptor-like kinase involved in the perception of the bacterial elicitor flagellin in Arabidopsis. Mol Cell 5:1003–1011

Gomez-Gomez L, Boller T (2002) Flagellin perception: a paradigm for innate immunity. Trends Plant Sci 7:251–256

Gou X, He K, Yang H, Yuan T, Lin H, Clouse SD, Li J (2010) Genome-wide cloning and sequence analysis of leucine-rich repeat receptor-like protein kinase genes in *Arabidopsis thaliana*. BMC Genomics 11:19

He K, Gou X, Yuan T, Lin H, Asami T, Yoshida S, Russell SD, Li J (2007) BAK1 and BKK1 regulate brassinosteroid-dependent growth and brassinosteroid-independent cell-death pathways. Curr Biol 17:1109–1115

Hecht V, Vielle-Calzada JP, Hartog MV, Schmidt ED, Boutilier K, Grossniklaus U, de Vries SC (2001) The Arabidopsis SOMATIC EMBRYOGENESIS RECEPTOR KINASE 1 gene is expressed in developing ovules and embryos and enhances embryogenic competence in culture. Plant Physiol 127:803–816

Heese A, Hann DR, Gimenez-Ibanez S, Jones AM, He K, Li J, Schroeder JI, Peck SC, Rathjen JP (2007) The receptor-like kinase SERK3/BAK1 is a central regulator of innate immunity in plants. Proc Natl Acad Sci USA 104:12217–12222

Horn MA, Walker JC (1994) Biochemical properties of the autophosphorylation of RLK5, a receptor-like protein kinase from *Arabidopsis thaliana*. Biochim Biophys Acta 1208: 65–74

Ichimura K, Casais C, Peck SC, Shinozaki K, Shirasu K (2006) MEKK1 is required for MPK4 activation and regulates tissue-specific and temperature-dependent cell death in Arabidopsis. J Biol Chem 281:36969–36976

Jinn TL, Stone JM, Walker JC (2000) HAESA, an Arabidopsis leucine-rich repeat receptor kinase, controls floral organ abscission. Genes Dev 14:108–117

Karlova R, Boeren S, Russinova E, Aker J, Vervoort J, de Vries S (2006) The Arabidopsis SOMATIC EMBRYOGENESIS RECEPTOR-LIKE KINASE1 protein complex includes BRASSINOSTEROID-INSENSITIVE1. Plant Cell 18:626–638

Kemmerling B, Schwedt A, Rodriguez P, Mazzotta S, Frank M, Qamar SA, Mengiste T, Betsuyaku S, Parker JE, Mussig C, Thomma BP, Albrecht C, de Vries SC, Hirt H, Nurnberger T (2007) The BRI1-Associated Kinase 1, BAK1, has a brassinolide-independent role in plant cell-death control. Curr Biol 17:1116–1122

Kim TW, Wang ZY (2010) Brassinosteroid signal transduction from receptor kinases to transcription factors. Annu Rev Plant Biol 61:681–704

Li J (2010) Multi-tasking of somatic embryogenesis receptor-like protein kinases. Curr Opin Plant Biol 13:509–514

Li J, Chory J (1997) A putative leucine-rich repeat receptor kinase involved in brassinosteroid signal transduction. Cell 90:929–938

Li J, Wen J, Lease KA, Doke JT, Tax FE, Walker JC (2002) BAK1, an Arabidopsis LRR receptor-like protein kinase, interacts with BRI1 and modulates brassinosteroid signaling. Cell 110: 213–222

Mittler R, Vanderauwera S, Gollery M, Van Breusegem F (2004) Reactive oxygen gene network of plants. Trends Plant Sci 9:490–498

Mou Z, Fan W, Dong X (2003) Inducers of plant systemic acquired resistance regulate NPR1 function through redox changes. Cell 113:935–944

Nam KH, Li J (2002) BRI1/BAK1, a receptor kinase pair mediating brassinosteroid signaling. Cell 110:203–212

Nodine MD, Yadegari R, Tax FE (2007) RPK1 and TOAD2 are two receptor-like kinases redundantly required for Arabidopsis embryonic pattern formation. Dev Cell 12:943–956

Oh MH, Wang X, Kota U, Goshe MB, Clouse SD, Huber SC (2009) Tyrosine phosphorylation of the BRI1 receptor kinase emerges as a component of brassinosteroid signaling in Arabidopsis. Proc Natl Acad Sci USA 106:658–663

Oh MH, Wang X, Wu X, Zhao Y, Clouse SD, Huber SC (2010) Autophosphorylation of Tyr-610 in the receptor kinase BAK1 plays a role in brassinosteroid signaling and basal defense gene expression. Proc Natl Acad Sci USA 107:17827–17832

Osakabe Y, Maruyama K, Seki M, Satou M, Shinozaki K, Yamaguchi-Shinozaki K (2005) Leucine-rich repeat receptor-like kinase1 is a key membrane-bound regulator of abscisic acid early signaling in Arabidopsis. Plant Cell 17:1105–1119

Pandey SP, Somssich IE (2009) The role of WRKY transcription factors in plant immunity. Plant Physiol 150:1648–1655

Rushton PJ, Somssich IE, Ringler P, Shen QJ (2010) WRKY transcription factors. Trends Plant Sci 15:247–258

Schmidt ED, Guzzo F, Toonen MA, de Vries SC (1997) A leucine-rich repeat containing receptor-like kinase marks somatic plant cells competent to form embryos. Development 124: 2049–2062

Schwessinger B, Roux M, Kadota Y, Ntoukakis V, Sklenar J, Jones A, Zipfel C (2011) Phosphorylation-dependent differential regulation of plant growth, cell death, and innate immunity by the regulatory receptor-like kinase BAK1. PLoS Genet 7:e1002046

Shah J (2003) The salicylic acid loop in plant defense. Curr Opin Plant Biol 6:365–371

Shirasu K, Lahaye T, Tan MW, Zhou F, Azevedo C, Schulze-Lefert P (1999) A novel class of eukaryotic zinc-binding proteins is required for disease resistance signaling in barley and development in *C. elegans*. Cell 99:355–366

Shiu SH, Bleecker AB (2001) Receptor-like kinases from Arabidopsis form a monophyletic gene family related to animal receptor kinases. Proc Natl Acad Sci USA 98:10763–10768

Spoel SH, Mou Z, Tada Y, Spivey NW, Genschik P, Dong X (2009) Proteasome-mediated turnover of the transcription coactivator NPR1 plays dual roles in regulating plant immunity. Cell 137:860–872

Suarez-Rodriguez MC, Adams-Phillips L, Liu YD, Wang HC, Su SH, Jester PJ, Zhang SQ, Bent AF, Krysan PJ (2007) MEKK1 is required for flg22-induced MPK4 activation in Arabidopsis plants. Plant Physiol 143:661–669

Thordal-Christensen H, Zhang Z, Wei Y, Collinge DB (1997) Subcellular localization of H_2O_2 in plants. H_2O_2 accumulation in papillae and hypersensitive response during the barley-powdery mildew interaction. Plant J 11:1187–1194

Torii KU, Mitsukawa N, Oosumi T, Matsuura Y, Yokoyama R, Whittier RF, Komeda Y (1996) The Arabidopsis *ERECTA* gene encodes a putative receptor protein kinase with extracellular leucine-rich repeats. Plant Cell 8:735–746

Tsuwamoto R, Fukuoka H, Takahata Y (2008) *GASSHO1* and *GASSHO2* encoding a putative leucine-rich repeat transmembrane-type receptor kinase are essential for the normal development of the epidermal surface in Arabidopsis embryos. Plant J 54:30–42

Wang X, Kota U, He K, Blackburn K, Li J, Goshe MB, Huber SC, Clouse SD (2008) Sequential transphosphorylation of the BRI1/BAK1 receptor kinase complex impacts early events in brassinosteroid signaling. Dev Cell 15:220–235

Wang Z, Meng P, Zhang X, Ren D, Yang S (2011) BON1 interacts with the protein kinases BIR1 and BAK1 in modulation of temperature-dependent plant growth and cell death in Arabidopsis. Plant J. doi:10.1111/j.1365-1313X.2011.04659.x

Zhao DZ, Wang GF, Speal B, Ma H (2002) The *EXCESS MICROSPOROCYTES1* gene encodes a putative leucine-rich repeat receptor protein kinase that controls somatic and reproductive cell fates in the Arabidopsis anther. Genes Dev 16:2021–2031

Zipfel C, Kunze G, Chinchilla D, Caniard A, Jones JD, Boller T, Felix G (2006) Perception of the bacterial PAMP EF-Tu by the receptor EFR restricts Agrobacterium-mediated transformation. Cell 125:749–760

Receptor Kinases Mediating Early Symbiotic Signalling

Esben Bjørn Madsen and Jens Stougaard

Abstract The intimate intercellular relation between plant and bacteria in root nodule symbiosis is established after specific recognition of the microsymbiont. This delicate process involves various cellular components including at least three receptor kinases in the model legume *Lotus japonicus*. Whereas the exact role of the SYMRK receptor is still unclear, the lipochitin-oligosaccharide receptors NFR1 and NFR5 have been shown to be major determinants in host recognition and the role of the extracellular LysM domains have been analysed. Biochemical analyses of the cytoplasmic domain of NFR5 reveal that NFR5 is a pseudokinase that lacks detectable kinase activity. However, recent data show that NFR1 can interact with and phosphorylate NFR5. The signalling mechanism of NFR1 and NFR5 therefore differs from that of previously described plant RLKs. Interestingly, NFR1 and the *Arabidopsis* chitin receptor CERK1 are highly similar and minor differences in the cytoplasmic domain determines the signalling potential.

1 Introduction

Symbiosis between plants and microorganisms is a common feature in the plant kingdom. In fact, fossil records date the origin of symbiosis between plants and fungi, mycorrhiza, back to the emergence of the first land plants, and arbuscular mycorrhiza is present in the majority of land plant species (Remy et al. 1994). In contrast to the widespread symbiotic relation with fungi, symbiosis with unicellular bacterial species such as *Bradyrhizobium*, *Mesorhizobium*, and *Rhizobium*, collectively named *Rhizobium* or rhizobia, is scarce in the plant kingdom and is with a single exception (*Parasponia*) only found in the legume family (Sprent 2007). The

E.B. Madsen • J. Stougaard (✉)
Center for Carbohydrate Recognition and Signalling, Department of Molecular Biology, Aarhus University, Aarhus, Denmark
e-mail: stougaard@mb.au.dk

F. Tax and B. Kemmerling (eds.), *Receptor-like Kinases in Plants*,
Signaling and Communication in Plants 13,
DOI 10.1007/978-3-642-23044-8_6,

interaction between legume plants and rhizobia is an important research area due to the environmental and economical benefits of symbiotic nitrogen fixation over the use of chemical fertilizers. A fundamental biological question is, which additional genetic components are required for legume plants to identify beneficial and compatible symbiotic partners among the millions of microbes found in the rhizosphere and to reorganize already differentiated tissue in order to create a new organ that can accommodate the microsymbiont. The development of nitrogen-fixing nodules is the result of a highly specific two-way signalling. The host roots secrete flavonoids into the rhizosphere where they, upon recognition by the rhizobial nodD protein, induce the transcription of the *nod* genes required for bacterial synthesis of the Nod factor. The Nod factor is highly morphogenic and correct perception by the legume root initiates signalling pathways, which can be observed and measured as morphological, physiological, biochemical and molecular events that lead to the development and infection of the nodule primodia (Ehrhardt et al. 1992, 1996; Hogslund et al. 2009; Lerouge et al. 1990; Miwa et al. 2006; Spaink et al. 1991). The Nod factor is a major determinant in microbe-host specificity and in the early nineties the Nod factor was purified and identified as a lipochitin-oligosaccharide (LCO) (Lerouge et al. 1990; Roche et al. 1991; Spaink et al. 1991). The identification and chemical characterization of the Nod factor combined with inoculation of rhizobial mutants led to the discovery that chemical modifications on both the reducing and non-reducing end of the LCO as well as the length and degree of saturation of the fatty acid are structural determinants for legume host recognition (D'Haeze and Holsters 2002; Lerouge et al. 1990; Rodpothong et al. 2009; Spaink et al. 1991).

2 NFR1 and NFR5, Two LysM Domain Containing Receptors

In the model legume *Lotus japonicus* (Handberg and Stougaard 1992), the bacterial LCO is perceived by at least two Nod-factor receptors called Nod Factor Receptor 1 (NFR1) and Nod Factor Receptor 5 (NFR5). The *Nfr1* and *Nfr5* genes were originally identified in a screen for nodulation deficient mutants. Detailed phenotypic characterization of *nfr1* and *nfr5* mutant plants revealed that they are blocked at one of the earliest steps in the signalling pathway leading to nodule development. In contrast to wild type plants, which show membrane depolarization, root hair deformation, extracellular and intracellular ion-fluxes and gene induction after application of purified LCO, *nfr5* and *nfr1* mutants remain unresponsive or show a severely attenuated response upon addition of LCO, respectively. *Nfr1* and *Nfr5* were cloned using map-based cloning techniques and were found to encode two proteins with predicted extracellular regions containing three LysM domains separated by cysteine pairs (Fig. 1) (Madsen et al. 2003; Radutoiu et al. 2003). LysM domains are frequently found in bacterial lysins, amidase and protease proteins, suggesting that LysM domains play a role in binding peptidoglycan, chitin and structurally related molecules (Eckert et al. 2006; Iizasa et al. 2010; Kaku et al.

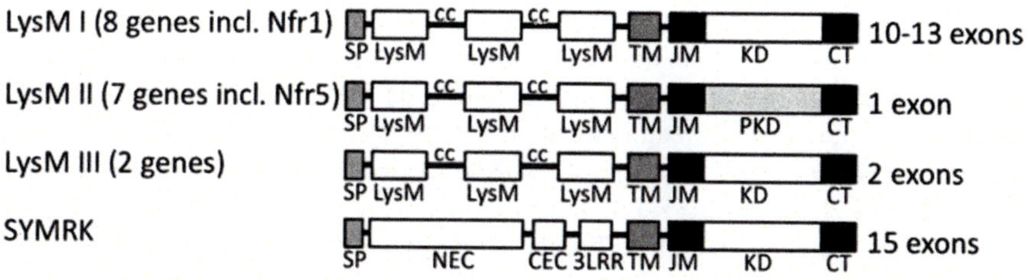

Fig. 1 Domain structure of *L. japonicus* LysM-containing receptors and SYMRK. Predicted protein domains are *SP* signal peptide, *LysM* LysM domain, *cc* cysteine pair, *NEC* N-terminal extracellular region, *CEC* conserved extracellular region preceeding LRRs, *LRR* leucine-rich repeats, *TM* transmembrane domain, *JM* juxta membrane domain, *KD* kinase domain, *PKD* pseudokinase domain and *CT* C-terminal tail

2006; Madsen et al. 2003; Radutoiu et al. 2003; Steen et al. 2003). As the Nod factor is an LCO, the LysM domain containing receptors NFR1 and NFR5 could potentially bind the Nod factor directly. The LysM domains are separated from cytoplasmic domains by single-pass transmembrane regions. The cytoplasmic domains of both NFR1 and NFR5 show sequence similarity to Ser/Thr kinases but where NFR1 carries all the conserved kinase subdomains NFR5 lacks several kinase subdomains (Madsen et al. 2003; Radutoiu et al. 2003). NFR5 orthologs have been identified in *Medicago truncatula* (NFP), *Pisum sativum* (Sym10) and *Glycine max* (NFR5) (Arrighi et al. 2006; Indrasumunar et al. 2010; Madsen et al. 2003). The *M. truncatula* LYK3 has been suggested as an LCO receptor involved in bacterial entry, but it is currently not clear whether LYK3 is the functional homolog of NFR1 (Limpens et al. 2003).

2.1 *NFR1 and NFR5 Localize to and Interact in the Plasma Membrane*

A strategic location for receptors involved in direct perception of extracellular signals is at the cell surface. Both NFR1 and NFR5 have predicted signal peptides and single pass transmembrane regions. Low expression levels and cleavage of epitope tags from NFR1 and NFR5 have so far prevented localization of NFR1 and NFR5 in legume roots. However, using transient expression of *Nfr1* and *Nfr5* in *Nicotiana benthamiana* or leek epidermal cells reveals that the Nod-factor receptors localize to the plasma membrane in heterologous tissue (Madsen et al. 2011). As *Nfr1* and *Nfr5* have similar mutant phenotypes, domain composition and localization and are predicted to bind the same ligand, it has long been suggested that NFR1 and NFR5 interact and form a Nod-factor receptor complex (Madsen et al. 2003; Radutoiu et al. 2003). The capacity of NFR1 and NFR5 for protein–protein interaction has been investigated using Bimolecular Fluoresecence Complementation (BiFC), in *N. benthamiana* and leek cells. The BiFC data show that both NFR1 and

NFR5 form homodimers, but when expressed together no BiFC signal could be detected. Instead cell death in the transformed tissue was observed within 2–3 days after transformation. The cell death phenomenon could be abrogated by substituting NFR1-T481 with alanine, creating a kinase dead version of NFR1 (see below). Co-expression of the kinase dead NFR1-T481A and NFR5 showed strong BiFC, indicating that NFR1 and NFR5 do interact in planta. This shows that the observed cell death is dependent on both NFR1 and NFR5 and requires NFR1 kinase activity (Madsen et al. 2011). Taken together these data indicate that NFR1 and NFR5 interact and in cooperation create a signal, which at least in heterologous tissue, causes cell death. The nature of this signal and downstream targets of NFR1 and NFR5 in *N. benthamiana* and leek are not yet known. It would be interesting to identify these targets and test whether corresponding components also interact with NFR1 and NFR5 in legume roots or the PUB1 E3 ubiquitin ligase recently found to interact with LCO receptors in *M. truncatula* (Mbengue et al. 2010) and *Lotus* (Samuelsen C. personal communication). Knock down experiments with *MtPUB1* and studies of TILLING mutants in *Lotus* indicate that *PUB1* is a negative regulator of nodulation (Mbengue et al. 2010, Samuelsen C. personal communication).

2.2 NFR1 Is an Active Protein Kinase and NFR5 a Pseudokinase

To better understand the signalling mechanism of NFR1 and NFR5, the cytoplasmic domains of *Nfr1* and *Nfr5* have been expressed in *E. coli* and analysed in detail (Madsen et al. 2011). The intracellular region of NFR1 contains a domain with similarity to Ser/Thr protein kinases and in vitro kinase assays with recombinant NFR1 protein have shown that the cytoplasmic domain of NFR1 is an active protein kinase, which is not only able to autophosphorylate but also to phosphorylate the cytoplasmic domain of NFR5 (Madsen et al. 2011). Tandem mass spectrometry analysis of the NFR1 cytoplasmic domain revealed that NFR1 is not a classic Ser/Thr kinase, but a dual specificity kinase as it, in addition to phosphorylating serine and threonine residues is capable of phosphorylating tyrosine residues (Madsen et al. 2011). Dual specificity of plant RLKs has recently also been described for BRI1 and BAK1 plus an additional six other RLKS (Oh et al. 2009).

The cytoplasmic domain of NFR5 (NFR5-CD) also shows similarity to protein kinases, but NFR5-CD lacks several conserved residues in kinase subdomain I, VII, VIII and IX and is classified as a pseudokinase (Boudeau et al. 2006). Kinase activity has been tested biochemically for both NFR5 and the *M. truncatula* ortholog NFP, which showed that NFP and NFR5 lack kinase activity in vitro (Arrighi et al. 2006; Madsen et al. 2011). The absence of detectable kinase activity in vitro does not, however, rule out that NFR5/NFP have kinase activity in planta where kinase substrates, ion content and membrane environment differ from the standard in vitro conditions. To test the in planta requirement for a putative NFR5 kinase activity, a *Nfr5* mutant construct was created, encoding a substitution (NFR5-K339E) of the conserved catalytic lysine residue in kinase subdomain II, which in classical protein kinases is directly involved in orientating ATP and essential for protein kinase activity. Confirming the in vitro results the mutagenized *Nfr5* construct complemented *nfr5* mutant plants. This strongly indicates that NFR5 does not require kinase activity for in planta Nod-factor signalling. Some pseudokinases function by being phosphorylated by interacting active protein kinases (Boudeau et al. 2006). This could also be the case for NFR5 as it can be phosphorylated by NFR1 in vitro. However, the biological importance of the NFR5-pS282 phosphorylation is not clear. Substitution of the NFR5 phosphorylation site residue with alanine did not alter the ability to rescue *nfr5* mutant plants. It is therefore currently an open question whether phosphorylation of NFR5 is involved in Nod-factor receptor signalling and what role the intracellular pseudokinase domain plays in Nod-factor signalling.

2.3 A Single NFR1 Phosphorylation Site Residue Is Essential for NFR1 in Planta Signalling

The biological role of NFR1 phosphorylation in Nod-factor signalling has been analysed by substituting NFR1 in vitro phosphorylation sites with non-phosphorylatable alanine and phenylalanine residues. By testing these mutagenized *Nfr1* constructs for functional complementation of *nfr1* mutant plants, the importance of the individual phosphorylation sites could be analysed. As observed for AtBRI1, substitution of phosphorylation site residues in the juxtamembrane and C-terminal tail had little or no effect on NFR1 function (Madsen et al. 2011; Wang et al. 2005a). Only mutagenesis of the conserved activation loop phosphorylation site residue T481 had a major effect on NFR1 function as it failed to complement *nfr1* mutant plants. In vitro kinase activity tests of recombinant NFR1-CD-T481A, revealed that T481 is essential for NFR1 substrate phosphorylation and autophosphorylation activity (Madsen et al. 2011). Therefore, the data indicate that NFR1 requires kinase activity for Nod-factor signalling in planta. This is in line with recent data showing that the putative *M. truncatula* ortholog LYK3 is a dual specificity kinase, which also requires kinase activity for in planta function (Klaus-Heisen et al. 2011).

2.4 NFR1 and NFR5 Are Major Determinants of Nod-Factor Recognition

The highly selective relationship between rhizobia and host plants has led to the definition of cross-inoculation groups. Among narrow host range interactions, *Sinorhizobium meliloti* (*S. meliloti*) and alfalfa and *M. truncatula*, belong to one cross-inoculation group, whereas *Mesorhizobium loti* (*M. loti*) and *Lotus* spp. like

L. japonicus, belong to another. Inoculation with rhizobia from another cross-inoculation group does not lead to nodulation. Direct evidence for a role of NFR1 and NFR5 in the legume host recognition was provided by expressing *Nfr1* and *Nfr5* from one inoculation group (*Lotus*) in a plant belonging to another inoculation group (*Medicago*) and analysis of the host specificity of the transgenic plant. *M. truncatula* plants transformed with either *Nfr1* or *Nfr5* alone were found to have an unaltered host specificity and were only able to nodulate with *S. meliloti* like untransformed *M. truncatula*. However, when both receptors were expressed in combination the *M. truncatula* plants were able to recognize not only *S. meliloti* but also *M. loti*, the symbiont of *L. japonicus*. The *M. truncatula Nfr1, Nfr5* plants formed infected nodules morphologically and anatomically similar to the ones developed with *S. meliloti* (Fig. 2) (Radutoiu et al. 2007). These findings indicate that NFR1 and NFR5 are major determinants of host specificity. However, nodules

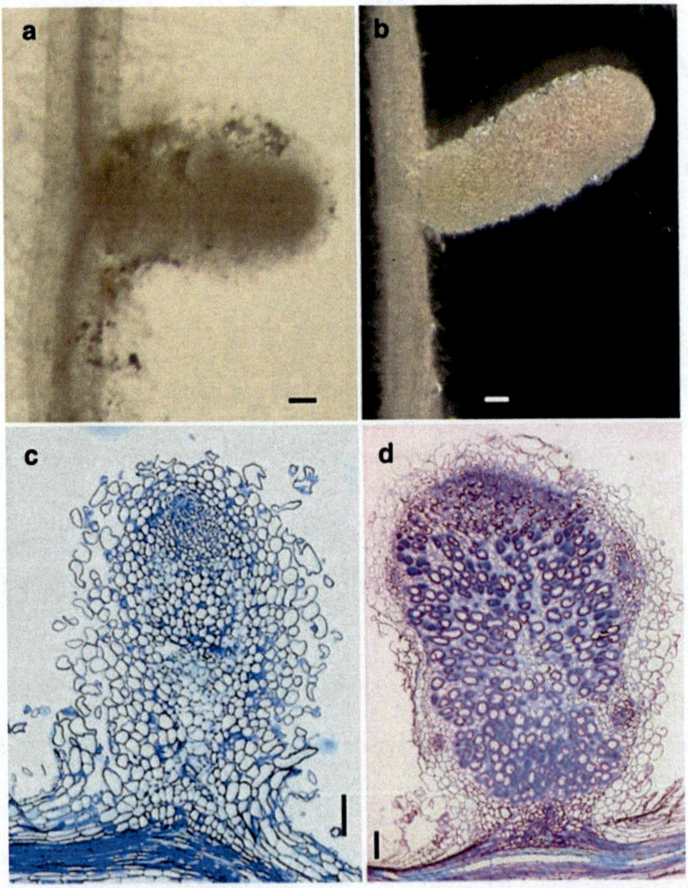

Fig. 2 *M. loti* nodules on *Nfr1, Nfr5* transgenic *M. truncatula* roots are morphologically (**a**) and anatomically(**c**) similar to *S. meliloti* nodules on untransformed *M. truncatula* roots (**b, d**)

on *M. truncatula Nfr1, Nfr5* plants inoculated with *M. loti* were uninfected, indicating that additional factors are involved in host recognition during the infection process.

Further analysis of NFR1 and NFR5's involvement in host recognition was analyzed by taking advantage of the differential host specificity between two *Lotus* species *L. japonicus* and *L. filicaulis*. The normal bacterial symbiont of pea, *Rhizobium leguminosarum* bv. *viciae* is not able to form nodules on *L. japonicus*. However, by genetically modifying the *Rhizobium leguminosarum* strain with the *f*lavonoid-*i*ndependent *t*ranscription *a*ctivator (FITA) *NodD* and the *NodZ* and *NolL* genes (DZL), the DZL strain constitutively produces a modified Nod factor with an acetylated fucosyl group at the reducing end of the Nod factor, thereby mimicking the Nod factor produced by *M. loti* (Pacios-Bras et al. 2003). This modification of the Nod factor permits *L. japonicus* but not *L. filicaulis* to recognize and form nodules with the DZL strain, indicating a difference in Nod-factor recognition between *L. japonicus* and *L. filicaulis* (Radutoiu et al. 2007).

By constructing chimeric Nod-factor receptors with extracellular domains from *L. filicaulis* NFR1 and NFR5 fused to the transmembrane and intracellular regions of *L. japonicus* NFR1 and NFR5 and testing the host specificity of the transformed roots, it could be shown that the extracellular domains of the NFRs were responsible for the change in host specificity. Subsequent substitutions of individual amino acids in the *L. filicaulis* extracellular domain with the amino acid found in *L. japonicus* NFR1 and NFR5, allowed the molecular basis for the difference in host specificity to be pinpointed to a single amino acid, NFR5-L/K118 located in the second LysM domain of NFR5. Mapping this specificity-determining residue to a homology model of the second NFR5 LysM domain, positions it at the entry/exit site of a potential Nod-factor binding groove (Fig. 3). Interestingly, the putative Nod-factor binding groove on the LysM domain is lined with aromatic residues, which could form stacking interactions with the sugar rings of the Nod factor

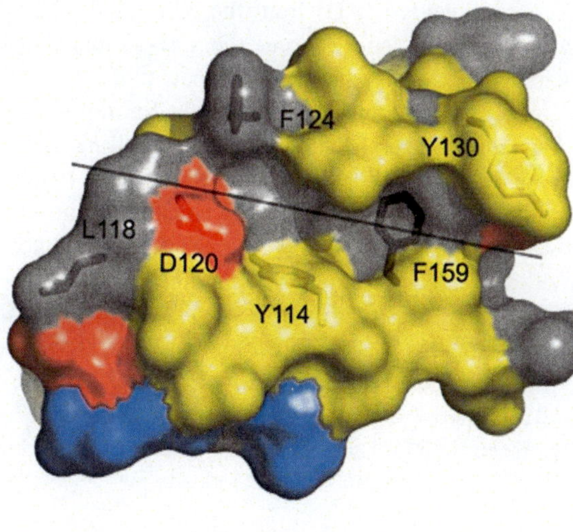

Fig. 3 Homology model of the wild type *L. japonicus* NFR5 LysM2 domain highlighting the possible Nod-factor binding groove and interacting amino acid positions including the leucine 118 at the exit/entry of the suggested Nod-factor binding groove. The model is coloured according to residue type. Hydrophobic residues are *grey*, polar residues *yellow*, acidic residues *red* and basic residues are *blue*

(Fig. 3) (Radutoiu et al. 2007). In conclusion, the molecular analysis of NFR1 and NFR5's role in legume recognition of rhizobia strongly suggest that NFR1 and NFR5 together or in parallel recognize the Nod factor and are major components in the host plants recognition.

3 Differences Between Symbiotic and Defence Signalling

Chitin is found abundantly in fungal cell walls and is a major microbe-associated molecular pattern (MAMP) that triggers plant innate immune responses (Boller and Felix 2009; Felix et al. 1993). By testing *Arabidopsis* LysM containing RLK knock-out mutants for lack of elicitor responses to treatment with chitin an Arabidopsis chitin receptor (CERK1) was identified (Miya et al. 2007; Wan et al. 2008). Sequence alignment with *Lotus* NFR1 reveals that NFR1 and CERK1 are highly similar. However, *nfr1* (and *nfr5*) mutants show a normal induction of chitooligo-saccharide-responsive genes upon chitin treatment, indicating that they are not affected in chitin signalling (Lohmann et al. 2010; Wan et al. 2008). In order to understand how the two highly identical receptors, NFR1 and CERK1 can give very contrasting signal outputs, either symbiosis or defence, a chimeric receptor containing the extracellular region of NFR1 fused to the intracellular region of CERK1 was constructed (Nakagawa et al. 2011). The NFR1-CERK1 chimeric receptor was unable to rescue nodulation deficient *nfr1-4* mutants, indicating that despite the receptor having the correct Nod-factor recognition LysM domains, the signalling by CERK1 intracellular domain differs from that required for Nod-factor signalling. The authors furthermore found that introducing the activation loop and the neighbouring EF/F loop from NFR1 into the chimeric receptor was sufficient to rescue the mutant phenotype, showing that only small changes to the CERK1 kinase allows it to produce a "symbiosis" signal output. Furthermore, adding only three amino acids (YAQ) from the NFR1 EF/F loop to the CERK1 intracellular domain was sufficient to rescue the *nfr1-4* mutant phenotype although with lower efficiency. Curiously, these data indicate that structural differences in or around the kinase activation loop are responsible for the ability of the NFR1 cytoplasmic domain to interact with downstream signalling elements in the symbiotic signal pathway. These findings challenge the general belief that protein kinase domains themselves are structurally highly conserved but structural differences including phosphorylation sites in the surrounding juxtamembrane and C-terminal tail provides the required specificity to the kinase by working as docking sites for interacting partners (Nühse et al. 2004; Wang et al. 2005b).

Lotus contains 4 LysM receptor kinase (*Lys*) genes, which belong to the same phylogenetic group as *Nfr1* and *CERK1* (Lohmann et al. 2010). Among these closely related *Lys* family members, the 'YAQ' sequence in the EF/F loop, is only present in NFR1, LYS6 and LYS7 and not found in LYS1 and LYS2. Chimeric receptors with the intracellular domain of LYS2, LYS6 and LYS7 fused to the extracellular domain of NFR1 confirmed the importance of the YAQ-motif as LYS2 failed to rescue the symbiotic phenotype of *nfr1-4*. In contrast, *nfr1-4* mutant plants transformed with the *Nfr1-Lys6 and Nfr1-Lys7* chimeric receptors were able to rescue the mutant phenotype, but the nodulation was delayed and most of the *Nfr1-Lys7* nodules did not contain endosymbiotic rhizobia (Nakagawa et al. 2011). The incomplete complementation with the cytoplasmic domains of LYS6 and LYS7 indicates that rhizobial recognition is taking place continuously in the infection process and involves different receptor complexes. An alternative explanation is that only a fully correct initial signal is capable of driving the infection process to completion. It is evident that further analysis of the individual *Lys* family members is required to fully understand the *Lys* genes role during rhizobial infection and defence signalling.

4 A Dual Role for the Symbiosis Receptor SYMRK

Forward genetic screens have identified, in addition to the Nod-factor receptors, a leucine-rich repeat (LRR) receptor kinase named SYMRK that is required for symbiosis with rhizobia. Genetic analysis indicates that *symrk* mutants are blocked further downstream of the signalling pathway than *nfr1* and *nfr5* mutants, as *symrk* mutants sense and respond to treatment with purified Nod factor (Radutoiu et al. 2003; Stracke et al. 2002). Furthermore, *symrk* mutants are not only defective in nodule formation but are also defective or delayed in mycorrhiza formation and lack the Ca^{2+}-spiking signature of the common symbiotic pathway required for both nodule and mycorrhiza formation (Kistner et al. 2005; Miwa et al. 2006; Wais et al. 2000). The domain structure with a large N-terminal region, a conserved region preceding three LRRs, a single-pass transmembrane region and a C-terminal kinase domain with biochemically proven protein kinase activity, indicates that SYMRK binds and perceives an extracellular ligand and transmits the signal across the plasma membrane to intercellular downstream components (Stracke et al. 2002; Yoshida and Parniske 2005). The identity of the SYMRK ligand and the role of SYMRK in symbiosis are, however, currently far from clear. A detailed phenotypic characterization of *symrk* and the corresponding *M. truncatula* mutant, *dmi2* showed that root hairs display an increased touch response compared to wild-type plants. This observation indicates that *SYMRK* in addition to its role in nodulation and mycorrhization might have a currently undescribed non-symbiotic function (Esseling et al. 2004). The function and potential ligand-binding activity of the SYMRK extracellular region has been analysed by testing *SYMRK* orthologs from several taxonomic clades for functional complementation of *symrk* mutants. Initially, *Lotus SYMRK* was tested for complementation of the *dmi2* mutant and *vice versa DMI2* for complementation of the *symrk* mutant. This lead to the discovery that despite *M. truncatula* belonging to a different cross-inoculation group than *L. japonicus*, both *SYMRK* and *DMI2* could fully complement *dmi2* and *symrk* mutants, respectively, for both nodulation and mycorrhiza formation. This indicates that SYMRK is not involved in determining legume host specificity (Markmann et al. 2008).

SYMRK orthologs from non-nodulating plants, which lack one LRR or one LRR and the large N-terminal extension, were all able to complement *Lotus symrk* mutants for mycorrhiza formation. Curiously, the rice *SYMRK* ortholog, which have a truncated extracellular domain and only two LRRs, was occasionally able to form small round nodules. This surprising finding indicates that not all three LRRs or the large N-terminal extension of *SYMRK* are essential elements of SYMRK for its function in symbiotic signalling (Markmann et al. 2008). As the extended extracellular domain has not been lost in evolution, it is tempting to speculate that it is involved in a currently unknown non-symbiotic function, and only the short extracellular region and intracellular part of SYMRK have been recruited for symbiosis signalling.

Further hints to the function of SYMRK in symbiosis came from an extensive genetic analysis in which the authors took advantage of a mutant encoding a constitutive active calcium calmodulin-dependent protein kinase (CCaMK), named *Spontaneous Nodule Formation 1* (*snf1*) to circumvent the requirement for Nod-factor signalling for nodule organogenesis (Madsen et al. 2010). This permitted an analysis where one can distinguish between the requirement of a given gene for infection thread formation and nodule organogenesis. In this manner, it was found that *snf1 symrk* double mutants are able to form infected nodules with *M. loti*. Infection threads were observed in the nodules but the formation was delayed and they occurred at a reduced number (Madsen et al. 2010). However, the finding that *snf1 symrk* double mutants are able to form infected nodules with a relative high efficiency shows that signal transduction through the SYMRK receptor is dispensable for root hair infection thread formation and bacterial release in *Lotus snf1* nodules. This is in contrast to results from two studies where *SYMRK* function was analyzed using *SYMRK* RNAi lines. Downregulation of either *MtDMI2* or *Sesbania rostrata SYMRK* led to serious defects in symbiosome formation. No or little release of bacteria from the infection threads was observed; instead the infection threads developed an unusual bag-like appearance, with protruding bulges and extensive branching. The abnormal infection thread growth and almost complete lack of symbiosomes in the infected tissue led to the conclusion that *SYMRK* plays a role in symbiosome formation, bacterial release from the infection threads or restriction of the infection thread growth (Capoen et al. 2005; Limpens et al. 2005). The discrepancies between the double mutant analysis and the RNAi lines could result from differences in the nodulation mechanism of *Lotus* and *Medicago* and *Sesbania*. Alternatively, the RNAi phenotype could be caused by a weak SYMRK signal early in the signalling pathway, which is insufficient to drive the nodulation process to completion.

5 The Elusive Role of the LYS Family Members

Besides the genetically defined *Nfr1*, *Nfr5* and *SYMRK* symbiotic receptor kinases, experimental evidence indicates that additional RLKs are involved in symbiosis. *snf1* mutants are able to form infected nodules with *M. loti* in a *nfr1* and *nfr5* mutant background, although at a reduced frequency. A detailed analysis of the infection process revealed that the infection did not occur through infection threads in the root hair, but instead infection threads initiated in cracks at the junction between the nodule and root. Once inside the root tissue transcellular infection threads develop and grow, leading to fully or partly infected nodules (Madsen et al. 2010). This shows that when the nodule primordia in *snf1* roots have formed, there is no strict requirement for *Nfr1* and *Nfr5* for crack entry. However, there is still a requirement for Nod-factor recognition for this type of rhizobial infection. When similar triple *nfr1 nfr5 snf1* mutants were inoculated with *M. loti nodC* or *M. loti nodA*, both unable to produce Nod factor, no transcellular infection threads were formed. Instead bacterial infections occur through intercellular epidermal entry where only single cells are infected. This indicates that besides *Nfr1* and *Nfr5*, which are required for the epidermal root hair infection process, one or several cortical Nod-factor receptors are involved in the subsequent infection of the dividing cortical tissue. Among the candidates for cortical Nod-factor receptors are the remaining members of the LYS protein family, which in *L. japonicus* consist of 17 members (Fig. 1).

The *Lys* gene family can be divided in 3 groups. The first group (LYS-I) contains *Lys* genes (*Lys1-7*), which have high similarity to *Nfr1* and contain 10–13 introns (Fig. 1). The second group (LYS-II) contains *Lys* genes (*Lys11-16*), which are most similar to *Nfr5* and all intron less. The last group (LYS-III) contains two members (*Lys20-21*), which are markedly different from both *Nfr1* and *Nfr5* and which both contain a single intron. Furthermore, the kinase domains of the LYS proteins in the LYS-I and LYS-III groups all contain the conserved kinase subdomains, whereas the LYS proteins in LYS-II group all lack or have substitutions in conserved residues in kinase subdomains. However, only LYS11 has an NFR5-like truncated activation loop. The role is currently unknown for most of the LysM-containing receptors. The likely *Medicago* ortholog of *Lys11*, *Lyr1* has been found to be three-, sixfold upregulated in mycorrhized roots (Gomez et al. 2009). Interestingly, simple lipochitin oligosaccharides have recently been identified as mycorrhiza signalling molecules, which can induce root hair deformation in *Vicia sativa*, and nodulin induction, lateral root formation and increase the number of mycorrhiza infection sites in *M. truncatula* (Maillet et al. 2011). Could *Lys11*/*Lyr1* have a role in recognition of mycorrhizal lipochitin similar to *Nfr5*'s role in Nod factor recognition?

Expression data show that nine of the *Lotus* Lys genes (*Lys2*, *Lys3*, *Lys5*, *Lys6*, *Lys12–16*) are significantly regulated after inoculation with rhizobia indicating a role in either symbiosis or MAMP perception (Lohmann et al. 2010). One group of Lys genes including *Lys2*, *Lys3*, *Lys13*, *Lys14* and *Lys16* are significantly upregulated, whereas another group of Lys genes are significantly downregulated upon inoculation (*Lys5*, *Lys6*, *Lys12*, *Lys15*). Furthermore, three genes, *Lys13*, *Lys14* and *Lys20* were found to be regulated by chitin in both shoot and root and to similar levels as known chitin-regulated genes, indicating a potential role in chitin signalling for this group (Lohmann et al. 2010).

6 Perspectives

The high similarity between the *Lotus* homologs of the *Arabidopsis* chitin receptor *CERK1* and the observation that several other *Lys* genes are upregulated by chitin makes it difficult to predict which *Lys* genes are involved in chitin perception in *Lotus*. Further complications may arise from the potential cross-talk and functional overlap between Nod-factor receptors, Myc-factor receptors and MAMP receptors. It is therefore increasingly evident that we are only beginning to understand the complexity of the signalling network of the symbiotic receptors. A better understanding of the LysM domains structure and binding properties would aid identifying the ligands and elucidate the role of the orphan LYS receptors. Furthermore, identifying the immediate downstream signalling components of the symbiotic receptors and obtaining insight into the molecular differences between receptor complexes that permit the interacting proteins to distinguish highly similar cytoplasmic domains will be interesting tasks for the future.

Acknowledgment Figures 2 and 3 were first published in Radutoiu et al. (2007) by Nature Publishing Group.

References

Arrighi JF, Barre A, Ben Amor B, Bersoult A, Soriano LC, Mirabella R, de Carvalho-Niebel F, Journet EP, Gherardi M, Huguet T, Geurts R, Denarie J, Rouge P, Gough C (2006) The Medicago truncatula lysin [corrected] motif-receptor-like kinase gene family includes NFP and new nodule-expressed genes. Plant Physiol 142:265–279

Boller T, Felix G (2009) A renaissance of elicitors: perception of microbe-associated molecular patterns and danger signals by pattern-recognition receptors. Annu Rev Plant Biol 60:379–406

Boudeau J, Miranda-Saavedra D, Barton GJ, Alessi DR (2006) Emerging roles of pseudokinases. Trends Cell Biol 16:443–452

Capoen W, Goormachtig S, De Rycke R, Schroeyers K, Holsters M (2005) SrSymRK, a plant receptor essential for symbiosome formation. Proc Natl Acad Sci USA 102:10369–10374

D'Haeze W, Holsters M (2002) Nod factor structures, responses, and perception during initiation of nodule development. Glycobiology 12:79R–105R

Eckert C, Lecerf M, Dubost L, Arthur M, Mesnage S (2006) Functional analysis of AtlA, the major N-acetylglucosaminidase of Enterococcus faecalis. J Bacteriol 188:8513–8519

Ehrhardt DW, Atkinson EM, Long SR (1992) Depolarization of alfalfa root hair membrane potential by Rhizobium meliloti Nod factors. Science 256:998–1000

Ehrhardt DW, Wais R, Long SR (1996) Calcium spiking in plant root hairs responding to Rhizobium nodulation signals. Cell 85:673–681

Esseling JJ, Lhuissier FG, Emons AM (2004) A nonsymbiotic root hair tip growth phenotype in NORK-mutated legumes: implications for nodulation factor-induced signaling and formation of a multifaceted root hair pocket for bacteria. Plant Cell 16:933–944

Felix G, Regenass M, Boller T (1993) Specific perception of subnanomolar concentrations of chitin fragments by tomato cells: induction of extracellular alkalinization, changes in protein phosphorylation, and establishment of a refractory state. Plant J 4:307–316

Gomez SK, Javot H, Deewatthanawong P, Torres-Jerez I, Tang Y, Blancaflor EB, Udvardi MK, Harrison MJ (2009) Medicago truncatula and Glomus intraradices gene expression in cortical cells harboring arbuscules in the arbuscular mycorrhizal symbiosis. BMC Plant Biol 9:10

Handberg K, Stougaard J (1992) Lotus japonicus, an autogamous, diploid legume species for classical and molecular genetics. Plant J 2:487–496

Hogslund N, Radutoiu S, Krusell L, Voroshilova V, Hannah MA, Goffard N, Sanchez DH, Lippold F, Ott T, Sato S, Tabata S, Liboriussen P, Lohmann GV, Schauser L, Weiller GF, Udvardi MK, Stougaard J (2009) Dissection of symbiosis and organ development by integrated transcriptome analysis of Lotus japonicus mutant and wild-type plants. PLoS One 4:e6556

Iizasa E, Mitsutomi M, Nagano Y (2010) Direct binding of a plant LysM receptor-like kinase, LysM RLK1/CERK1, to chitin in vitro. J Biol Chem 285:2996–3004

Indrasumunar A, Kereszt A, Searle I, Miyagi M, Li D, Nguyen CD, Men A, Carroll BJ, Gresshoff PM (2010) Inactivation of duplicated nod factor receptor 5 (NFR5) genes in recessive loss-of-function non-nodulation mutants of allotetraploid soybean (Glycine max L. Merr.). Plant Cell Physiol 51:201–214

Kaku H, Nishizawa Y, Ishii-Minami N, Akimoto-Tomiyama C, Dohmae N, Takio K, Minami E, Shibuya N (2006) Plant cells recognize chitin fragments for defense signaling through a plasma membrane receptor. Proc Natl Acad Sci USA 103:11086–11091

Kistner C, Winzer T, Pitzschke A, Mulder L, Sato S, Kaneko T, Tabata S, Sandal N, Stougaard J, Webb KJ, Szczyglowski K, Parniske M (2005) Seven Lotus japonicus genes required for transcriptional reprogramming of the root during fungal and bacterial symbiosis. Plant Cell 17:2217–2229

Klaus-Heisen D, Nurisso A, Pietraszewska-Bogiel A, Mbengue M, Camut S, Timmers T, Pichereaux C, Rossignol M, Gadella TW, Imberty A, Lefebvre B, Cullimore JV (2011) Structure-function similarities between a plant receptor-like kinase and the human interleukin-1 receptor-associated kinase-4. J Biol Chem 286:11202–11210

Lerouge P, Roche P, Faucher C, Maillet F, Truchet G, Prome JC, Denarie J (1990) Symbiotic host-specificity of Rhizobium meliloti is determined by a sulphated and acylated glucosamine oligosaccharide signal. Nature 344:781–784

Limpens E, Franken C, Smit P, Willemse J, Bisseling T, Geurts R (2003) LysM domain receptor kinases regulating rhizobial Nod factor-induced infection. Science 302:630–633

Limpens E, Mirabella R, Fedorova E, Franken C, Franssen H, Bisseling T, Geurts R (2005) Formation of organelle-like N2-fixing symbiosomes in legume root nodules is controlled by DMI2. Proc Natl Acad Sci USA 102:10375–10380

Lohmann GV, Shimoda Y, Nielsen MW, Jorgensen FG, Grossmann C, Sandal N, Sorensen K, Thirup S, Madsen LH, Tabata S, Sato S, Stougaard J, Radutoiu S (2010) Evolution and regulation of the Lotus japonicus LysM receptor gene family. Mol Plant Microbe Interact 23:510–521

Madsen EB, Madsen LH, Radutoiu S, Olbryt M, Rakwalska M, Szczyglowski K, Sato S, Kaneko T, Tabata S, Sandal N, Stougaard J (2003) A receptor kinase gene of the LysM type is involved in legume perception of rhizobial signals. Nature 425:637–640

Madsen LH, Tirichine L, Jurkiewicz A, Sullivan JT, Heckmann AB, Bek AS, Ronson CW, James EK, Stougaard J (2010) The molecular network governing nodule organogenesis and infection in the model legume Lotus japonicus. Nat Commun 1:1–12

Madsen EB, Meritxell A, Grossmann C, Ye J, Vieweg S, Broghammer A, Krusell L, Radutoiu S, Jensen ON, Stougaard J, Parniske M (2011) Autophosphorylation is essential for the in vivo function of the Lotus japonicus Nod factor receptor 1 and receptor-mediated signalling in cooperation with Nod factor receptor 5. Plant J 65(3):404–417. doi:10.1111/j.1365-313X.2010.04431.x

Maillet F, Poinsot V, Andre O, Puech-Pages V, Haouy A, Gueunier M, Cromer L, Giraudet D, Formey D, Niebel A, Martinez EA, Driguez H, Becard G, Denarie J (2011) Fungal lipochitooligosaccharide symbiotic signals in arbuscular mycorrhiza. Nature 469:58–63

Markmann K, Giczey G, Parniske M (2008) Functional adaptation of a plant receptor-kinase paved the way for the evolution of intracellular root symbioses with bacteria. PLoS Biol 6:e68

Mbengue M, Camut S, de Carvalho-Niebel F, Deslandes L, Froidure S, Klaus-Heisen D, Moreau S, Rivas S, Timmers T, Herve C, Cullimore J, Lefebvre B (2010) The Medicago truncatula E3 ubiquitin ligase PUB1 interacts with the LYK3 symbiotic receptor and negatively regulates infection and nodulation. Plant Cell 22:3474–3488

Miwa H, Sun J, Oldroyd GE, Downie JA (2006) Analysis of Nod-factor-induced calcium signaling in root hairs of symbiotically defective mutants of Lotus japonicus. Mol Plant Microbe Interact 19:914–923

Miya A, Albert P, Shinya T, Desaki Y, Ichimura K, Shirasu K, Narusaka Y, Kawakami N, Kaku H, Shibuya N (2007) CERK1, a LysM receptor kinase, is essential for chitin elicitor signaling in Arabidopsis. Proc Natl Acad Sci USA 104:19613–19618

Nakagawa T, Kaku H, Shimoda Y, Sugiyama A, Shimamura M, Takanashi K, Yazaki K, Aoki T, Shibuya N, Kouchi H (2011) From defense to symbiosis: limited alterations in the kinase domain of LysM receptor-like kinases are crucial for evolution of legume-Rhizobium symbiosis. Plant J 65:169–180

Nuhse TS, Stensballe A, Jensen ON, Peck SC (2004) Phosphoproteomics of the Arabidopsis plasma membrane and a new phosphorylation site database. Plant Cell 16:2394–2405

Oh MH, Wang X, Kota U, Goshe MB, Clouse SD, Huber SC (2009) Tyrosine phosphorylation of the BRI1 receptor kinase emerges as a component of brassinosteroid signaling in Arabidopsis. Proc Natl Acad Sci USA 106:658–663

Pacios-Bras C, Schlaman HR, Boot K, Admiraal P, Langerak JM, Stougaard J, Spaink HP (2003) Auxin distribution in Lotus japonicus during root nodule development. Plant Mol Biol 52:1169–1180

Radutoiu S, Madsen LH, Madsen EB, Felle HH, Umehara Y, Gronlund M, Sato S, Nakamura Y, Tabata S, Sandal N, Stougaard J (2003) Plant recognition of symbiotic bacteria requires two LysM receptor-like kinases. Nature 425:585–592

Radutoiu S, Madsen LH, Madsen EB, Jurkiewicz A, Fukai E, Quistgaard EM, Albrektsen AS, James EK, Thirup S, Stougaard J (2007) LysM domains mediate lipochitin-oligosaccharide recognition and Nfr genes extend the symbiotic host range. EMBO J 26:3923–3935

Remy W, Taylor TN, Hass H, Kerp H (1994) Four hundred-million-year-old vesicular arbuscular mycorrhizae. Proc Natl Acad Sci USA 91:11841–11843

Roche P, Debelle F, Maillet F, Lerouge P, Faucher C, Truchet G, Denarie J, Prome JC (1991) Molecular basis of symbiotic host specificity in Rhizobium meliloti: nodH and nodPQ genes encode the sulfation of lipo-oligosaccharide signals. Cell 67:1131–1143

Rodpothong P, Sullivan J, Songsrirote K, Sumpton D, Cheung K, Thomas-Oates J, Radutoiu S, Stougaard J, Ronson CW (2009) Nodulation gene mutants of Mesorhizobium loti R7A-nodZ and nolL mutants have host-specific phenotypes on Lotus spp. Mol Plant Microbe Interact 22:1546–1554

Spaink HP, Sheeley DM, van Brussel AA, Glushka J, York WS, Tak T, Geiger O, Kennedy EP, Reinhold VN, Lugtenberg BJ (1991) A novel highly unsaturated fatty acid moiety of lipo-oligosaccharide signals determines host specificity of Rhizobium. Nature 354:125–130

Sprent JI (2007) Evolving ideas of legume evolution and diversity: a taxonomic perspective on the occurrence of nodulation. New Phytol 174:11–25

Steen A, Buist G, Leenhouts KJ, El Khattabi M, Grijpstra F, Zomer AL, Venema G, Kuipers OP, Kok J (2003) Cell wall attachment of a widely distributed peptidoglycan binding domain is hindered by cell wall constituents. J Biol Chem 278:23874–23881

Stracke S, Kistner C, Yoshida S, Mulder L, Sato S, Kaneko T, Tabata S, Sandal N, Stougaard J, Szczyglowski K, Parniske M (2002) A plant receptor-like kinase required for both bacterial and fungal symbiosis. Nature 417:959–962

Wais RJ, Galera C, Oldroyd G, Catoira R, Penmetsa RV, Cook D, Gough C, Denarie J, Long SR (2000) Genetic analysis of calcium spiking responses in nodulation mutants of Medicago truncatula. Proc Natl Acad Sci USA 97:13407–13412

Wan J, Zhang XC, Neece D, Ramonell KM, Clough S, Kim SY, Stacey MG, Stacey G (2008) A LysM receptor-like kinase plays a critical role in chitin signaling and fungal resistance in Arabidopsis. Plant Cell 20:471–481

Wang X, Goshe MB, Soderblom EJ, Phinney BS, Kuchar JA, Li J, Asami T, Yoshida S, Huber SC, Clouse SD (2005a) Identification and functional analysis of in vivo phosphorylation sites of the Arabidopsis BRASSINOSTEROID-INSENSITIVE1 receptor kinase. Plant Cell 17:1685–1703

Wang X, Li X, Meisenhelder J, Hunter T, Yoshida S, Asami T, Chory J (2005b) Autoregulation and homodimerization are involved in the activation of the plant steroid receptor BRI1. Dev Cell 8:855–865

Yoshida S, Parniske M (2005) Regulation of plant symbiosis receptor kinase through serine and threonine phosphorylation. J Biol Chem 280:9203–9209

The Cell Wall-Associated Kinases, WAKs, Regulate Cell Expansion and the Stress Response

Bruce D. Kohorn and Susan L. Kohorn

Abstract The plant cell wall is secreted and assembled by cells to provide structure and shape, and thereby helps to determine the form of a plant organ. Control of the synthesis and directional enlargement of the wall is therefore crucial for plant development, but the wall also serves as a first defense against common plant stresses such as pathogens and physical wounding. There is now substantial evidence to suggest that the Cell Wall Associated Kinases, WAKs, are pectin receptors required for both normal cell elongation and for an induced stress response.

1 Introduction

Until recently, common perceptions in cell signaling centered on linear transduction pathways that depended upon ligands activating receptors in the plasma membrane, cytoplasm, or nucleus. The resulting cascade of events was thought to promote a change in gene expression or enzyme regulation that was directly attributable to that one ligand. This understandably naive view has helped to focus efforts and has indeed resulted in great strides in understanding these pathways. However, it has become very clear that signaling pathways form an integrated network, involving multiple coordinated ligands, receptors, and outputs. An excellent example of this is the way in which plants build, modify, and regulate their extracellular matrix, or cell wall. Indeed, mutations in receptors initially identified to regulate gametogenesis turn out to have dramatic effects on vegetative cell wall dynamics, and conversely disruptions in cell wall biosynthesis enzymes cause perturbations in normal developmental patterns governed by numerous receptors. The root of this relationship lies in the role cell walls play in plant development, but also in the amazing interconnectedness of these pathways.

B.D. Kohorn (✉) • S.L. Kohorn
Biology Department, Bowdoin College, Brunswick, ME 04011, USA
e-mail: bkohorn@bowdoin.edu; skohorn@bowdoin.edu

F. Tax and B. Kemmerling (eds.), *Receptor-like Kinases in Plants*,
Signaling and Communication in Plants 13,
DOI 10.1007/978-3-642-23044-8_7, © Springer-Verlag Berlin Heidelberg 2012

1.1 The Plant Cell Wall

Cell walls differ greatly between species and cell types, but the primary wall or region that is first laid down seems to have a similar basic underlying architecture (Fig. 1) (Keegstra 2010; Kohorn 2000; Seifert and Blaukopf 2010). A rosette of plasma membrane cellulose synthases extrudes cellulose polymers into the extracellular space, resulting in intertwined cellulose fibers of varying complexity (Arioli et al. 1998; Keegstra 2010; Mutwil et al. 2008; Taylor 2008). The sugar building blocks are directed to the rosette from the cytoplasm via sucrose synthase, and indeed alterations in sugar metabolism and partitioning within and between cells would be expected to have direct consequences on cellulose synthesis (Seifert 2004; Seifert and Blaukopf 2010; Winter and Huber 2000). Cellulose synthase is associated with the cytoskeleton, which helps in its directional synthesis, and as such alterations to actin or tubulin dynamics would also be expected to affect cell wall biology and perhaps vice versa (Crowell et al. 2010; Szymanski 2009).

The cell wall contains a number of other sugar-based polymers, such as hemicelluose (Scheller and Ulvskov 2010) and pectin (Harholt et al. 2010; Mohnen 2008) both of which are synthesized in the golgi and secreted via vesicles (Fig. 1). Thus, any regulation of the golgi, its enzyme content and activity, or of secretory processes including plasma membrane fusion dynamics, will have direct effects on cell wall architecture and composition. Thus, it is not surprising that screens for mutants in developmental processes have turned up numerous alleles of cell wall

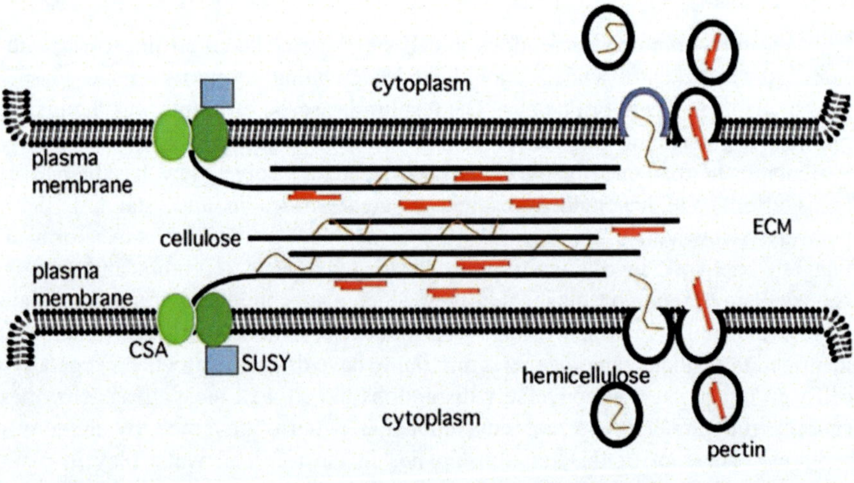

Fig. 1 Cartoon of cell wall assembly and structure between two cells. Cellulose is synthesized and secreted by a rosette of cellulose synthases (CSAs) on the plasma membrane, and forms a crystalline microfibrils (*black cables*; cellulose). Cellulose synthase binds cytoplasmic sucrose synthase (SUSY) that provides the sugars for polymer synthesis. Pectin (*red bars*) and hemicellulose (*orange wiggles*) are synthesized in and secreted through the golgi and secretory vesicles. Reprinted with permission from (Kohorn 2000)

biosynthesis genes, and conversely mutations in cell wall aspects reveal alleles in genes normally associated with a variety of metabolic and developmental pathways (Seifert and Blaukopf 2010).

Perhaps useful but less usual is to remember that the cell wall lies between adjacent cells (Fig. 1), and that contributions by one cell could greatly affect its neighbor, an attribute commonly accepted by metazoan biologists, but not discussed sufficiently in the plant world, with the exception of those studying pollen tube growth through the stigma and style (Bosch and Hepler 2005; Krichevsky et al. 2007; Winship et al. 2010). What is readily recognized is the importance for the structural integrity of the wall such that the turgor-driven pressure from the contained cell is maintained. Regulation of the direction and time in which the wall is synthesized and expanded therefore dictates cell size and shape and organ characteristics.

The cell wall has a large array of diverse proteins that are anchored in the membrane and extend into the matrix, or are situated in the matrix (Anderson et al. 2001; Keegstra 2010; Kohorn 2000; Seifert and Blaukopf 2010). There are numerous proteins whose functions have been defined independently of the cell wall, but subsequently are realized to have a role in cell wall dynamics and makeup. This chapter will include and briefly review examples that help us to understand why there is an emerging picture of a network of signals that shapes both plant development and the cell wall and will then focus on one family, the Wall Associated Kinases (WAKs) that are now known to provide a direct signaling link from the cell wall to regulate development and the pathogen response.

1.2 Receptors Associated with Cell Walls

There are a growing number of plasma membrane receptors that have been identified by genetic screens for a variety of developmental phenotypes, and many turn out to have some relation to cell wall biology. The converse is true where mutations and their suppressors that were thought to be peculiar to the cell wall, also turn out to have larger physiological relevance. These were thoroughly and thoughtfully reviewed recently (Seifert and Blaukopf 2010) and will not be discussed extensively, except to emphasize how receptors and cell wall biosynthesis machinery are quite interconnected.

One of the best examples of the complexity of relationships that governs cell wall biology is the isolation of the *THESEUS* gene (*THE1*) (Hematy and Hofte 2008; Hematy et al. 2007), and whose discovery lead the understanding of a larger family that includes *FERONIA (FER)*, *HERKULES (HERK)*, and *ANXUR* (Guo et al. 2009a, b; Hematy and Hofte 2008; Miyazaki et al. 2009). *THE1* was isolated as a suppressor of a cellulose synthase mutation that led to a loss of cell elongation. *FER* was discovered as a mutation effecting gametogenesis, but is now realized to influence many other aspects of vegetative growth as it is required for cell elongation. The family has been termed CrRLK1L (for its founding members first identified in

identified in this yeast selection, there were some interesting lessons. The selection was established to identify proteases if the protease cleavage site was defined. By expressing plant cDNAs in a yeast also expressing a GAL4 fusion with the protease target site imbedded, protease activity cleaved the GAL4 rendering it inactive; yeast could be selected for loss of GAL4 activity and thus the responsible cDNA recovered from a library (Smith and Kohorn 1991). The protease target used contained a phosphorylation substrate site and this apparently allowed the WAK kinase domain to inactivate GAL4. Most interesting was that polyclonal antibodies subsequently raised to the WAK kinase domain reacted with WAKs, but one monoclonal serum only identified the correct kinase for this phosphorylation substrate, and these are the Thylakoid-Associated Kinases or TAKs that regulate light harvesting in the chloroplast (Snyders and Kohorn 1999, 2001). TAKs represent the only TGF-like receptor family in the plant kingdom.

3 Defining a WAK

What distinguished the WAKs from the few other identified plant receptors at the time was the combination of three observations. First, WAKs could only be extracted from cells by boiling in detergent and reductant and appeared crosslinked to insoluble material. Second, electron micrographs with specific WAK-kinase antiserum showed WAK protein in the cell wall and on the plasma membrane (Fig. 2). Yet topology experiments with proteases and western blots showed that the kinase domain was cytoplasmic, and the receptor indeed traversed the membrane to extend the EGF domains into the extracellular space (He et al. 1996). It was by these criteria that the protein was named WAK. A third peculiarity of WAKs was that this receptor defined a class of kinase with multiple EGF repeats. To date, the S locus SRK (Naithani et al. 2007) and a vacuolar sorting receptor (Cao et al. 2000) are the only other examples of plant proteins with an EGF repeat. Subsequent genome analysis, before the Arabidopsis genome was available, led to the identification of five WAKs all tightly clustered in tandem in a 30 kb locus on chromosome 1 (He et al. 1999) (Fig. 3). Their kinase domains are 85% identical, and the extracellular regions show up to 65% identity. However, they all contain the same conserved spacing of Cyteine residues, the hallmark of the EGF repeat of metazoans. Mammalian EGFs have been shown to create disulfide bonds in a consistent (1,3,2–4,5–6) pattern where the cysteines that form these bonds are positioned close together and create a distinctive β-5- sheet and β-hairpin turn (Li and Montelione 1995). Folding and bonding assays reveal that all EGF cysteines participate in disulfide-bonding activity, but a distinct disulfide-bonding pattern is not always necessary for EGF activity. The EGF disulfide bonding is quite plastic in that the EGF backbone can support several different cysteine-bonding patterns without a significant change to the EGF shape and function in mice (Sampoli Benitez and Komives 2000). EGF-like domains in metazoan protein receptors are

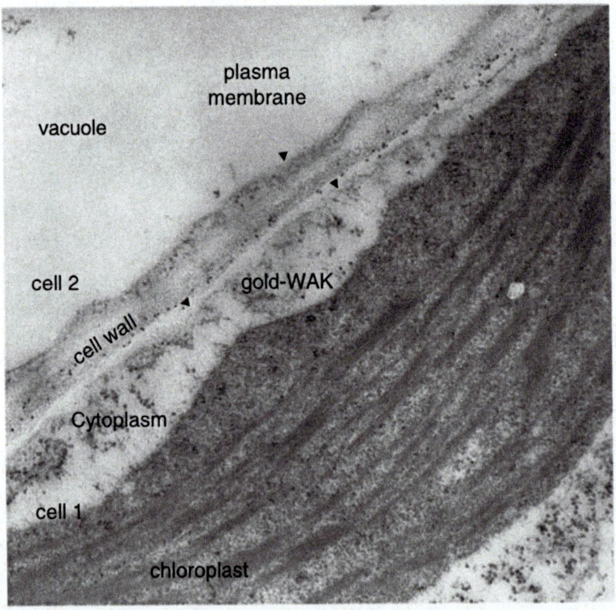

Fig. 2 Immunocytochemical localization of Wak1 by electron microscopy. Leaf sections reacted with anti-Wak1 serum, and gold-conjugated secondary antibodies show WAK present in the plasma membrane and cell wall. Reprinted with permission from (He et al. 1996)

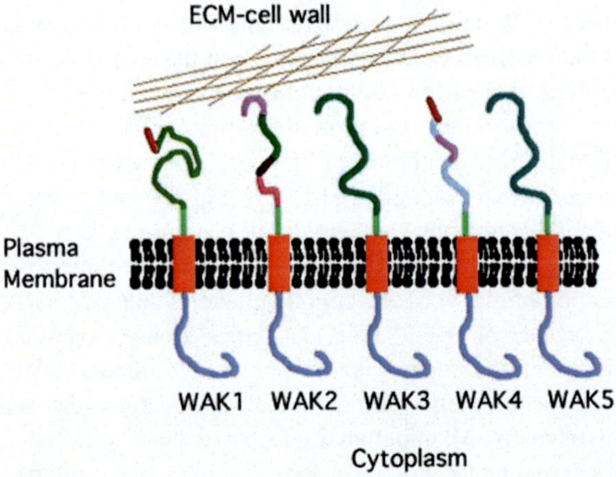

Fig. 3 The five WAKs have near identical cytoplasmic kinases, traverse the plasma membrane to bind pectin in the cell wall with divergent extracellular domains. Colors in the extracellular domains indicate distinct EGF-containing domains that differ between the WAK isoforms

responsible for calcium binding and receptor activity, but the role that the EGFs play in WAK function has yet to be reported.

In Arabidopsis, there are at least 21 *WAK*-like or *WAKL* genes, identified by their sequence similarity to *WAKs* (Verica et al. 2003; Verica and He 2002). The expansion and size of this family indicates their importance, but there is insufficient evidence to conclude that the WAKL proteins are wall associated, and indeed the

TAK kinase domain has almost the same amino acid identity as WAKs. The *WAKLs* instead may describe a more general class of plant receptor that simply contains EGF repeats, and whose additional domains and perhaps function has diverged from the *WAKs*. This remains to be determined, although recent mutations in *WAKLs* point to a role in pathogenesis and metal tolerance (Hou et al. 2005).

Initially, loss of function alleles of individual *WAKs* provided no phenotype in Arabidopsis (Wagner and Kohorn 2001), and since the genes are tightly clustered, single null alleles could not be readily combined. Attempts to identify a *WAK* locus deletion were unsuccessful. Inducible antisense RNA directed to the five member family did result in a 50% reduction in protein levels as assayed by the WAK kinase serum, and this lead to dwarf plants. The affected leaves showed a smaller cell size, rather than fewer cells, and it was concluded that WAKs are required for cell elongation (Anderson et al. 2001; Lally et al. 2001; Wagner and Kohorn 2001). More recent work identified a more subtle phenotype for a *WAK2* null allele, *wak2-1*, that caused a loss of cell expansion in roots, but only under limiting sugar and salt conditions (Kohorn et al. 2006b). Individual loss-of-function alleles in any of the four other WAKs do not result in an obvious phenotype. Enzyme analysis of tissue from *wak2-1* null mutants indicated that there was a reduction in the activity of a vacuolar invertase, which cleaves sucrose into glucose and fructose thereby increasing the solute concentration and consequently can lead to an increase in turgor. The enzyme activity reduction was accompanied by a similar reduction in invertase RNA levels, and thus WAKs appear to be regulated in part at the transcriptional level. This finding provided a theoretical mechanism by which WAKs could cause cell expansion through turgor control. It is noteworthy that the phenotype could be rescued by the repartitioning of sugar sinks through the expression of sucrose phosphate synthase, providing yet another example of the complex interactions between metabolism, the cell wall, and development (Kohorn et al. 2006b). However, it remained unclear how the WAKs could be activated to drive this possible change in turgor.

4 WAKs Bind to Pectin

The initial observation that WAKs are wall associated and crosslinked led to an analysis where it was shown that pectinase treatments but not other cell wall degrading enzymes, were able to release WAKs from the insoluble material. Further analysis revealed a pectin epitope remaining on WAKs though Western blotting, suggesting a covalent binding to a pectic fragment (Anderson et al. 2001; Wagner and Kohorn 2001). When the extracellular domain of WAK1 and 2 were expressed in yeast and purified, they bound to purified pectin in vitro. WAKs had a higher affinity for de-esterified pectin than to esterified molecules that did not have as negative a charge. Moreover, short pectin fragments of degree of polymerization (dp) 9-15 effectively competed with longer pectins for WAK binding (Decreux and Messiaen 2005). Mutation of the positively charged residues in WAK1 to neutral amino acids lead to a loss of binding to de-esterified pectin suggesting that the interaction was in part charge-based (Decreux et al. 2006). These in vitro results provided confirmation of a WAK- pectin association, but also created a puzzle as to why the WAK isolated from plants after pectinase treatment had a covalent attachment to pectin. Moreover, these results raised the question as to what type of pectin WAK was binding to and was perhaps responsive to in plants. In vitro, both WAK1 and 2 bind to a variety of pectins including polymers of homogalacturonan, Oligogalacturonides (OGs), and to Rhamnogalacturonans I and II (Kohorn et al. 2009). The common feature of these molecules is the galacturonic acid backbone, and this is therefore the predicted target of WAKs. Since WAKs bind to pectin in a calcium-dependent manner and EGF repeat structure and function involves calcium, it is possible that the EGF repeats in the ECM domain facilitate pectin binding, yet this remains to be determined in plants.

The biological activity of short pectin fragment, or Oligogalacturonides (OGs) has been established for many years, and there is a stronger link to defense and stress responses than to developmental processes (Mohnen 2008; Willats et al. 2001; Yamazaki et al. 1983). Pathogen invasion, physical wounding, and herbivory have the ability to generate OGs in the cell wall, and it has been suggested by many that there was a specific receptor for these OGs. WAKs may indeed be this receptor, but how they distinguish OGs from nascent pectins is not clear. The in vitro data raise several other interesting questions. Pectin can be de-esterified once secreted into the cell wall by a family of methyl esterases, revealing a negative charge that permits a calcium-induced crosslinking with other pectins, and perhaps other carbohydrates and proteins in the wall. The best-understood example of this regulation is in growing pollen tubes, where pectin is de-esterified along the sides of the pollen tube after it has been deposited into a more fluid and growing tip (Bosch and Hepler 2005; Krichevsky et al. 2007). Numerous reports have also documented an unequal distribution of esterified and de-esterified pectins in a variety of cell all types, and correlations between the degree of esterification and cell enlargement have been documented (Caffall and Mohnen 2009; Harholt et al. 2010; Mohnen 2008; Willats et al. 2001; Wolf et al. 2009). How the differential binding of WAKs to esterified and de-esterified pectin relates to the unequal distribution has yet to be explored. WAKs in most cells appear to be uniformly distributed (He et al. 1996), with the exception of the emerging pollen tube where WAK epitopes are only detected near the tip (Fig. 4). WAKs are present in the growing pollen tube and their relationship to pectin deposition and directional growth needs more exploration.

More recent work provides a biological relevance for the pectin-binding activity of WAKs. Pectin treatment of protoplasts leads to the induction and repression of hundreds of genes involved in cell wall biogenesis and stress responses, and this is dramatically altered in cells lacking WAK2 (Kohorn et al 2006b, 2009). Using an Invertase promoter-RFP reporter to monitor invertase expression in protoplasts, it was shown that pectin can activate invertase expression, but not in a *wak2-1* null. While there may be additional proteins in the membrane that mediate the pectin

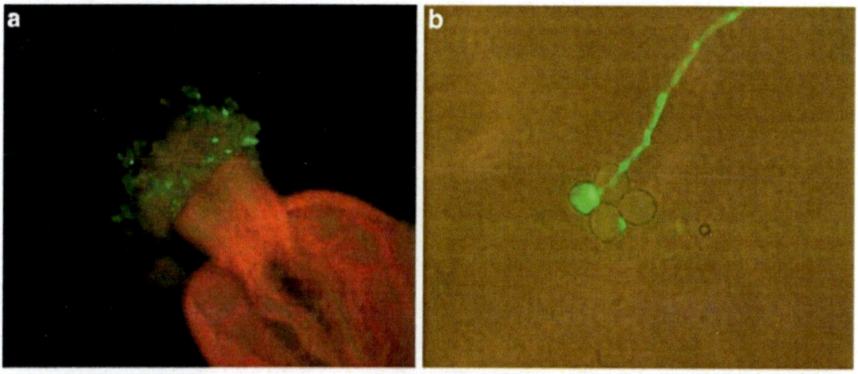

Fig. 4 WAK serum identifies WAK in pollen deposited on an Arabidopsis stigma (**a**) and in germinating pollen tubes in vitro (**b**). Fluorescent secondary antibody (*green*) was detected by confocal microscopy (**a**) or epifluorescent microscopy (**b**)

response, there appears to be much support for the claim that at least WAK1 and 2 are pectin receptors. This suggestion received more support from experiments where the WAK1 extracellular domain was fused to an unrelated cytoplasmic kinase that had a known and recordable downstream pathway (Brutus et al. 2010). The pathway downstream of the fusion kinase was activated by OGs when transiently expressed in leaves. While these experiments support the idea that WAKs are a pectin receptor, more detailed structural information and kinase activation studies are needed to make this claim conclusive. These reports do not address the question of how WAKs might distinguish longer pectin polymers present in the cell wall that they may already be bound to, from newly generated and bioactive OGs. Indeed, as discussed below, WAKs appear to have a role in pathogenesis and the stress response in addition to normal development. Intriguingly, this bifunctional role mirrors the binding to several types of pectin, but this relationship requires further analysis.

5 WAK and Pectin Biogenesis

To follow the biogenesis of WAKs and to determine where they might first bind to pectin in the cell, a WAK1-GFP fusion was followed over time in protoplasts, cells stripped of their walls (Kohorn et al. 2006a). To date attempts at detecting WAK-GFP fusions in whole plants have not been successful. These WAK1-GFP-expressing protoplasts begin regenerating their wall within 24 h. It was noted that the migration of WAK1-GFP to the cell surface is far slower than that of a cell surface receptor not associated with the cell wall, as shown in Fig. 5. The rate of this migration is increased dramatically in a *mur1* mutant that blocks the addition of fucose side chains to pectin and other unidentified molecules. The migration is also

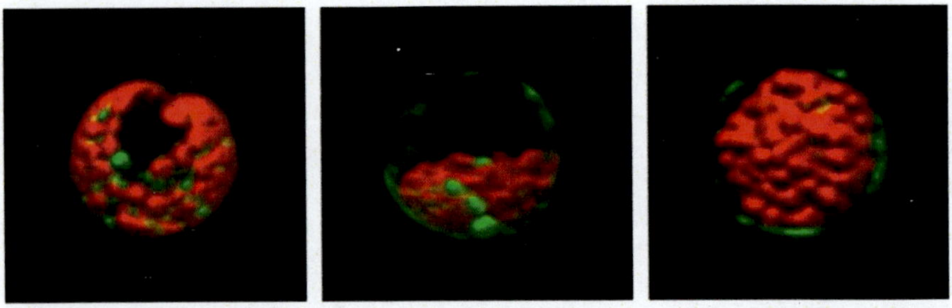

Fig. 5 Movement of WAKs to the cell surface is delayed by crosslinking in the Golgi. Three-dimensional reconstruction by confocal microscopy of WAK-GFP (*green*) expressed in protoplasts containing red chloroplasts. Images were taken after 1, 2 or 5 days of expression. See (Kohorn et al. 2006a) for details

dependent upon cellulose synthesis on the plasma membrane. These studies noted that WAK is crosslinked into a detergent-insoluble complex within the cytoplasmic compartment before it appears on the cell surface, and this is independent of fucose modification or cellulose synthesis. Thus, the assembly and crosslinking of WAKs may begin at an early stage within a cytoplasmic compartment rather than in the cell wall itself, and is coordinated with synthesis of surface cellulose.

6 Other WAK Ligands

Pectin may not be the only WAK ligand. A yeast two-hybrid assay revealed that Glycine-Rich Protein 3 (GRP3) interacts with the extracellular portion of WAK, in a domain distinct from the EGF repeats (Park et al. 2001). GRP3 interacts with WAK1 but not WAK2 or 4 in the yeast two-hybrid assay (Anderson et al. 2001). Secreted GRPs form a large family in Arabidopsis and it is possible that different family members interact with different WAKs, but this too needs to be explored (Anderson et al. 2001; Mangeon et al. 2010; Mousavi and Hotta 2005). Some suggest that GRPs are structural proteins (Ringli et al. 2001), but their function really remains undefined. Fractionation experiments under conditions that attempt to stabilize protein–protein interactions identify a 450 kDa complex that contains WAK1 and GRP3, yet this result is surprising given that WAK1 is normally not detected as a water-soluble protein, and further analysis is needed to confirm these results (Park et al. 2001). Blue native gel analysis of protein complexes fails to identify this WAK–GRP complex, and indeed the relationship between GRP, pectin, and WAKs warrants more exploration.

7 WAKs and the Stress Response

While experiments point to a role of WAKs in cell expansion during plant growth there are several lines of evidence that WAKs and WAKLs are also involved in the response to pathogen and stress. WAK expression is induced by wounding, pathogen infection, and by many stresses such as ozone and heavy metals (Anderson et al. 2001; He et al. 1998; Kohorn 2001; Sivaguru et al. 2003; Wagner and Kohorn 2001). Expression, however, is late in the response and does not necessarily show a mechanistic relationship. Dominant alleles of *WAK1* that lacked the extracellular domain did cause a resistance to high levels of salicilate, and antisense *WAK1* plants had increased sensitivity (He et al. 1998). While it was not clear how WAKs were involved, the work did establish a connection with pathogenesis, and reinforced the need to explore the role that pathogen induced OGs might have in regulating WAKs. More recently, a dominant allele of *WAK2*, *WAK2cTAP*, was observed to cause ectopic lesions, curling of leaves, and stunted growth, all hallmarks of a pathogen effect, yet in the absence of pathogen (Fig. 6) (Kohorn et al. 2009). *WAK2cTAP* alleles lacking the extracellular or kinase domains do not induce these phenotypes indicating that active receptor function is required. It is not clear why the *WAK2cTAP* allele is dominant, and additional genetic analysis is needed to understand how this allele induces the stress response. Secondly, OGs were shown to stimulate downstream targets of a WAK-kinase chimera in a heterologous transient expression assay, further evidence that OGs can regulate WAKs (Brutus et al. 2010). There is additional convincing evidence that the WAKL family plays a significant role in the pathogen response.

In Arabidopsis, there are over 21 *WAK*-like genes (*WAKLs*): this family appears to have gone through a further large expansion in family size in crop plants such as rice, leading to the suggestion of some role in pathogen resistance (Verica and He 2002; Zhang et al. 2005). Dominant mutant alleles of *WAKLs* in Arabidopsis provide resistance to Fusarium, and a rice *WAK* allele is less sensitive to rice

Fig. 6 The *WAK2cTAP* allele induces necrotic lesions, dwarfing, and leaf curling

blast disease (Diener and Ausubel 2005; Li et al. 2009). It would be helpful to establish whether the rice genes are *WAKLs* or *WAKs*, and indeed whether WAKLs are wall associated or are only related to WAKs as a result of their EGF and kinase domains.

8 WAKs and MAPKs

Mitogen-Activated Protein Kinases (MAPKs) form a major signaling link between cell surface receptors and both transcriptional and enzyme regulation in eukaryotes, and Arabidopsis MAPK6 and MAPK3 have important roles in development and the response to stress and pathogens (Andreasson and Ellis 2010; Colcombet and Hirt 2008). Evidence points to a role of MAPKs in WAK signaling. Plants homozygous for the null allele *wak2-1* showed a reduction in the activation of MAPK3 relative to wild type and in protoplasts, MAPK3 activity is elevated in cells treated with long polymers of pectin (Kohorn et al 2009). Plants expressing the *WAK2cTAP*-dominant allele in a $mapk3^{-/-}$ background had more severe growth defects than *WAK2cTAP* alone, supporting the concept that MAPK3 is required for downstream WAK2 signaling (Kohorn et al 2009). OGs are known to activate MAPKs (Andreasson and Ellis 2010; Moscatiello et al 2006), but the effect of OGs on WAK-dependent MAPK activity has yet to be reported, and this knowledge may help to distinguish the activity stimulated by long pectin polymers and OGs. Experiments to date have also failed to distinguish whether WAK2 regulates MAPK6 and indeed other MAPKs. Importantly, MAPK3 and 6 have been implicated in the response to pathogens and stress, and are known to play a role in pectin- and OG-induced plant responses (Andreasson and Ellis 2010; Colcombet and Hirt 2008). Thus, WAKs are required not only during normal cellular development but also during the pathogen and stress response, and this dual role remains unexplained (Fig. 7).

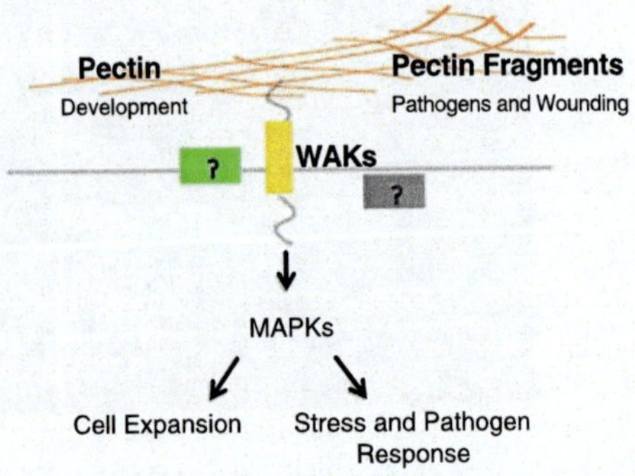

Fig. 7 Model for WAK regulation by pectin for both cell expansion and the stress response

9 Summary

The WAKs are like many other plant receptors in that they regulate some aspect of cell division or growth, but also affect aspects of the cell wall. But WAKs remain unique in that they are known to bind directly to a major structural component of the cell wall, pectin, which appears to alter a WAK-dependent signaling pathway regulating cell expansion. That WAKs are also involved in the pathogen and the stress responses can be rationalized through an understanding of how crucial pectin is to these biological events. To solve the puzzle of how WAKs are activated or inhibited by the various forms of pectin will require structural analysis of the proteins and carbohydrates, and an understanding of the nature and spatial distribution of pectin in cell walls. Given the growing number of examples of receptor interactions and convergence of their stimulated pathways, it would not be surprising to find a connection between WAKs and other classes of receptor kinases. More attention is needed in this rapidly growing and exciting field. What is clear, however, is that the many receptor kinases, including WAKs, form a signaling network that is connected by the many hundreds of developmental, metabolic, and environmental cues that shape plant form and function, many of which in some way involve the cell wall.

References

Abrash EB, Bergmann DC (2010) Regional specification of stomatal production by the putative ligand CHALLAH. Development 137:447–455

Anderson CM, Wagner TA, Perret M, He ZH, He D, Kohorn BD (2001) WAKs: cell wall-associated kinases linking the cytoplasm to the extracellular matrix. Plant Mol Biol 47:197–206

Andreasson E, Ellis B (2010) Convergence and specificity in the Arabidopsis MAPK nexus. Trends Plant Sci 15:106–113

Arioli T, Peng L, Betzner AS, Burn J, Wittke W, Herth W, Camilleri C, Hofte H, Plazinski J, Birch R et al (1998) Molecular analysis of cellulose biosynthesis in Arabidopsis. Science 279:717–720

Bai L, Zhang G, Zhou Y, Zhang Z, Wang W, Du Y, Wu Z, Song CP (2009) Plasma membrane-associated proline-rich extensin-like receptor kinase 4, a novel regulator of Ca signalling, is required for abscisic acid responses in *Arabidopsis thaliana*. Plant J 60:314–327

Boisson-Dernier A, Roy S, Kritsas K, Grobei MA, Jaciubek M, Schroeder JI, Grossniklaus U (2009) Disruption of the pollen-expressed FERONIA homologs ANXUR1 and ANXUR2 triggers pollen tube discharge. Development 136:3279–3288

Boller T, He SY (2009) Innate immunity in plants: an arms race between pattern recognition receptors in plants and effectors in microbial pathogens. Science 324:742–744

Bosch M, Hepler PK (2005) Pectin methylesterases and pectin dynamics in pollen tubes. Plant Cell 17:3219–3226

Brutus A, Sicilia F, Macone A, Cervone F, De Lorenzo G (2010) A domain swap approach reveals a role of the plant wall-associated kinase 1 (WAK1) as a receptor of oligogalacturonides. Proc Natl Acad Sci USA 107:9452–9457

Caffall KH, Mohnen D (2009) The structure, function, and biosynthesis of plant cell wall pectic polysaccharides. Carbohydr Res 344:1879–1900

Cao X, Rogers SW, Butler J, Beevers L, Rogers JC (2000) Structural requirements for ligand binding by a probable plant vacuolar sorting receptor. Plant Cell 12:493–506

Chinchilla D, Bauer Z, Regenass M, Boller T, Felix G (2006) The Arabidopsis receptor kinase FLS2 binds flg22 and determines the specificity of flagellin perception. Plant Cell 18:465–476

Colcombet J, Hirt H (2008) Arabidopsis MAPKs: a complex signalling network involved in multiple biological processes. Biochem J 413:217–226

Crowell EF, Gonneau M, Vernhettes S, Hofte H (2010) Regulation of anisotropic cell expansion in higher plants. C R Biol 333:320–324

Decreux A, Messiaen J (2005) Wall-associated kinase WAK1 interacts with cell wall pectins in a calcium-induced conformation. Plant Cell Physiol 46:268–278

Decreux A, Thomas A, Spies B, Brasseur R, Van Cutsem P, Messiaen J (2006) In vitro characterization of the homogalacturonan-binding domain of the wall-associated kinase WAK1 using site-directed mutagenesis. Phytochemistry 67:1068–1079

Diener AC, Ausubel FM (2005) RESISTANCE TO FUSARIUM OXYSPORUM 1, a dominant Arabidopsis disease-resistance gene, is not race specific. Genetics 171:305–321

Escobar-Restrepo JM, Huck N, Kessler S, Gagliardini V, Gheyselinck J, Yang WC, Grossniklaus U (2007) The FERONIA receptor-like kinase mediates male-female interactions during pollen tube reception. Science 317:656–660

Guo H, Li L, Ye H, Yu X, Algreen A, Yin Y (2009a) Three related receptor-like kinases are required for optimal cell elongation in *Arabidopsis thaliana*. Proc Natl Acad Sci USA 106:7648–7653

Guo H, Ye H, Li L, Yin Y (2009b) A family of receptor-like kinases are regulated by BES1 and involved in plant growth in *Arabidopsis thaliana*. Plant Signal Behav 4:784–786

Haffani YZ, Silva-Gagliardi NF, Sewter SK, Grace Aldea M, Zhao Z, Nakhamchik A, Cameron RK, Goring DR (2006) Altered expression of PERK receptor kinases in Arabidopsis leads to changes in growth and floral organ formation. Plant Signal Behav 1:251–260

Hara K, Yokoo T, Kajita R, Onishi T, Yahata S, Peterson KM, Torii KU, Kakimoto T (2009) Epidermal cell density is autoregulated via a secretory peptide, EPIDERMAL PATTERNING FACTOR 2 in Arabidopsis leaves. Plant Cell Physiol 50:1019–1031

Harholt J, Suttangkakul A, Vibe Scheller H (2010) Biosynthesis of pectin. Plant Physiol 153:384–395

He ZH, Fujiki M, Kohorn BD (1996) A cell wall-associated, receptor-like protein kinase. J Biol Chem 271:19789–19793

He ZH, He D, Kohorn BD (1998) Requirement for the induced expression of a cell wall associated receptor kinase for survival during the pathogen response. Plant J 14:55–63

He ZH, Cheeseman I, He D, Kohorn BD (1999) A cluster of five cell wall-associated receptor kinase genes, Wak1-5, are expressed in specific organs of Arabidopsis. Plant Mol Biol 39:1189–1196

Hematy K, Hofte H (2008) Novel receptor kinases involved in growth regulation. Curr Opin Plant Biol 11:321–328

Hematy K, Sado PE, Van Tuinen A, Rochange S, Desnos T, Balzergue S, Pelletier S, Renou JP, Hofte H (2007) A receptor-like kinase mediates the response of Arabidopsis cells to the inhibition of cellulose synthesis. Curr Biol 17:922–931

Hematy K, Cherk C, Somerville S (2009) Host-pathogen warfare at the plant cell wall. Curr Opin Plant Biol 12:406–413

Hou X, Tong H, Selby J, Dewitt J, Peng X, He ZH (2005) Involvement of a cell wall-associated kinase, WAKL4, in Arabidopsis mineral responses. Plant Physiol 139:1704–1716

Keegstra K (2010) Plant cell walls. Plant Physiol 154:483–486

Kohorn BD (2000) Plasma membrane-cell wall contacts. Plant Physiol 124:31–38

Kohorn BD (2001) WAKs; cell wall associated kinases. Curr Opin Cell Biol 13:529–533

Kohorn BD, Lane S, Smith TA (1992) An Arabidopsis serine/threonine kinase homologue with an epidermal growth factor repeat selected in yeast for its specificity for a thylakoid membrane protein. Proc Natl Acad Sci USA 89:10989–10992

Kohorn BD, Kobayashi M, Johansen S, Friedman HP, Fischer A, Byers N (2006a) Wall-associated kinase 1 (WAK1) is crosslinked in endomembranes, and transport to the cell surface requires correct cell-wall synthesis. J Cell Sci 119:2282–2290

Kohorn BD, Kobayashi M, Johansen S, Riese J, Huang LF, Koch K, Fu S, Dotson A, Byers N (2006b) An Arabidopsis cell wall-associated kinase required for invertase activity and cell growth. Plant J 46:307–316

Kohorn BD, Johansen S, Shishido A, Todorova T, Martinez R, Defeo E, Obregon P (2009) Pectin activation of MAP kinase and gene expression is WAK2 dependent. Plant J 60:974–982

Krichevsky A, Kozlovsky SV, Tian GW, Chen MH, Zaltsman A, Citovsky V (2007) How pollen tubes grow. Dev Biol 303:405–420

Lally D, Ingmire P, Tong HY, He ZH (2001) Antisense expression of a cell wall-associated protein kinase, WAK4, inhibits cell elongation and alters morphology. Plant Cell 13:1317–1331

Li YC, Montelione GT (1995) Human type-alpha transforming growth factor undergoes slow conformational exchange between multiple backbone conformations as characterized by nitrogen-15 relaxation measurements. Biochemistry 34:2408–2423

Li H, Zhou SY, Zhao WS, Su SC, Peng YL (2009) A novel wall-associated receptor-like protein kinase gene, OsWAK1, plays important roles in rice blast disease resistance. Plant Mol Biol 69:337–346

Mangeon A, Junqueira RM, Sachetto-Martins G (2010) Functional diversity of the plant glycine-rich proteins superfamily. Plant Signal Behav 5:99–104

Miyazaki S, Murata T, Sakurai-Ozato N, Kubo M, Demura T, Fukuda H, Hasebe M (2009) ANXUR1 and 2, sister genes to FERONIA/SIRENE, are male factors for coordinated fertilization. Curr Biol 19:1327–1331

Mohnen D (2008) Pectin structure and biosynthesis. Curr Opin Plant Biol 11:266–277

Moscatiello R, Mariani P, Sanders D, Maathuis FJ (2006) Transcriptional analysis of calcium-dependent and calcium-independent signalling pathways induced by oligogalacturonides. J Exp Bot 57:2847–2865

Mousavi A, Hotta Y (2005) Glycine-rich proteins: a class of novel proteins. Appl Biochem Biotechnol 120:169–174

Mutwil M, Debolt S, Persson S (2008) Cellulose synthesis: a complex complex. Curr Opin Plant Biol 11:252–257

Naithani S, Chookajorn T, Ripoll DR, Nasrallah JB (2007) Structural modules for receptor dimerization in the S-locus receptor kinase extracellular domain. Proc Natl Acad Sci USA 104:12211–12216

Nakhamchik A, Zhao Z, Provart NJ, Shiu SH, Keatley SK, Cameron RK, Goring DR (2004) A comprehensive expression analysis of the Arabidopsis proline-rich extensin-like receptor kinase gene family using bioinformatic and experimental approaches. Plant Cell Physiol 45:1875–1881

Park AR, Cho SK, Yun UJ, Jin MY, Lee SH, Sachetto-Martins G, Park OK (2001) Interaction of the Arabidopsis receptor protein kinase Wak1 with a glycine-rich protein, AtGRP-3. J Biol Chem 276:26688–26693

Rae PM, Kohorn BD, Wade RP (1980) The 10 kb Drosophila virilis 28S rDNA intervening sequence is flanked by a direct repeat of 14 base pairs of coding sequence. Nucleic Acids Res 8:3491–3504

Ringli C, Keller B, Ryser U (2001) Glycine-rich proteins as structural components of plant cell walls. Cell Mol Life Sci 58:1430–1441

Sampoli Benitez BA, Komives EA (2000) Disulfide bond plasticity in epidermal growth factor. Proteins 40:168–174

Sanchez-Rodriguez C, Estevez JM, Llorente F, Hernandez-Blanco C, Jorda L, Pagan I, Berrocal M, Marco Y, Somerville S, Molina A (2009) The ERECTA receptor-like kinase regulates cell wall-mediated resistance to pathogens in *Arabidopsis thaliana*. Mol Plant Microbe Interact 22:953–963

Scheller HV, Ulvskov P (2010) Hemicelluloses. Annu Rev Plant Biol 61:263–289

Seifert GJ (2004) Nucleotide sugar interconversions and cell wall biosynthesis: how to bring the inside to the outside. Curr Opin Plant Biol 7:277–284

Seifert GJ, Blaukopf C (2010) Irritable walls: the plant extracellular matrix and signaling. Plant Physiol 153:467–478

Shpak ED, Berthiaume CT, Hill EJ, Torii KU (2004) Synergistic interaction of three ERECTA-family receptor-like kinases controls Arabidopsis organ growth and flower development by promoting cell proliferation. Development 131:1491–1501

Sivaguru M, Ezaki B, He ZH, Tong H, Osawa H, Baluska F, Volkmann D, Matsumoto H (2003) Aluminum-induced gene expression and protein localization of a cell wall-associated receptor kinase in Arabidopsis. Plant Physiol 132:2256–2266

Smith TA, Kohorn BD (1991) Direct selection for sequences encoding proteases of known specificity. Proc Natl Acad Sci USA 88:5159–5162

Snyders S, Kohorn BD (1999) TAKs, thylakoid membrane protein kinases associated with energy transduction. J Biol Chem 274:9137–9140

Snyders S, Kohorn BD (2001) Disruption of thylakoid-associated kinase 1 leads to alteration of light harvesting in Arabidopsis. J Biol Chem 276:32169–32176

Szymanski DB (2009) Plant cells taking shape: new insights into cytoplasmic control. Curr Opin Plant Biol 12:735–744

Taylor NG (2008) Cellulose biosynthesis and deposition in higher plants. New Phytol 178:239–252

van Zanten M, Snoek LB, Proveniers MC, Peeters AJ (2009) The many functions of ERECTA. Trends Plant Sci 14:214–218

Verica JA, He ZH (2002) The cell wall-associated kinase (WAK) and WAK-like kinase gene family. Plant Physiol 129:455–459

Verica JA, Chae L, Tong H, Ingmire P, He ZH (2003) Tissue-specific and developmentally regulated expression of a cluster of tandemly arrayed cell wall-associated kinase-like kinase genes in Arabidopsis. Plant Physiol 133:1732–1746

Wagner TA, Kohorn BD (2001) Wall-associated kinases are expressed throughout plant development and are required for cell expansion. Plant Cell 13:303–318

Willats WG, McCartney L, Mackie W, Knox JP (2001) Pectin: cell biology and prospects for functional analysis. Plant Mol Biol 47:9–27

Winship LJ, Obermeyer G, Geitmann A, Hepler PK (2010) Under pressure, cell walls set the pace. Trends Plant Sci 15:363–369

Winter H, Huber SC (2000) Regulation of sucrose metabolism in higher plants: localization and regulation of activity of key enzymes. Crit Rev Biochem Mol Biol 35:253–289

Wolf S, Mouille G, Pelloux J (2009) Homogalacturonan methyl-esterification and plant development. Mol Plant 2:851–860

Xu SL, Rahman A, Baskin TI, Kieber JJ (2008) Two leucine-rich repeat receptor kinases mediate signaling, linking cell wall biosynthesis and ACC synthase in Arabidopsis. Plant Cell 20:3065–3079

Yamazaki N, Fry SC, Darvill AG, Albersheim P (1983) HopPathogen interactions: XXIV. Fragments isolated from suspension-cultured sycamore cell walls inhibit the ability of the cells to incorporate [C]leucine into proteins. Plant Physiol 72:864–869

Zhang S, Chen C, Li L, Meng L, Singh J, Jiang N, Deng XW, He ZH, Lemaux PG (2005) Evolutionary expansion, gene structure, and expression of the rice wall-associated kinase gene family. Plant Physiol 139:1107–1124

The Regulation of Pollen–Pistil Interactions by Receptor-Like Kinases

Emily Indriolo and Daphne R. Goring

Abstract Receptor-Like Kinases (RLKs) have been discovered to regulate several signaling events during pollen–pistil interactions, with functions uncovered in both the pollen tube and the pistil. The earliest role is for *S* Receptor Kinase (SRK) in the rapid activation of the Brassicaceae self-incompatible response in the stigma following the perception of self-pollen. The next role is in the communication of a growing pollen tube with the transmitting tract of the *Solanum lycopersicum* pistils through the Pollen Receptor Kinases (LePRKs). Once the pollen tube has reached the ovule, it is fundamentally important for the pollen tube to have the correct timing to cease growth and release the sperm cells for the successful double fertilization of the egg cell and central cell. In *Arabidopsis thaliana*, this event is mediated by RLKs on both the female (FERONIA/SIRENE) and male (ANXUR1 and ANXUR2) side. Thus, RLKs play several fundamental roles in the successful creation of the next generation. This chapter will review the functions of these receptor kinases as well as other players that work alongside with these RLKs in regulating these precise steps during pollen–pistil interactions.

E. Indriolo
Department of Cell and Systems Biology, University of Toronto, Toronto, ON, Canada M5S 3B2,
e-mail: emily.indriolo@utoronto.ca

D.R. Goring (✉)
Department of Cell and Systems Biology, University of Toronto, Toronto, ON, Canada M5S 3B2,

Centre for the Analysis of Genome Evolution and Function, University of Toronto, Toronto, ON, Canada M5S 3B2,
e-mail: d.goring@utoronto.ca

F. Tax and B. Kemmerling (eds.), *Receptor-like Kinases in Plants*,
Signaling and Communication in Plants 13,
DOI 10.1007/978-3-642-23044-8_8,

1 Introduction

Cell–cell communication is an integral aspect of pollen–pistil interactions: from the moment that pollen lands on the pistil surface, to the growth of the pollen tube toward the ovule to deliver sperm cells for fertilization. Pollen–pistil interactions describe the interplay and signaling between a pollen grain and its subsequent pollen tube with the stigma, the transmitting tract of the style, and the ovary. This entire process requires constant communication between the pollen and pistil for successful and optimal fertilization. Examples of RLKs involved in this process have been identified at three different stages of pollen–pistil interactions: (1) the recognition and rejection of self-incompatible pollen grains in the Brassicaceae (*Brassica* spp. and *Arabidopsis lyrata*); (2) the growth of a pollen tube through the stigma as studied primarily in *Solanum lycopersicum* (tomato); and (3) the proper delivery of the sperm cells to the ovule for fertilization of the egg cell as studied in *A. thaliana*. This chapter describes the specific roles of each these RLKs and how they function to regulate a particular stage during pollen–pistil interactions. While beyond the scope of this chapter, it is interesting to note that RLKs have also been discovered in the regulation of floral, anther, and ovule developmental processes (reviewed in De Smet et al. 2009; Zhao 2009; Feng and Dickinson 2010; Fulton et al. 2010).

2 The *S* Receptor Kinase in Early Pollen–Stigma Interactions in the Brassicaceae

The most well-characterized role of RLKs in pollen–pistil interactions is in the mechanism of self-incompatibility as studied in *Brassica* spp. and *A. lyrata*. Self-incompatibility is a trait present in many flowering plants that allows these plants to prevent inbreeding. This mechanism is defined by the rejection of self-pollen, and thus, only allowing for the acceptance of nonself-pollen from other plants. This complex signaling event is initiated by the arrival of a "self" pollen grain on the stigmatic papilla of the pistil (Fig. 1). As the Brassicaceae have dry-type stigmas, the desiccated pollen grain must normally receive water from the stigmatic papilla to hydrate (Heslop-Harrison 1979). If a pollen grain is identified as self-incompatible, it will be rejected by blocking the transfer of water. Only compatible pollen grains will be provided water from the stigmatic papilla for hydration and germination. Similarly, loosening of the stigmatic surface is required for pollen tubes to penetrate into the stigma and subsequently grow down into the transmitting tissue, and this step is also blocked in the self-incompatibility response.

In *Brassica* spp., the recognition event between a pollen grain and the stigmatic papilla was found to be driven by the stigma-specific *S* Receptor Kinase (SRK) (Stein et al. 1991; Goring and Rothstein 1992; Takasaki et al. 2000; Silva et al. 2001) and the pollen coat protein, S-locus Protein 11/S-locus Cysteine-Rich Protein (SP11/SCR), which is the ligand of SRK (Schopfer et al. 1999; Kachroo et al. 2001;

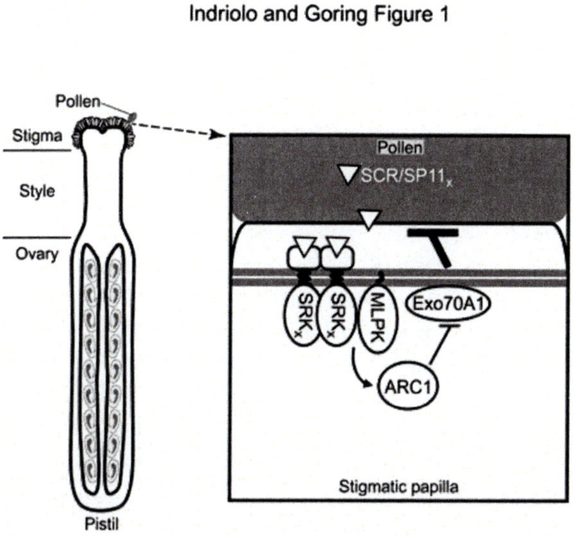

Fig. 1 Model of the *Brassica* self-incompatibility signaling pathway. *Left*: Illustration of a longitudinal cross-section of a *Brassica* pistil. *Right*: The self-incompatibility response is initiated when a pollen grain originates from a plant sharing the same S-haplotype as the pistil (e.g., S_x-haplotype). The SCR/SP11 ligand found on the surface of the pollen grain is recognized by the *S* Receptor Kinase (SRK) present at the surface of the plasma membrane as a homodimer. In cooperation with the *M* Locus Protein Kinase (MLPK), SRK is proposed to "activate" the ARC1 E3 ubiquitin ligase so that ARC1 will now promote the ubiquitination of its target protein, Exo70A1. Ubiquitination is predicted to either inhibit Exo70A1's activity or targets Exo70A1 for degradation, thus inhibiting the assembly of the exocyst complex at the plasma membrane. With Exo70A1 no longer present at the pollen contact site, the delivery of secretory vesicles is blocked, and factors required for pollen grain hydration and pollen tube penetration are not delivered; thus causing pollen rejection. Please see text for further details

Takayama et al. 2000, 2001). Related SRK and SCR genes were subsequently found to regulate the self-incompatibility trait in *A. lyrata* (Kusaba et al. 2001; Schierup et al. 2001). The genes encoding the SRK and SP11/SCR proteins are tightly linked to the *S* locus region and are highly polymorphic, allowing for genetic diversity and preventing inbreeding (reviewed in Charlesworth and Charleswoth 1987, 2009; Barrett 2002; Fujimoto and Nishio 2007).

Binding of SP11/SCR to SRK is proposed to activate signaling events in the stigmatic papilla to cause rejection of the self-incompatible pollen (Fig. 1). This response is thought to be localized to the pollen contact site as a compatible pollen grain and an incompatible pollen grain can both be accepted and rejected on the same papilla (Dickinson 1995). In addition to the initial ligand–receptor interaction of SP11/SCR and SRK, there are several other proteins involved in regulating the rejection of self-incompatible pollen. There are two known positive regulators of this process; the E3 ubiquitin ligase, ARM-Repeat Containing 1 (ARC1) (Gu et al. 1998; Stone et al. 1999, 2003) and M-locus Protein Kinase (MLPK) (Murase et al. 2004; Kakita et al. 2007a, b) as well as the negative regulator of SI, Thioredoxin

H-like 1 (THL1) (Bower et al. 1996; Cabrillac et al. 2001; Haffani et al. 2004). Additionally, several proteins, Kinase-Associated Protein Phosphatase (KAPP), Sorting Nexin 1 (SNX1), and calmodulin have been shown to interact with the kinase domain of SRK, but the role of their interactions in regards to self-incompatibility are less clear (Vanoosthuyse et al. 2003). Recent evidence has linked the negative regulation of vesicle trafficking via the exocyst complex to the blockage of pollen grain hydration in a self-incompatibility signaling event (Samuel et al. 2009).

SRK is a stigma-specific gene that is expressed in flower buds and peaks at maturity (Stein et al. 1991; Goring and Rothstein 1992; Delorme et al. 1995; Kusaba et al. 2001). SRK mRNA is subject to alternative splicing with most of the transcripts uncharacterized (Stein et al. 1991; Delorme et al. 1995). The full-length protein consists of a transit signal peptide, the extracellular domain for ligand binding, transmembrane domain, juxtamembrane domain and the kinase domain (Stein et al. 1991; Goring and Rothstein 1992; Giranton et al. 1995). Besides the full-length transcript, one of the alternative splice variants is of the extracellular domain alone, known as eSRK, and it is glycosylated (Giranton et al. 1995; Takayama et al. 2001; Shimosato et al. 2007). There is another splice variant form of SRK that is a result of posttranslational processing of SRK, known as tSRK (Shimosato et al. 2007). tSRK has the extracellular portion of the receptor and the transmembrane domain (Shimosato et al. 2007).

SRK is localized to the plasma membrane in order to perceive its ligand SP11/SCR, but SRK is hypothesized to be synthesized at the endoplasmic reticulum, as it has a signal transit peptide for localization at the plasma membrane (Stein et al. 1991, 1996; Goring and Rothstein 1992; Delorme et al. 1995). SRK is predicted to reach the plasma membrane by way of the Golgi and the trans-Golgi network to the endosomes and then to the plasma membrane (Ivanov and Gaude 2009a, b). At the plasma membrane, SRK is locally distributed at various points but not throughout the entire membrane, resulting in "patches" of SRK (Ivanov and Gaude 2009a). At the endosome, SRK was found to be quite abundant and SRK is co-localized with THL1, an inhibitor of its activity (Cabrillac et al. 2001; Ivanov and Gaude 2009a). It is hypothesized that if THL1 were not present to inhibit SRK, the high quantity of SRK in the endosome would autoactivate itself leading to incorrect signaling and the rejection of a pollen grain before a signal is even received (Giranton et al. 2000; Ivanov and Gaude 2009a). The interaction of the SRK kinase domain with Sorting Nexin 1 (SNX1) may be related to its endosomal localization. Previously, it has been shown in yeast and mammalian systems that sorting nexins play a role in the recycling and degradation of endosomal receptors (Carlton and Cullen 2005). Given the fact that SRK can be found at the endosome, SNX1 may play a role in regulating the levels of SRK at the plasma membrane and endosome.

When SRK is localized to the plasma membrane, it forms homodimers (and not heterodimers e.g., SRK_8 with SRK_8 or SRK_9 with SRK_9). The formation of these homodimers is proposed to be spontaneous and not ligand-dependent (Giranton et al. 2000; Shimosato et al. 2007). SRK is unable to bind SP11/SCR unless the receptor is found in its homodimeric state (Giranton et al. 2000;

Shimosato et al. 2007). Following pollination, it is unknown how SP11/SCR is able to go from the pollen coat through the papillar cell wall to reach the SRK receptor, but it is hypothesized to involve unidentified pollen coat proteins (Iwano et al. 2003). The binding of SP11/SCR to SRK is mediated by the hypervariable region of the extracellular domain of SRK. Studies have shown that the SRK homodimers will bind the corresponding *S*-haplotype-specific SP11/SCR most strongly (e.g., SRK_8 binds $SP11/SCR_8$), but SRK homodimers will also more weakly bind SP11/SCR proteins from other *S*-haplotypes (Kemp and Doughty 2007; Naithani et al. 2007; Shimosato et al. 2007). It has been proposed that co-dominance, nonlinear dominant–recessive relationships, and mutual weakening could be a result of the formation of SRK heterodimers (from two different *S*-haplotypes), but these have yet to be studied in depth (Naithani et al. 2007). The molecular control of dominant-recessive relationships for pollen recessive alleles has been recently uncovered and turns out to be at the level of gene expression rather than receptor–ligand interactions (Shiba et al. 2002, 2006; Tarutani et al. 2010). The *SP11 methylation inducer* (*SMI*) locus, also linked to the *S* locus region, was discovered to encode a small noncoding RNA which promotes the methylation of the SP11/SCR promoter from the pollen-recessive *S*-haplotype. *SMI,* originating from the dominant *S*-haplotype, silences the expression of the pollen-recessive SP11/SCR gene, and thus, giving rise to pollen-recessive phenotype (Tarutani et al. 2010).

Once SP11/SCR has bound SRK, the signaling cascade is activated and pollen rejection occurs. SRK propagates the signal rapidly across the plasma membrane by transphosphorylation that in turn is able to activate the downstream interactors (Giranton et al. 2000; Stone et al. 2003; Samuel et al. 2008). Another kinase necessary for the successful rejection of self-pollen is MLPK and without it, *Brassica* are unable to reject self-pollen (Murase et al. 2004). MLPK belongs to the receptor-like cytoplasmic kinase subfamily and is localized to the plasma membrane (Murase et al. 2004). MLPK mRNA is expressed as two different transcripts, known as *MLPKf1* and *MLPKf2*, with different start sites (Kakita et al. 2007a). *MLPKf1* encodes MLPK with an N-terminal myristoylation site for plasma membrane localization, while *MLPKf2* encode MLPK with an N-terminal hydrophobic region for plasma membrane localization. Both forms are expressed in the stigma and are able to complement *mlpk* mutant plants, implying that there is some complexity to the regulation of this protein (Kakita et al. 2007a). SRK can interact with MLPK at the plasma membrane, albeit as a more transient relationship. MLPK may also be interacting with the inactive SRK complex prior to SP11/SCR binding (Kakita et al. 2007a, b).

ARC1 is an SRK-interactor that was originally isolated through a yeast two-hybrid screen of a *B. napus* pistil cDNA library and binds very well to the phosphorylated SRK kinase domain through its C-terminal ARM repeats (Gu et al. 1998). The ARC1–SRK interaction was also confirmed for the *B. oleracea* versions of these proteins (Vanoosthuyse et al. 2003; Niu et al. 2009). ARC1 can be phosphorylated in vitro by both SRK (poorly) and MLPK (well), indicating that perhaps ARC1 is recruited to the SRK-MLPK complex at the plasma membrane for phosphorylation by MLPK (Gu et al. 1998; Samuel et al. 2008). In *B. napus*, the expression of ARC1 is stigma-specific and peaks at flower opening (Gu et al. 1998). When the expression of ARC1 is knocked down via antisense ARC1, the resulting phenotype is the breakdown of self-incompatibility as self-incompatible pollen tubes are able to penetrate into the pistil and set seed occurs. Therefore, ARC1 is a positive regulator of self-incompatibility and functions downstream of SRK (Stone et al. 1999). ARC1 is an E3 ubiquitin ligase with several different domains: the U-box domain that has been shown to bind the E2 conjugating enzyme in a closely related protein, AtPUB14 (Stone et al. 2003; Andersen et al. 2004); a conserved U-box N-terminal domain that is found in a subset of plant U-box proteins and may participate in protein–protein interactions (Stone et al. 2003); and the C-terminal ARM repeat domain that was shown to interact with the SRK kinase domain (Gu et al. 1998). While ARC1 is expected to interact with SRK at the plasma membrane, it also contains a functional nuclear localization signal and has been shown to shuttle between the cytoplasm and nucleus when transiently expressed in BY2 cells (Stone et al. 2003). The biological reason for ARC1's ability to shuttle throughout the cell is not understood and instead poses many more questions as to what this means in regards to ARC1's cellular function in the self-incompatibility pathway.

With ARC1 potentially being phosphorylated by SRK and MLPK in vivo, the question arises as to the role of phosphorylation on ARC1's function during the self-incompatibility response. Previous transient expression studies in BY2 cells found that ARC1 on its own shuttles between the nucleus and cytosol, and is localized to both compartments. When an active SRK or MLPK kinase domains are co-expressed with ARC1, ARC1 it is prevented from shuttling to the nucleus suggesting that phosphorylation plays a role in shifting the subcellular localization of ARC1 (Stone et al. 2003; Samuel et al. 2008). In the cytosol, ARC1 becomes localized to the proteasomes in the presence of the active kinases which raises the question of whether phosphorylation is also regulating ARC1's stability (or localization to target substrates for degradation). Interestingly, the ARM-repeat protein, β-catenin, shuttles between different compartments of the animal cell, and phosphorylation plays an important role in its regulation. For example, phosphorylation of cytosolic-localized β-catenin by casein kinase Iα and glycogen synthase kinase-3 results in the degradation of β-catenin while the phosphorylation by C-Jun N-terminal kinase-2 results in the nuclear localization of β-catenin from the cytosol (Liu et al. 2002; Wu et al. 2008). As well, β-catenin has been found to have stronger affinities to some of its binding partners when the partner protein is phosphorylated (Daugherty and Gottardi 2007). Similarly, the phosphorylation-dependent interaction of ARC1 to SRK suggests that ARC1 may have a stronger affinity for the phosphorylated SRK kinase domain (Gu et al. 1998). Finally, phosphorylation may change the binding affinity of ARC1 for a target protein for ubiquitination, in this case perhaps Exo70A1 that is found at the plasma membrane (Samuel et al. 2009).

The role of ARC1 in self-incompatibility is to ubiquitinate target proteins for degradation by the 26S proteasome, and one of the targets has recently been identified as Exo70A1 (Stone et al. 2003; Samuel et al. 2009). Exo70A1 is a subunit of the evolutionarily conserved exocyst complex, which mediates polarized vesicle secretion to the plasma membrane (Boyd et al. 2004; Synek et al. 2006). In *A. thaliana*, RFP tagged Exo70A1 is shown to move from the Golgi to the plasma membrane right before the stigmatic papillae are set to receive pollen, following flower-opening. When Exo70A1 expression is supressed in the stigma of transgenic Exo70A1-RNAi *B. napus* plants, compatible pollen grains were unable to hydrate properly and pollen tubes were blocked from penetrating the stigmatic surface, thus preventing seed set. This suggests that Exo70A1 as part of the exocyst is required to deliver secretory vesicles to the stigmatic papillar plasma membrane at the pollen contact site for delivering cargo required for these processes. In contrast, when Exo70A1 is overexpressed in *B. napus* stigmas, self-incompatible pollen grains were able to hydrate, pollen tubes penetrated the stigmatic surface, and seeds were produced, thus overcoming self-incompatibility (Samuel et al. 2009). This phenotype is hypothesized to be a result of the increased amounts of Exo70A1 at the plasma membrane, resulting from overexpression, which is able to overcome ARC1's ubiquitination activity during the self-incompatibility response. Alternatively, the RFP tag on the Exo70A1 protein could be interfering with ARC1 binding while not affecting Exo70A1 activity (as stigma-expressed RFP:Exo70A1 was able to complement the stigmatic defect of the *A. thaliana exo70A1* mutant) (Samuel et al. 2009). With both of these observations, it is proposed that ARC1 targets Exo70A1 for ubiquitination and degradation to block the delivery of secretory vesicles by the exocyst complex resulting in the rejection of self-incompatible pollen (Fig 1). While the binding site of Exo70A1 on ARC1 is unknown, it would shed light into the function of ARC1 with its unique localization pattern as well as its activation by SRK via phosphorylation.

As stated previously, KAPP and calmodulin can bind to the kinase domain of SRK, but their roles in self-incompatibility are unclear. KAPP was identified based on sequence homology to the previously characterized *A. thaliana* KAPP, while calmodulin 1 and calmodulin 2 were identified through the screening a *B. oleracea* pistil cDNA yeast two-hybrid library (Vanoosthuyse et al. 2003). KAPP was shown to interact with SRK in the yeast two-hybrid system and could be phosphorylated by SRK. KAPP was also able to dephosphorylate SRK, which would return SRK to an inactive state, and is consistent with its putative function of down-regulating activated receptor kinases (Vanoosthuyse et al. 2003; Johnson and Ingram 2005). Previously, *A. thaliana* KAPP was found to act downstream of other *A. thaliana* receptor kinases as a negative regulator (Stone et al. 1994, 1998), possibly by the down-regulation of receptor kinases by endocytosis (Shah et al. 2002). Both of the calmodulin proteins 1 and 2 were shown to bind the kinase domain of SRK, however, the binding was not specific to SRK nor was there an effect on SRK kinase activity. Thus, the calmodulin proteins may also have a more general function in the regulation of receptor kinases though the nature of this function is unclear (Vanoosthuyse et al. 2003). Finally, there are other cellular changes that have been observed as part of the *Brassica* self-incompatibility response such as a reorganization of the actin cytoskeleton and disruption of the vacuolar network (Iwano et al. 2007). How these events are connected to SRK, MLPK, ARC1 or the other potential signaling proteins in the self-incompatibility response is not known.

3 *Solanum lycopersicum* Pollen Receptor Kinases in the Regulation of Pollen Tube Growth Through the Transmitting Tissue

After a compatible pollen grain has landed on the stigma, pollen hydration and germination follows, and a pollen tube emerges to penetrate the stigmatic surface and grow down the style to the ovary. While the pollen tube grows through the transmitting tract of the style and ovary, there is a constant exchange of information between the pollen tube and the pistil in order for the pollen tube to reach the ovules. The role of RLKs in the communication between the pollen tube and the pistil is most well characterized in *S. lycopersium* (tomato). The system has revealed more about pollen tube growth through the stigma into the style as the pollen is much more accessible in comparison to the model plant, *A. thaliana*. In *S. lycopersium*, three pollen RLKs were identified from mature tomato anther cDNAs: Pollen Receptor Kinase (LePRK) 1, 2 and 3. The predicted structure of the LePRK proteins consists of a signal peptide at the most N-terminal portion of the protein followed by an leucine-rich repeat extracellular domain, transmembrane domain, and a C-terminal kinase domain (Muschietti et al. 1998; Kim et al. 2002).

LePRK1 and 2 were found to be present in a protein complex in mature pollen grains and localized to the plasma membrane of growing pollen tubes (Muschietti et al. 1998; Wengier et al. 2003). As well, LePRK2 is phosphorylated in the pollen tube membrane and requires phosphorylation for interacting with the LePRK complex. LePRK2 has been shown to be a positive regulator of pollen germination and pollen tube growth through the analysis of antisense transgenic tomato plants (Muschietti et al. 1998; Zhang et al. 2008). The LePRK2 antisense transgenic pollen grains showed reduced rates of germination and less pollen tube growth when compared to wild-type pollen (Zhang et al. 2008). One major morphological difference between wild-type and antisense LePRK2 pollen tubes is the mislocalization of the vacuole in the LePRK2 antisense transgenic pollen tubes. Unlike wild-type pollen tubes where the large vacuole is at the rear of the tip with thin vacuoles near the tip, the LePRK2 antisense transgenic pollen tubes had large vacuoles near the pollen tube tip implying that there was a problem with the proper localization of subcellular components that are necessary for normal pollen tube growth (Zhang et al. 2008). Calcium has been shown to be required for the growth of pollen germination and pollen tube growth, and the proper concentration of exogenous calcium allows for normal growth of pollen tubes in vitro (Steer 1989). While in vitro-grown wild-type pollen tubes responded to exogenous calcium levels in a dose-dependent manner, the antisense LePRK2 pollen tubes showed no difference (Zhang et al. 2008). Therefore, LePRK2 may play a role in calcium sensing in the pistil as a pollen tube grows to the ovary.

The LePRK1/2 complex can bind several different factors, as the pollen tubes grow into the stigma and through the style towards the ovules. In the pollen, there is the Late Anther Tomato 52 (LAT52), SHY, and Kinase Partner Protein (KPP); while in the pistil, there is the tomato Stigma1 (LeSTIG1) and the Style Interactor for LePRKs (STIL) (Tang et al. 2002, 2004; Guyon et al. 2004; Wengier et al. 2010; Kaothien et al. 2005). Each factor interacts with the LePRK complex in a directional and developmental sequence tied to the germination and growth of the pollen tube through the pistil (Fig. 2). The first protein that LePRK2 is bound to is LAT52, a ~20 kDa cysteine rich protein, which forms a complex with the extracellular domain on the surface of the mature pollen grain (Tang et al. 2002). When the pollen grain lands on the surface of the stigma, the LAT52 complex dissociates from the extracellular binding domain, and LeSTIG1, a cysteine rich protein of ~15 kDa will bind to the LePRK2 through the same extracellular binding domain as LAT52 (Tang et al. 2004). As the pollen tube begins to penetrate the stigma and into the style the LeSTIG1 protein is still bound to LePRK2, and in vitro LeSTIG1 has been shown to promote pollen tube growth (Tang et al. 2004). It has been hypothesized that there may be another ligand that binds LePRK2 when the pollen tube reaches the ovary, but that proposed ligand has yet to be identified.

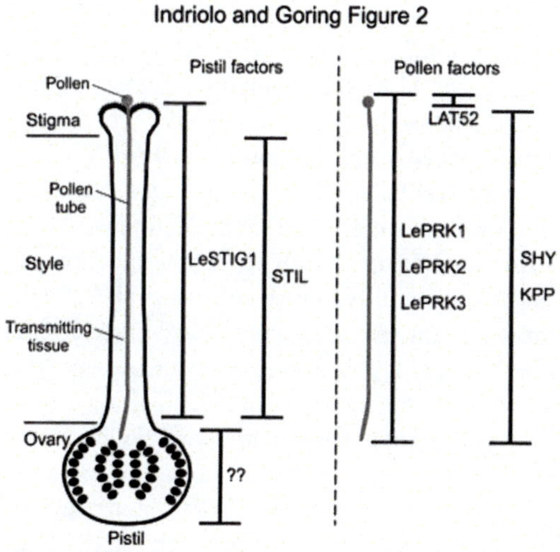

Fig. 2 Model of pistil and pollen factors involved in pollen tube growth through the tomato pistil. *Left*: Illustration of a longitudinal cross-section of a tomato pistil. *Right*: Both pistil and pollen factors have been identified for the process of pollen tube growth through the pistil to the ovary. The *lines* indicate the range of their activities (for example, LeSTIG1 is a pistil factor active in both the stigma and style). LePRK1, 2, and 3 are RLKs expressed in the pollen tube, and both pistil factors have been shown to bind to LePRK2. It is hypothesized there is an additional factor present in the ovary to act on the RLKs (denoted as ??). Other pollen factors include LAT52, SHY and KPP. Please see text for further details

SHY is a gene that was identified in a screen for genes that were up-regulated during pollen germination and encodes a ~35 kDa LRR protein that is predicted to be secreted to the outside of the pollen grain (Guyon et al. 2000). SHY was shown to bind to LePRK2 in yeast two-hybrid screens and also bound to LePRK2 in pull-down assays (Tang et al. 2002, 2004). When the expression of SHY was knocked down by antisense in pollen grains, the pollen tubes showed abnormal growth and germination in vitro as well as in pollen–pistil interactions (Guyon et al. 2004). The SHY-antisense pollen tubes were also unable to fertilize the ovules as they only grew to the top of the ovary but were unable to make it any further (Guyon et al. 2004). Clearly, SHY is required for the successful delivery of the sperm cells to the ovary, but how it is interacting with LePRK2 through its extracellular domain is unclear. The last known interactor of LePRK2 is STIL, a small 3.5 kDa compound, that is found in style extracts and promotes pollen tube growth in vitro (Wengier et al. 2003, 2010). STIL is resistant to heat, acid, base, DTT, and proteases, and its molecular structure is currently unknown. STIL can promote the dephosphorylation of LePRK2 as well as the disassociation of LePRK1 and LePRK2 when added as a part of a style extract to pollen microsomes (Wengier et al. 2003, 2010).

The only cytoplasmic interactor that has been identified to date is KPP, and it was pulled out of a yeast two-hybrid assay with the cytoplasmic domains of LePRK1 and 2 (Kaothien et al. 2005). KPP was found to interact with both the phosphorylated and mutant (unphosphorylated) forms of the LePRK1 and 2 cytoplasmic domains indicating that phosphorylation is not required for KPP binding (Kaothien et al. 2005). When the interaction of KPP with LePRK1 and 2 were examined in pollen, only LePRK2 bound KPP, which is consistent with all other known interactors of the LePRK complex. Interestingly, when KPP is overexpressed in pollen tubes in vitro and in vivo, the resulting phenotype is depolarized pollen tube growth and balloon shaped tips (isotropic growth). This balloon tip phenotype is hypothesized to be a result of the de-localization of the endogenous KPP or a result of excess KPP in the cytoplasm of which either would result in a more diffuse area of tip growth versus the inherent aspect of tip growth being driven by a localized small region (Kaothien et al. 2005).

The KPP protein is a plant specific guanine nucleotide exchange factor (GEF) for Rop GTPases (Berken et al. 2005; Zhang and McCormick 2007). The process of pollen tube growth involves extensive vesicle trafficking and actin cytoskeleton rearrangements in the pollen tube, and a number of pollen factors involved in the regulation of these processes have been identified including Rop GTPases (reviewed in Yang 2008; Yalovsky et al. 2008; Zhang and McCormick 2010; Zonia 2010). GEFs are responsible for stimulating small GTPases to exchange GDP to GTP, and thus, switch from an inactive to an active form of the GTPase. Thus, the activation of the Rop GTPase by KPP and LePRK2 is proposed to establish the cell polarity of the pollen tube tip for proper growth (Zhang and McCormick 2007; Lee et al. 2008).

The *A. thaliana* KPP homologue, AtRopGEF12, has also been characterized along with the LePRK2 homologue, AtPRK2a (Zhang and McCormick 2007). An interaction between AtRopGEF12 and AtPRK2a, via the cytoplasmic domain of

AtPRK2a and the C-terminal end of AtRopGEF12, was demonstrated using the split ubiquitin system (Zhang and McCormick 2007). While overexpressing the full length AtRopGEF12 had little effect on pollen tubes growing in vitro, the expression of the C-terminal domain was found to alter pollen tube growth (Zhang and McCormick 2007). Interestingly, when AtRopGEF12 was co-expressed with AtPRK2a, isotropic growth (swelling of the pollen tube tip) was observed. This is similar phenotype to that observed in the overexpression of a constitutively active Rop GTPase, and thus, linked AtPRK2a-AtRopGEF12 as positive regulators of Rops in regulating pollen tube tip growth (Zhang and McCormick 2007).

From all known research into this stage of pollen–pistil interactions, two major foci of LePRK2 can be generated (Fig. 2). The first is directed at the ligands and extracellular interactors of LePRK2 including LAT52, LeSTIG, SHY and STIL while the second is directed on the cytoplasmic side of pollen tube growth with LePRK2/AtPRK2a and the KPP/AtRopGEF12 in their proposed roles in regulating Rop GTPases. Future research in this field should focus on integrating all of the different signals from the exterior and interior of the growing pollen tube to determine how they are related. It also would be interesting to further determine how this process is controlled in pollen tubes for other plant species outside of the Solanaceae.

4 Role of RLKs in Pollen Tube Delivery and Sperm Release in *Arabidopsis thaliana*

After the pollen tube has navigated its way through the style to the ovary, the next major step for the pollen tube is to reach an ovule for fertilization. The pollen tube moves from the transmitting tract to the funiculus and then to the micropyle (Fig. 3). Various guidance cues from the pistil are responsible for guiding the pollen tube to the ovule (reviewed in Chapman and Goring 2010). Once entering the micropyle, the final step is for the pollen tube to be recognized by the female gametophyte where it enters one of the two synergid cells. After pollen tube reception, the synergid cell degenerates, and the pollen tube ruptures to discharge the two sperm cells which fertilize the egg cell and central cell. Recently, RLKs have been identified on both the female and male side in the role of pollen tube delivery and sperm release in *A. thaliana*.

A screen to find plants which had defects with pollen tube delivery and rupture led to the identification of the *feronia* (*fer*), also known as *sirène* (*ser*), mutant which showed an inability of the pollen tubes to rupture. Instead, when a pollen tube entered the *fer* micropyle, it would continue to grow, coiling inside the female gametophyte (Huck et al. 2003; Rotman et al. 2003). The FER gene was determined to encode a RLK that belonged to the CrRLK1-like family of RLKs (Escobar-Restrepo et al. 2007). CrRLK1was first identified as a novel RLK in *Catharanthus roseus* with no identifiable motifs in the extracellular domain, but otherwise a

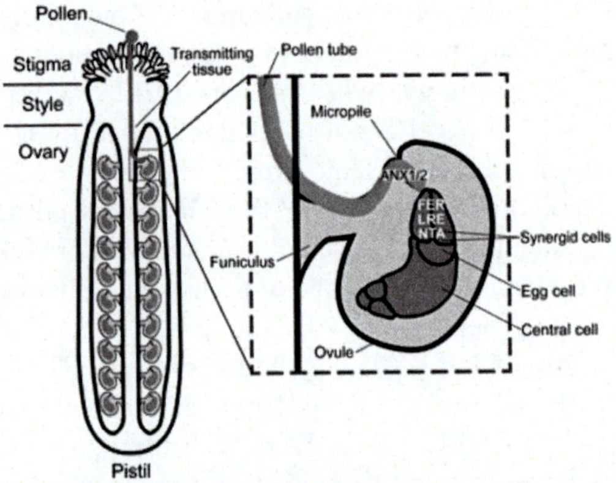

Fig. 3 Model of factors regulating the timing of pollen tube rupture and release of sperm cells for fertilization. *Left*: Illustration of a longitudinal cross-section of an *A. thaliana* pistil. *Right*: The pollen-specific RLKs, ANX1/2, are found at the tip of the growing pollen tube. The closely related RLK, FER, is localized to the most apical portion of the synergid cells in the ovule, where contact with the pollen tube tip will occur. Other factors localized with FER in the synergid cells include LRE and NTA. All of these proteins are critical to the process of pollen tube rupture and sperm cell release, once the pollen tube has entered a synergid cell; thus, allowing for successful fertilization of the egg cell and central cell. Please see text for further details

typical RLK structure of an N-terminal transit peptide domain, extracellular domain, transmembrane domain and a C-terminal kinase domain (Schulze-Muth et al. 1996; Shiu and Bleecker 2003). FER is expressed in the vegetative tissue of the mature *A. thaliana* plant and is specifically expressed in the synergid cells within the ovule, but not in the male gametophyte (Escobar-Restrepo et al. 2007). When the FER protein was fused to GFP, the fusion protein was able to rescue the mutant plants and restore the phenotype of normal pollen tube rupture and fertilization. The FER GFP fusion was localized to the synergid cell plasma membrane as predicted for signal perception and subsequent signal transduction. Thus, FER's role in the female gametophyte would be to communicate with the pollen tube to stop, rupture, and release the sperm cells. Interestingly, Escobar-Restrepo et al. (2007) also found that when pollen grains from two other members of the Brassicaceae (*A. lyrata* and *C. flexuosa*) were applied to wild-type *A. thaliana* pistils, the pollen tubes were observed to appear phenotypically similar to that of the *fer* mutant plants with pollen tubes unable to stop or release their sperm. The authors hypothesized that the signal for the arrest and rupture of pollen tubes was not perceived due to divergent ligand-FER receptor systems in these species. Consistent with this, a comparison of the extracellular domain sequences of the FER homologues from *A. lyrata* and *C. flexuosa* with the *A. thaliana* domain showed that the extracellular domain sequences were diverging at a faster rate than the remainder of the protein.

With FER's role to communicate to the growing pollen tube from the female gametophyte, two different groups used male gametophyte-specific expression patterns to search for RLKs that would play a similar role in the pollen tube. This led to the discovery of the *Anxur1* and *Anxur2* (*ANX1*, *ANX2*) genes which encode RLKs belonging to the same CrRLK1-like family (Miyazaki et al. 2009; Boisson-Dernier et al. 2009). The ANX1/2 proteins are localized to the cell periphery of the growing pollen tube tips, and appeared to be associated with vesicles involved in the exocytic and endocytic cycle of tip growth (Boisson-Dernier et al. 2009). While the single *anx1* and *anx2* mutants had a wild-type phenotype, the double *anx1;anx2* mutant pollen tubes showed spontaneous discharge of the pollen tip and sperm release, with the pollen tubes rupturing predominately in the style and rarely reaching the top of the ovary (Miyazaki et al. 2009; Boisson-Dernier et al. 2009). Therefore, ANX1 and ANX2 are required for the stability of pollen tube growth and the proper timing of the release of the sperm cells (Miyazaki et al. 2009; Boisson-Dernier et al. 2009). With ANX1 and ANX2 being closely related to FER, it was proposed that ANX1/2 may function complementary to FER in recognizing a signal at the pollen tube tip to cease pollen tube growth and release the sperm cells into the female gametophyte. Thus, this model proposes that these closely related receptors work together on both the female and male side to assure successful fertilization.

Screens for other mutants displaying the pollen tube overgrowth phenotype in the female gametophyte have uncovered two other components that may function with FER in the synergid cells: LORELEI (LRE) and NORTIA (NTA) (Capron et al. 2008; Kessler et al. 2010). *LRE* encodes a glycosylphosphatidylinositol (GPI)-anchored protein with a transit signal peptide on the N-terminal region, a C-terminal GPI anchor and central novel domain that is hypothesized to be involved in the activity of the protein (Capron et al. 2008). *LRE* was found to be expressed predominantly in the synergids, but only when these cells were fully differentiated (Capron et al. 2008). While not a lot is known about GPI-anchored proteins in *A. thaliana*, they appear to have a wide range of functions, including the regulation of gametogenesis, polarized growth and pathogen responses (reviewed in Seifert and Roberts 2007). GPI-anchored proteins have been found to be localized to lipid rafts in the plasma membrane (van Meer 2002), and LRE has been proposed to modulate FER's localization or function as part of this proposed localization to lipid rafts (Capron et al. 2008). Alternatively, LRE has been proposed to perceive the arrival of the pollen tube on the outside of the synergid cells where the GPI anchor could be cleaved leading to a response to the pollen tube (Capron et al. 2008). LRE is also expressed in egg cell, and more recent work has uncovered an additional role for LRE in egg cell fertilization following sperm cell release (Capron et al. 2008; Tsukamoto et al. 2010).

NTA was found to belong to MLO family of multiple membrane-spanning proteins, which were first identified by loss-of-function mutants exhibiting resistance to powdery mildew fungi (Kessler et al. 2010). Similarly to FER and LRE, NTA was found to be specifically expressed in the synergid cell within the ovule. Interestingly, with the connection of *mlo* mutants to powdery mildew resistance, the authors infected *fer* mutants with powdery mildew and found that the *fer* mutants displayed a similar resistance to host cell invasion by the fungi. At the micropylar end of the synergid cell (where the pollen tube enters) is a membrane-rich region called the filiform apparatus. FER is localized to the filiform apparatus while NTA is redistributed to this structure with pollen tube arrival in a FER-dependent manner. MLO proteins have been proposed to have a role in vesicle trafficking to the plasma membrane, and so NTA is thought to deliver NTA-associated vesicles to the filiform apparatus, perhaps to secrete cargo for pollen tube rupture (Kessler et al. 2010; Govers and Angenent 2010).

Interestingly, FER was recently identified as an interactor of ROPGEF1 in a yeast-two hybrid screen. FER was linked to ROPGEF1 in regulating root hair growth in *A. thaliana*, and consistent with this, the *fer* mutant was found to have defective root hairs (collapsed or abnormal in shape) (Duan et al. 2010). The FER-ROPGEF1 complex was then linked to the activation of a Rop GTPase and the subsequent ROS production and calcium influx to drive root hair growth (Duan et al. 2010). There are interesting parallels between FER-ROPGEF1 driving root hair growth and the tomato LePRK2-KPP driving pollen tube tip growth (described above). Based on these results, Kanaoka and Torii (2010) proposed a model where *A. thaliana* ANX1/2 are using a similar pathway to drive pollen tube growth until the pollen tube reaches the FER expressing synergid cell. At this point, FER is proposed to compete away the ANX1/2 ligands causing the pollen tube to rupture from loss of ANX1/2 signaling. One issue that needs to be reconciled with this model is the connection of FER and the potential delivery of NTA associated vesicles which are also proposed to secrete cargo to block ANX1/2 signaling on the pollen tube. A second issue is the observed differences in ROS production between the roots and leaves of *fer* mutants. Duan et al. (2010) found that the *fer* mutant root and root hairs had decreased levels of ROS compared to wild-type roots (consistent with the model of ROS production in root hair growth). However, Kessler et al. (2010) found that the *fer* mutant leaves had increased accumulation of hydrogen peroxide and spontaneous cell death which suggests that ROS production may be regulated by other factors in the leaves. Thus, FER may function through different signaling pathways in the different tissues, and it may be difficult to have one general model fit its different roles.

5 Conclusions

From the initial event of a pollen grain landing on the pistil to the pollen tube growing and reaching an ovule for double fertilization, RLKs have been uncovered as important players in several different stages. These include SRK as a key player in the recognition and rejection of self-pollen on the stigmas in the Brassicaceae; the tomato LePRK proteins in pollen tube growth, and the *A. thaliana* FER and ANX1/2 proteins in controlling the timing of pollen tube rupture and release of sperm cells for fertilization. While each of these unique events has been characterized in a particular species or system, it does allow for us to begin developing a model for RLK-mediated events in pollen–pistil interactions. As research continues, the next steps will lie in continuing to map the signals perceived by RLKs and the downstream events that they initiate, and to determine how conserved the LePRK, FER and ANX1/2 functions are in other plant families. Finally, given that RLKs constitute the largest family of signaling proteins in plant genomes (Shiu and Bleecker 2003); there are likely other new roles yet to be discovered for RLKs in mediating successful fertilization.

References

Andersen P, Kragelund BB, Olsen AN, Larsen FH, Chua NH, Poulsen FM, Skriver K (2004) Structure and biochemical function of a prototypical *Arabidopsis* U-box domain. J Biol Chem 279:40053–40061

Barrett SC (2002) Sexual interference of the floral kind. Heredity 88:154–159

Berken A, Thomas C, Wittinghofer A (2005) A new family of RhoGEFs activates the Rop molecular switch in plants. Nature 436:1176–1180

Boisson-Dernier A, Roy S, Kritsas K, Grobei MA, Jaciubek M, Schroeder JI, Grossniklaus U (2009) Disruption of the pollen-expressed FERONIA homologs ANXUR1 and ANXUR2 triggers pollen tube discharge. Development 136:3279–3288

Bower MS, Matias DD, Fernandes-Carvalho E, Mazzurco M, Gu T, Rothstein SJ, Goring DR (1996) Two members of the thioredoxin-h family interact with the kinase domain of a *Brassica* *S* locus receptor kinase. Plant Cell 8:1641–1650

Boyd C, Hughes T, Pypaert M, Novick P (2004) Vesicles carry most exocyst subunits to exocytic sites marked by the remaining two subunits, Sec3p and Exo70p. J Cell Biol 167:889–901

Cabrillac D, Cock JM, Dumas C, Gaude T (2001) The S-locus receptor kinase is inhibited by thioredoxins and activated by pollen coat proteins. Nature 410:220–223

Capron A, Gourgues M, Neiva LS, Faure JE, Berger F, Pagnussat G, Krishnan A, Alvarez-Mejia C, Vielle-Calzada JP, Lee YR, Liu B, Sundaresan V (2008) Maternal control of male-gamete delivery in *Arabidopsis* involves a putative GPI-anchored protein encoded by the LORELEI gene. Plant Cell 20:3038–3049

Carlton JG, Cullen PJ (2005) Sorting nexins. Curr Biol 15:R819–R820

Chapman LA, Goring DR (2010) Pollen-pistil interactions regulating successful fertilization in the Brassicaceae. J Exp Bot 61:1987–1999

Charlesworth B, Charlesworth D (2009) Darwin and Genetics. Genetics 183:757–766

Charlesworth D, Charleswoth B (1987) Inbreeding depression and its evolutionary consequences. Annu Rev Ecol Syst 18:237–268

Daugherty RL, Gottardi CJ (2007) Phospho-regulation of β-catenin adhesion and signaling functions. Physiology (Bethesda) 22:303–309

De Smet I, Voss U, Jürgens G, Beeckman T (2009) Receptor-like kinases shape the plant. Nat Cell Biol 11:1166–1173

Delorme V, Giranton JL, Hatzfeld Y, Friry A, Heizmann P, Ariza MJ, Dumas C, Gaude T, Cock JM (1995) Characterization of the S locus genes, SLG and SRK, of the *Brassica* S3 haplotype: identification of a membrane-localized protein encoded by the S locus receptor kinase gene. Plant J 7:429–440

Dickinson HG (1995) Dry stigmas, water and self-incompatibility in *Brassica*. Sex Plant Reprod 8:1–10

Duan Q, Kita D, Li C, Cheung AY, Wu HM (2010) FERONIA receptor-like kinase regulates RHO GTPase signaling of root hair development. Proc Natl Acad Sci USA 107:17821–17826

Escobar-Restrepo J, Huck N, Kessler S, Gagliardini V, Gheyselinck J, Yang W, Grossniklaus U (2007) The FERONIA receptor-like kinase mediates male-female interactions during pollen tube reception. Science 317:656–660

Feng X, Dickinson HG (2010) Cell-cell interactions during patterning of the *Arabidopsis* anther. Biochem Soc Trans 38:571–576

Fujimoto R, Nishio T (2007) Self-Incompatibility. In: Gupta SK (ed) Advances in Botanical Research: Incorporating Advances in Plant Pathology. Advances in Botanical Research 45:139–154, DOI: 10.1016/S0065-2296(07)45005-4

Fulton L, Vaddepalli P, Yadav RK, Batoux M, Schneitz K (2010) Inter-cell-layer signalling during *Arabidopsis* ovule development mediated by the receptor-like kinase STRUBBELIG. Biochem Soc Trans 38:583–587

Giranton JL, Ariza MJ, Dumas C, Cock JM, Gaude T (1995) The S locus receptor kinase gene encodes a soluble glycoprotein corresponding to the SRK extracellular domain in *Brassica oleracea*. Plant J 8:827–834

Giranton JL, Dumas C, Cock JM, Gaude T (2000) The integral membrane S-locus receptor kinase of *Brassica* has serine/threonine kinase activity in a membranous environment and spontaneously forms oligomers in planta. Proc Natl Acad Sci USA 97:3759–3764

Goring DR, Rothstein SJ (1992) The S-locus receptor kinase gene in a self-incompatible *Brassica napus* line encodes a functional serine/threonine kinase. Plant Cell 4:1273–1281

Govers F, Angenent GC (2010) Plant science. Fertility goddesses as Trojan horses. Science 330:922–923

Gu T, Mazzurco M, Sulaman W, Matias DD, Goring DR (1998) Binding of an arm repeat protein to the kinase domain of the S-locus receptor kinase. Proc Natl Acad Sci USA 95:382–387

Guyon V, Astwood J, Garner E, Dunker A, Taylor L (2000) Isolation and characterization of cDNAs expressed in the early stages of flavonol-induced pollen germination in Petunia. Plant Physiol 123:699–710

Guyon V, Tang W, Monti M, Raiola A, De Lorenzo G, McCormick S, Taylor L (2004) Antisense phenotypes reveal a role for SHY, a pollen-specific leucine-rich repeat protein, in pollen tube growth. Plant J 39:643–654

Haffani Y, Gaude T, Cock J, Goring D (2004) Antisense suppression of thioredoxin h mRNA in *Brassica napus* cv. Westar pistils causes a low level constitutive pollen rejection response. Plant Mol Biol 55:619–630

Heslop-Harrison J (1979) An interpretation of the hydrodynamics of pollen. Am J Bot 66:737–743

Huck N, Moore J, Federer M, Grossniklaus U (2003) The *Arabidopsis* mutant feronia disrupts the female gametophytic control of pollen tube reception. Development 130:2149–2159

Ivanov R, Gaude T (2009a) Endocytosis and endosomal regulation of the S-receptor kinase during the self-incompatibility response in *Brassica oleracea*. Plant Cell 21:2107–2117

Ivanov R, Gaude T (2009b) *Brassica* self-incompatibility: a glimpse below the surface. Plant Signal Behav 4:996–998

Iwano M, Shiba H, Funato M, Shimosato H, Takayama S, Isogai A (2003) Immunohistochemical studies on translocation of pollen S-haplotype determinant in self-incompatibility of *Brassica rapa*. Plant Cell Physiol 44:428–436

Iwano M, Shiba H, Matoba K, Miwa T, Funato M, Entani T, Nakayama P, Shimosato H, Takaoka A, Isogai A, Takayama S (2007) Actin dynamics in papilla cells of *Brassica rapa* during self- and cross-pollination. Plant Physiol 144:72–81

Johnson KL, Ingram GC (2005) Sending the right signals: regulating receptor kinase activity. Curr Opin Plant Biol 8:648–656

Kachroo A, Schopfer CR, Nasrallah ME, Nasrallah JB (2001) Allele-specific receptor-ligand interactions in *Brassica* self-incompatibility. Science 293:1824–1826

Kakita M, Murase K, Iwano M, Matsumoto T, Watanabe M, Shiba H, Isogai A, Takayama S (2007a) Two distinct forms of M-locus protein kinase localize to the plasma membrane and interact directly with S-locus receptor kinase to transduce self-incompatibility signaling in *Brassica rapa*. Plant Cell 19:3961–3973

Kakita M, Shimosato H, Murase K, Isogai A, Takayama S (2007b) Direct interaction between the S-locus receptor kinase and M-locus protein kinase involved in *Brassica* self-incompatibility signaling. Plant Biotechnol 24:185–190

Kanaoka MM, Torii KU (2010) FERONIA as an upstream receptor kinase for polar cell growth in plants. Proc Natl Acad Sci USA 107:17461–17462

Kaothien P, Ok S, Shuai B, Wengier D, Cotter R, Kelley D, Kiriakopolos S, Muschietti J, McCormick S (2005) Kinase partner protein interacts with the LePRK1 and LePRK2 receptor kinases and plays a role in polarized pollen tube growth. Plant J 42:492–503

Kemp BP, Doughty J (2007) S cysteine-rich (SCR) binding domain analysis of the *Brassica* self-incompatibility S-locus receptor kinase. New Phytol 175:619–629

Kessler SA, Shimosato-Asano H, Keinath NF, Wuest SE, Ingram G, Panstruga R, Grossniklaus U (2010) Conserved molecular components for pollen tube reception and fungal invasion. Science 330:968–971

Kim HU, Cotter R, Johnson S, Senda M, Dodds P, Kulikauska R, Tang W, Ezcura I, Herzmark P, McCormick S (2002) New pollen-specific receptor kinases identified in tomato, maize and *Arabidopsis*: the tomato kinases show overlapping but distinct localization patterns on pollen tubes. Plant Mol Biol 50:1–16

Kusaba M, Dwyer K, Hendershot J, Vrebalov J, Nasrallah JB, Nasrallah ME (2001) Self-incompatibility in the genus *Arabidopsis*: characterization of the *S* locus in the outcrossing *A. lyrata* and its autogamous relative *A. thaliana*. Plant Cell 13:627–643

Lee Y, Szumlanski A, Nielsen E, Yang Z (2008) Rho-GTPase-dependent filamentous actin dynamics coordinate vesicle targeting and exocytosis during tip growth. J Cell Biol 181:1155–1168

Liu C, Li Y, Semenov M, Han C, Baeg GH, Tan Y, Zhang Z, Lin X, He X (2002) Control of β-catenin phosphorylation/degradation by a dual-kinase mechanism. Cell 108:837–847

Miyazaki S, Murata T, Sakurai-Ozato N, Kubo M, Demura T, Fukuda H, Hasebe M (2009) ANXUR1 and 2, sister genes to FERONIA/SIRENE, are male factors for coordinated fertilization. Curr Biol 19:1327–1331

Murase K, Shiba H, Iwano M, Che FS, Watanabe M, Isogai A, Takayama S (2004) A membrane-anchored protein kinase involved in *Brassica* self-incompatibility signaling. Science 303:1516–1519

Muschietti J, Eyal Y, McCormick S (1998) Pollen tube localization implies a role in pollen-pistil interactions for the tomato receptor-like protein kinases LePRK1 and LePRK2. Plant Cell 10:319–330

Naithani S, Chookajorn T, Ripoll DR, Nasrallah JB (2007) Structural modules for receptor dimerization in the S-locus receptor kinase extracellular domain. Proc Natl Acad Sci USA 104:12211–12216

Niu Y, Wang ZM, Gao QG, Song M, Wang XJ, Zhu LQ (2009) Expression of ARC1 in vitro and test of interaction between ARC1 and SRK from *Brassica oleracea* L. in signal transduction pathway of self-incompatibility. Acta Agron Sin 35:1202–1208

Rotman N, Rozier F, Boavida L, Dumas C, Berger F, Faure J (2003) Female control of male gamete delivery during fertilization in *Arabidopsis thaliana*. Curr Biol 13:432–436

Samuel MA, Mudgil Y, Salt JN, Delmas F, Ramachandran S, Chilelli A, Goring DR (2008) Interactions between the S-domain receptor kinases and AtPUB-ARM E3 ubiquitin ligases suggest a conserved signaling pathway in *Arabidopsis*. Plant Physiol 147:2084–2095

Samuel MA, Chong YT, Haasen KE, Aldea-Brydges MG, Stone SL, Goring DR (2009) Cellular pathways regulating responses to compatible and self-incompatible pollen in *Brassica* and *Arabidopsis* stigmas intersect at Exo70A1, a putative component of the exocyst complex. Plant Cell 21:2655–2671

Schierup MH, Mable BK, Awadalla P, Charlesworth D (2001) Identification and characterization of a polymorphic receptor kinase gene linked to the self-incompatibility locus of *Arabidopsis lyrata*. Genetics 158:387–399

Schopfer CR, Nasrallah ME, Nasrallah JB (1999) The male determinant of self-incompatibility in Brassica. Science 286:1697–1700

Schulze-Muth P, Irmler S, Schröder G, Schröder J (1996) Novel type of receptor-like protein kinase from a higher plant (Catharanthus roseus). cDNA, gene, intramolecular autophosphorylation, and identification of a threonine important for auto- and substrate phosphorylation. J Biol Chem 271:26684–26689

Seifert GJ, Roberts K (2007) The biology of arabinogalactan proteins. Annu Rev Plant Biol 58:137–161

Shah K, Russinova E, Gadella TW Jr, Willemse J, De Vries SC (2002) The *Arabidopsis* kinase-associated protein phosphatase controls internalization of the somatic embryogenesis receptor kinase 1. Genes Dev 16:1707–1720

Shiba H, Iwano M, Entani T, Ishimoto K, Shimosato H, Che FS, Satta Y, Ito A, Takada Y, Watanabe M, Isogai A, Takayama S (2002) The dominance of alleles controlling self-incompatibility in *Brassica* pollen is regulated at the RNA level. Plant Cell 14:491–504

Shiba H, Kakizaki T, Iwano M, Tarutani Y, Watanabe M, Isogai A, Takayama S (2006) Dominance relationships between self-incompatibility alleles controlled by DNA methylation. Nat Genet 38:297–299

Shimosato H, Yokota N, Shiba H, Iwano M, Entani T, Che FS, Watanabe M, Isogai A, Takayama S (2007) Characterization of the SP11/SCR high-affinity binding site involved in self/nonself recognition in *Brassica* self-incompatibility. Plant Cell 19:107–117

Shiu SH, Bleecker AB (2003) Expansion of the receptor-like kinase/Pelle gene family and receptor-like proteins in *Arabidopsis*. Plant Physiol 132:530–543

Silva NF, Stone SL, Christie LN, Sulaman W, Nazarian KA, Burnett LA, Arnoldo MA, Rothstein SJ, Goring DR (2001) Expression of the S receptor kinase in self-compatible Brassica napus cv. Westar leads to the allele-specific rejection of self-incompatible *Brassica napus* pollen. Mol Genet Genomics 265:552–559

Steer M (1989) Calcium control of pollen-tube tip growth. Biol Bull 176:18–20

Stein JC, Howlett B, Boyes DC, Nasrallah ME, Nasrallah JB (1991) Molecular cloning of a putative receptor protein kinase gene encoded at the self-incompatibility locus of *Brassica* oleracea. Proc Natl Acad Sci USA 88:8816–8820

Stein JC, Dixit R, Nasrallah ME, Nasrallah JB (1996) SRK, the stigma-specific S locus receptor kinase of Brassica, is targeted to the plasma membrane in transgenic tobacco. Plant Cell 8:429–445

Stone JM, Collinge MA, Smith RD, Horn MA, Walker JC (1994) Interaction of a protein phosphatase with an *Arabidopsis* serine-threonine receptor kinase. Science 266:793–795

Stone JM, Trotochaud AE, Walker JC, Clark SE (1998) Control of meristem development by CLAVATA1 receptor kinase and kinase-associated protein phosphatase interactions. Plant Physiol 117:1217–1225

Stone SL, Arnoldo M, Goring DR (1999) A breakdown of *Brassica* self-incompatibility in ARC1 antisense transgenic plants. Science 286:1729–1731

Stone SL, Anderson EM, Mullen RT, Goring DR (2003) ARC1 is an E3 ubiquitin ligase and promotes the ubiquitination of proteins during the rejection of self-incompatible *Brassica* pollen. Plant Cell 15:885–898

Synek L, Schlager N, Elias M, Quentin M, Hauser MT, Zarsky V (2006) AtEXO70A1, a member of a family of putative exocyst subunits specifically expanded in land plants, is important for polar growth and plant development. Plant J 48:54–72

Takasaki T, Hatakeyama K, Suzuki G, Watanabe M, Isogai A, Hinata K (2000) The S receptor kinase determines self-incompatibility in *Brassica* stigma. Nature 403:913–916

Takayama S, Shiba H, Iwano M, Shimosato H, Che FS, Kai N, Watanabe M, Suzuki G, Hinata K, Isogai A (2000) The pollen determinant of self-incompatibility in *Brassica campestris*. Proc Natl Acad Sci USA 97:1920–1925

Takayama S, Shimosato H, Shiba H, Funato M, Che FS, Watanabe M, Iwano M, Isogai A (2001) Direct ligand-receptor complex interaction controls *Brassica* self-incompatibility. Nature 413:534–538

Tang W, Ezcurra I, Muschietti J, McCormick S (2002) A cysteine-rich extracellular protein, LAT52, interacts with the extracellular domain of the pollen receptor kinase LePRK2. Plant Cell 14:2277–2287

Tang W, Kelley D, Ezcurra I, Cotter R, McCormick S (2004) LeSTIG1, an extracellular binding partner for the pollen receptor kinases LePRK1 and LePRK2, promotes pollen tube growth in vitro. Plant J 39:343–353

Tarutani Y, Shiba H, Iwano M, Kakizaki T, Suzuki G, Watanabe M, Isogai A, Takayama S (2010) Trans-acting small RNA determines dominance relationships in *Brassica* self-incompatibility. Nature 466:983–986

Tsukamoto T, Qin Y, Huang Y, Dunatunga D, Palanivelu R (2010) A role for LORELEI, a putative glycosylphosphatidylinositol-anchored protein, in *Arabidopsis thaliana* double fertilization and early seed development. Plant J 62:571–588

van Meer G (2002) Cell biology. The different hues of lipid rafts. Science 296:855–857

Vanoosthuyse V, Tichtinsky G, Dumas C, Gaude T, Cock JM (2003) Interaction of calmodulin, a sorting nexin and kinase-associated protein phosphatase with the *Brassica oleracea S* locus receptor kinase. Plant Physiol 133:919–929

Wengier D, Valsecchi I, Cabanas ML, Tang WH, McCormick S, Muschietti J (2003) The receptor kinases LePRK1 and LePRK2 associate in pollen and when expressed in yeast, but dissociate in the presence of style extract. Proc Natl Acad Sci USA 100:6860–6865

Wengier D, Mazzella M, Salem T, McCormick S, Muschietti J (2010) STIL, a peculiar molecule from styles, specifically dephosphorylates the pollen receptor kinase LePRK2 and stimulates pollen tube growth in vitro. BMC Plant Biol 10:33

Wu X, Tu X, Joeng KS, Hilton MJ, Williams DA, Long F (2008) Rac1 activation controls nuclear localization of β-catenin during canonical Wnt signaling. Cell 133:340–353

Yalovsky S, Bloch D, Sorek N, Kost B (2008) Regulation of membrane trafficking, cytoskeleton dynamics, and cell polarity by ROP/RAC GTPases. Plant Physiol 147:1527–1543

Yang Z (2008) Cell polarity signaling in *Arabidopsis*. Annu Rev Cell Dev Biol 24:551–575

Zhang Y, McCormick S (2007) A distinct mechanism regulating a pollen-specific guanine nucleotide exchange factor for the small GTPase Rop in *Arabidopsis thaliana*. Proc Natl Acad Sci USA 104:18830–18835

Zhang Y, McCormick S (2010) The regulation of vesicle trafficking by small GTPases and phospholipids during pollen tube growth. Sex Plant Reprod 23:87–93

Zhang D, Wengier D, Shuai B, Gui C, Muschietti J, McCormick S, Tang W (2008) The pollen receptor kinase LePRK2 mediates growth-promoting signals and positively regulates pollen germination and tube growth. Plant Physiol 148:1368–1379

Zhao D (2009) Control of anther cell differentiation: a teamwork of receptor-like kinases. Sex Plant Reprod 22:221–228

Zonia L (2010) Spatial and temporal integration of signalling networks regulating pollen tube growth. J Exp Bot 61:1939–1957

Receptor Kinase Interactions: Complexity of Signalling

Milena Roux and Cyril Zipfel

Abstract Receptor-like kinases (RLKs) are signal detection systems used for myriad aspects of plant life, including growth, development and defense. RLKs are regulated by differential phosphorylation controlled by associated kinases and phosphatases. These regulators and other interactors control receptor localization and abundance to carefully balance signal perception and changes in downstream gene expression. RLK-mediated signalling is mediated initially by phosphorylation events occurring between proteins present in receptor complexes. The following chapter is a summary of our current knowledge on plant RLK signalling at the plasma membrane, highlighting some of the common principles evident in RLK complex composition.

Abbreviations

AHA	*Arabidopsis* H^+-ATPase
ARC1	ARM-repeat containing 1
BAK1	BRI1-associated RK 1
BAM	Barely any meristem
BIK1	Botrytis-induced kinase 1
BKI1	BRI1 kinase inhibitor 1
BRI1	Brassinosteroid insensitive 1
BSK	BR signalling kinase
CaM	Calmodulin
CLV	Clavata
CRN	Coryne

M. Roux
University of Copenhagen, Ole Maaløes Vej 5, Copenhagen 2200, Denmark

C. Zipfel (✉)
The Sainsbury Laboratory, Norwich Research Park, NR4 7UH, Norwich, UK
e-mail: cyril.zipfel@tsl.ac.uk

F. Tax and B. Kemmerling (eds.), *Receptor-like Kinases in Plants*,
Signaling and Communication in Plants 13,
DOI 10.1007/978-3-642-23044-8_9,

CZ	Central zone
DRM	Detergent-resistant membrane
Exo70A1	Exocyst subunit
FER	Feronia
HERK	Hercules
KAPP	Kinase-associated protein phosphatases
MLPK	M-locus protein kinase
OC	Organizing centre
PAMP	Pathogen-associated molecular pattern
PBL	PBS1-like
PZ	Peripheral zone
ROP	Rho GTPase
RPK2	Receptor-like protein kinase 2
RZ	Rib zone
SCD1	Stomatal cytokinesis defective 1
SNX1	Sorting nexin 1
SRK	S-locus protein kinase
THL1	Thioredoxin-h-like 1
TRIP-1	TGFbeta-interacting protein 1
TTL	Transthyretin-like

1 General Introduction

In diverse systems, receptor proteins engage in interactions by forming homo- or heteromers in order to co-ordinate signalling responses, often through reciprocal phosphorylation events (Schlessinger 2002). Mammalian growth factor receptors, one class of receptor tyrosine kinase (RTK), are composed of a glycosylated ligand-binding extracellular domain, a single transmembrane region and a cytoplasmic domain with catalytic tyrosine kinase activity (Schlessinger 1988). RTKs can be envisaged as membrane-associated allosteric enzymes, where ligand binding and enzyme activity are separated by their topology in the plasma membrane (Ullrich and Schlessinger 1990). Thus, receptor activation by ligand binding must somehow be translated across this membrane barrier into an altered signalling state of the intracellular domain. Receptor oligomerization, which occurs universally among growth factor receptors, is such a mechanism for ligand-induced activation (Schlessinger 1988; Williams 1989). Ligand binding stabilizes interactions between receptor molecules, leading to trans-phosphorylation, and this combines to create positive feedback, with enhanced ligand binding and kinase activity (Ullrich and Schlessinger 1990). For example, insulin receptor activation loop auto-phosphorylation increases its catalytic efficiency up to 200-fold (Cobb et al. 1989). Without specific stimuli, receptor kinases are auto-inhibited by intra-molecular interactions specific to each type of receptor, and the release of this inhibition is achieved when kinase domain phosphorylation releases the active site into a conformation suitable for phosphotransfer. Interestingly, auto-phosphorylation occurs in *trans* and in a particular sequence, each subsequent phosphorylation event further destabilizes the auto-inhibitory interactions (Lemmon and Schlessinger 2010). Once auto- and trans-phosphorylation occur, the phosphorylated receptor becomes a site for assembly of signalling protein complexes.

Plant receptor-like kinases (RLKs) share a similar domain-organization to mammalian growth factor receptors, although plant RLKs possess an intracellular Ser/Thr kinase domain that is more closely related to the *Drosophila* Pelle kinases and human IRAKs (Interleukin-1 receptor-associated kinase) than RTKs (Shiu and Bleecker 2001). Despite this distinction, plant RLKs, such as brassinosteroid insensitive 1 (BRI1) and BRI1-associated kinase 1 (BAK1), were recently shown to also auto-phosphorylate on tyrosine residues, revealing that plant RLKs are dual-specificity kinases (Oh et al. 2009, 2010). It is clear that receptor complex formation is also the key to plant RLK signalling; although nowhere near as much mechanistic detail is yet available for plant RLKs, which number over >600 in *Arabidopsis* and >1,000 in rice (Shiu and Bleecker 2001; Shiu et al. 2004; Zhang et al. 2010b).

In this chapter, we will discuss examples of RLK complexes in different plant signalling pathways, positive and negative regulators of RLKs and downstream signalling, and highlight some of the questions remaining to be answered.

2 Complexes Involved in Plant Growth and Development

2.1 *Brassinosteroid Signalling Complexes*

The major receptor for the growth-promoting plant steroid hormones brassinosteroids (BR) is the RLK brassinosteroid insensitive 1 (BRI1) (Li and Chory 1997; Wang et al. 2001; Kinoshita et al. 2005). Originally, *bri1* was identified as a BR-non-responsive dwarf mutant (Clouse et al. 1996). BRI1 has an extracellular domain composed of 25 leucine-rich repeats (LRRs), interrupted by an island domain between LRR 20 and 21 (Li and Chory 1997; Wang et al. 2001). This is connected to a cytoplasmic Ser/Thr kinase domain by a transmembrane and a juxtamembrane domain. The island domain and its flanking 21st LRR were shown to be required for BR binding (Kinoshita et al. 2005). BRI1 activity is negatively regulated by its C-terminus, which is a site of auto- and trans-phosphorylation (Wang et al. 2005). It is postulated that phosphorylation in this region causes conformational changes that release the auto-inhibition of the C-terminal tail (Wang et al. 2005).

The LRR-RLK BRI1-associated receptor kinase (BAK1) was identified as an interactor of BRI1 in a yeast-two hybrid (Y2H) screen and in an activation-tagging screen for a suppressor of the weak *bri1-5* allele (Li et al. 2002; Nam and Li 2002). BAK1 belongs to a sub-class of the subfamily II of LRR-RLKs, referred to as the SERK family based on sequence homology with the carrot LRR-RLK SERK protein involved in somatic embryogenesis (Hecht et al. 2001). Similar to BRI1,

BAK1 possesses an active intracellular Ser/Thr kinase domain (Li et al. 2002; Nam and Li 2002), but only 5 extracellular LRRs. Importantly, BAK1 does not appear to play a direct role in BR perception, as BR binding is maintained in *bak1* mutants (Kinoshita et al. 2005). Nonetheless, the importance of BAK1 for BRI1 function has been illustrated in several ways. First, *BAK1* over-expression results in a phenotype similar to *BRI1* over-expression (Nam and Li 2002), while *bak1* knock-out mutants resemble weak *bri1* alleles such as *bri1-301* that carries a G989I mutation in the kinase domain (Li et al. 2002). The fact that *bak1* mutants only resemble weak *bri1* alleles is explained by the redundancy between BAK1 and other SERKs (discussed later). Furthermore, the *bri1-301* dwarf phenotype was enhanced by crossing in the *bak1* mutation (Nam and Li 2002), while over-expression of *BAK1* partially suppresses *bri1-5* phenotypes (Li et al. 2002).

While BRI1 homodimers can mediate a basal level of BR signalling in the absence of SERKs (Wang et al. 2008), the amplitude of signalling is enhanced with their co-operation. BRI1 auto-phosphorylates its activation loop after ligand binding. In parallel, BAK1 associates with BRI1 and its activation loop residues become phosphorylated. The activated BAK1 then trans-phosphorylates BRI1 on intracellular juxtamembrane and C-terminal residues to enhance its activity (Wang et al. 2008). BRI1 also interacts with the BAK1 paralogs SERK1 and SERK4/BAK1-like (BKK1) that act redundantly with BAK1 in BR signalling (Karlova et al. 2006; He et al. 2007; Albrecht et al. 2008; Jeong et al. 2010).

De-phosphorylation is a widely studied mechanism for regulation of receptor kinases. In mammals, the dual-specificity MAP kinase phosphatases serve as feedback control regulators of innate immunity during infection (Wang and Liu 2007). In plants, PP2Cs are responsible for de-phosphorylation of Ser/Thr kinases. The kinase-associated protein phosphatase (KAPP) is capable of interacting with BRI1, BAK1 and SERK1 (Shah et al. 2002; Ding et al. 2007), and may function as a negative regulator of BR signalling (Ding et al. 2007). Indeed, *kapp bri1-5* double-mutants display mildly enhanced BR sensitivity compared to *bri1-5* (Ding et al. 2007). However, the mechanism of action of KAPP in BRI1 de-phosphorylation remains to be investigated.

Another negative regulator of BRI1 signalling, BRI1 kinase inhibitor 1 (BKI1), was identified as a BRI1-interacting protein by Y2H using the BRI1 intracellular domain (Wang and Chory 2006). Notably, BKI1 did not interact with the BRI1 paralogs BRL1 and BRL3, or with BAK1. RNA interference of *BKI1* enhances hypocotyl elongation, while over-expression causes a dwarf phenotype and modifies BR-regulated gene expression (Wang and Chory 2006). Importantly, BKI1 is detected at the plasma membrane but undergoes cytosolic relocalization upon BR treatment (Wang and Chory 2006). Furthermore, BKI1 associates more strongly with a kinase-inactive BRI1 than with the kinase-active WT BRI1 in Y2H, and BKI1 is a substrate for BRI1. Interestingly, persistent localization of BKI1 to the plasma membrane, imposed by the addition of an artificial myristoylation site, amplifies the dwarf phenotype seen in BKI1-overexpressing plants (Wang and Chory 2006). Notably, BKI1 appears to inhibit BRI1-BAK1 interaction in vitro and the release of BKI1 is required for signalling to be activated. These results together with the current model of phosphorylation events occurring between BRI1 and BAK1 suggest that upon BR perception, BKI1 phosphorylation by BRI1 causes its release from the complex and allows BRI1 to interact with BAK1 leading to full signalling capacity (Fig. 1). However, it is still unclear if BAK1 recruitment induces BKI1 release, or whether BKI1 release is a prerequisite for BAK1 heteromerization. These data suggest a regulatory mechanism where either BKI1 is a negative regulator of BRI1 under low-BR conditions, or BKI1 could act as a positive regulator of downstream signalling upon release from the plasma membrane.

BRI1-interacting proteins that are phosphorylated by BRI1 include transthyretin-like (TTL) protein, TGF-β receptor interacting protein 1 (TRIP-1) and BR signalling kinases (BSKs) (discussed below; Fig. 1; Table 1). TTL1, a plasma-membrane-localized protein, was shown to interact with the BRI1 kinase domain in yeast, and to be phosphorylated by BRI1 in vitro (Nam and Li 2004). Interestingly, TTL binds kinase-dead BRI1 with less affinity than the wild-type (Nam and Li 2004), although the interaction in vivo has not been characterized. TTL may regulate BR signalling through its interaction with the BRI1 kinase domain, though it is not known how, or whether, this interaction influences downstream BR-responsive gene expression. However, a role for TTL as a negative regulator of BR signalling is suggested by the enhanced BR-sensitivity and enhanced growth observed in *ttl* mutants, in contrast to the *bri1-301*-like dwarf phenotype caused by *TTL* over-expression (Nam and Li 2004). The authors suggest that PM-localization of TTL may be required for its function, which may be regulation of BR-mediated cell elongation. However, presently there are no data to support this hypothesis, and it remains to be seen whether biological activity of TTL or its phosphorylation status is affected by exogenous hormone application.

In a different approach, the similarity between plant and animal RLK signalling was exploited to identify further BRI1-interacting proteins. Extensive analysis to understand the regulation of BRI1 function has revealed several parallels between these receptors and transforming growth factor-β (TGF-β) receptor (Kim and Wang 2010). In mammals, TGF-β polypeptides mediate various developmental processes by binding to surface receptor complexes consisting of type I and type II TGF-β receptor kinases (Massagué 1998; Shi and Massagué 2003). TGF-β I receptor homodimers hetero-oligomerize with TGF-β II homodimers. The ligand TGF-β binds to TGF-β II to drive its interaction with and phosphorylation of TGF-β I receptors. This results in recruitment of downstream signalling components and consequently signal transduction (Lutz and Knaus 2002). The WD domain protein TGF-β receptor interacting protein 1 (TRIP-1) is a cytoplasmic substrate of type II TGF-β receptor (Chen et al. 1995; Choy and Derynck 1998) and also plays an independent role as a subunit of the eukaryotic translation initiation factor (eIF3) complex (Asano et al. 2000). In *Arabidopsis* and bean, antisense suppression of *TRIP-1* homologs causes dwarfism, delayed senescence and bushiness, resembling BR-deficient/insensitive mutants (Jiang and Clouse 2001). These data provided the first hint at a potential BR-related function for TRIP-1 in plants. Kinase assays and phosphoamino acid analysis in vitro revealed phosphorylation of TRIP-1 by BRI1 kinase activity (Ehsan et al. 2005). This was supplemented by mass spectrometry

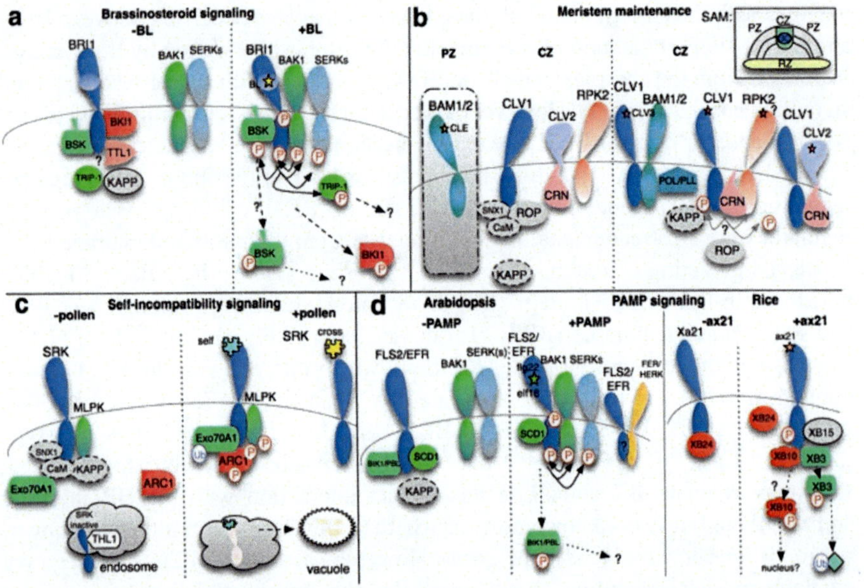

Fig. 1 Models of selected plant RLK complexes. (**a**) Brassinosteroid signaling complexes. *Left panel*: prior to BL perception, BRI1 is associated with BKI1 and TTL (*red*), *Right panel*: Following BL perception, BKI1 dissociation may allow BRI1 to interact with positive regulators. BRI1 is activated by autophosphorylation and BAK1-mediated transphosphorylation. SERKs act with BAK1 to regulate BRI1 activity. BSKs and TRIP-1, both positive regulators of BRI1, are phosphorylated by BRI1 and released to act downstream. (**b**) Meristem maintenance complexes. *Inset* shows schematic representation of shoot apical meristem (SAM) with central zone (CZ) containing the organizing centre (OC), where CLV1 is expressed; peripheral zone (PZ) where BAMs are expressed; and rib zone (RZ). In the peripheral zone, BAMs are thought to sequester CLE peptides. In the central zone, CLV1 binds CLV3 peptide and is associated with CLV2 and RPK2, via CRN. CLV1 is also associated with BAMs in the central zone. CLV2 also binds CLV3 and in association with CRN homodimers also transduces CLV3 signal. ROP GTPase was identified in CLV1 complexes but its biological function in this pathway remains uncharacterized. (**c**) Self-incompatibility RLK complexes. SRK is associated with negative regulator TTL in endosomes. Following receptor activation by self-pollen (*blue puzzle*), SRK is targeted to vacuolar degradation. SRK phosphorylates MLPK and MLPK also phosphorylates ARC1, and ARC1 ubiquitinates Exo70A1, preventing delivery of vesicles containing factors required for pollen tube formation. In the presence of cross-pollen (*yellow puzzle*), SRK is not activated and Exo70A1 delivers its cargo for pollen reception. (**d**) PAMP perception-related RLK complexes. *In Arabidopsis*: FLS2 is associated with SCD1 prior to flg22 elicitation. Following PAMP addition, BAK1 rapidly associates with FLS2 and EFR, and this is accompanied by auto-/trans-phosphorylation of the receptors. BIK1 and related RLCKs are also constitutively associated with EFR and FLS2 and released following phosphorylation, possibly for downstream signaling activation. FLS2 was found in flg22-induced DRMs containing other RLKs such as FER and HERK, as well as AHAs. *In rice*: Xa21 is associated with the ATPase XB24 in the absence of Xoo; XB24 promotes Xa21 auto-phosphorylation and is dissociated following ax21 binding. Xa21 phosphorylates downstream components such as XB3 and XB10. PP2C XB15 dephosphorylates and inhibits Xa21 activity. Phosphorylation is indicated by *red circles* enclosing P; ubiquitination is indicated by *purple circles* enclosing Ub. *Question marks* indicate missing data. Interactors identified in vitro only are marked by *dotted outlines*. Note that for simplicity the stoichiometry of complexes is not reflected in the figure

Table 1 Common classes of plant RLK-interactors

Receptor	BRI1	FLS2	Xa21	CLV1, CLV2	SRK1
Biological function	Growth cell elongation	Innate immunity	Innate immunity	Apical meristem maintenance	Self-incompatibilty
Ligand	BR[a–c]	flg22[d,e]	ax21[f]	CLV3[g–j]	SCR/SP11[k,l]
Regulatory RLK/RLP	BAK1[m,n]	BAK1[o,p,q]		BAM1/2[r,s]	
	SERK1[t]	BKK1[u]		RPK2[v,w]	
	BKK1[x–aa]	SERK1/SERK2?[u]			
Cytoplasmic kinase	BSKs[ab]	BIK1/PBLs[ac,ad]		CRN[ba,bb,bc]	MLPK[af,ag]
ATPase			XB24[ah]		
E3 ligase			XB3[ai]		ARC1[aj,ak]
Phosphatase	KAPP[al]	KAPP[am]	XB15[an]	KAPP[ao,ap]	KAPP[aq]
				POL/PLL1[ar,as]	
Transcription factor			XB10[at]		
Other	TTL[au]	SCD1[av]		SNX1[aq]	Exo70A1[aw]
	TRIP-1[ax]			CaM[aq]	SNX1[aq]
	BKI1[ay]			Rho GTPase[ao]	CaM[aq]
					THL[k,l,af,ag,aj,ak,aw,az]

References: [a]Li and Chory (1997), [b]Wang et al. (2001), [c]Kinoshita et al. (2005), [d]Gómez-Gómez and Boller (2000), [e]Chinchilla et al. (2006), [f]Lee et al. (2009), [g]Ogawa et al. (2008), [h]Jeong et al. (1999), [i]Fiers et al. (2005), [j]Wang et al. (2010), [k]Cabrillac et al. (2001), [l]Takayama et al. (2001), [m]Li et al. (2002), [n]Nam and Li (2002), [o]Chinchilla et al. (2007), [p]Heese et al. (2007), [q]Schulze et al. (2010), [r]deYoung et al. (2006), [s]Guo et al. (2010), [t]Shah et al. (2002), [u]Roux et al. (in revision), [v]Kinoshita et al. (2010), [w]Betsuyaku et al. (2010), [x]Karlova et al. (2006), [y]He et al. (2007), [z]Albrecht et al. (2008), [aa]Jeong et al. (2010), [ab]Tang et al. (2008), [ac]Lu et al. (2010), [ad]Zhang et al. (2010a), [ae]Miwa et al. (2008), [af]Murase et al. (2004), [ag]Kakita et al. (2007a), [ah]Chen et al. (2010b), [ai]Wang et al. (2006), [aj]Gu et al. (1998), [ak]Stone et al. (2003), [al]Ding et al. (2007), [am]Gómez-Gómez et al. (2001), [an]Park et al. (2008), [ao]Trotochaud et al. (1999), [ap]Zhao et al. (2011), [aq]Vanoosthuyse et al. (2003), [ar]Song et al. (2006), [as]Muller et al. (2008), [at]Peng et al. (2008), [au]Nam and Li (2004), [av]Korasick et al. (2010), [aw]Samuel et al. (2009), [ax]Ehsan et al. (2005), [ay]Wang and Chory (2006), [az]Bower et al. (1996), [ba]Muller et al. (2008), [bb]Bleckmann et al. (2010), [bc]Zhu et al. (2010)

analysis, which led to the identification of three specific TRIP-1 phosphosites (T14, T89, T197/S198). Furthermore, TRIP-1 and BRI1 were shown to interact in planta in a series of comprehensive co-immunoprecipitation experiments (Ehsan et al. 2005). These data suggest a role for TRIP-1 as a cytoplasmic substrate of BRI1, possibly in BR-mediated translational regulation; however the molecular mechanism of TRIP-1 in BR signalling awaits investigation.

In a proteomic approach towards identification of BR signalling components, two-dimensional difference gel electrophoresis (2D-DIGE) was used to study proteins with altered migration upon BL application, using plasma membrane protein derived from *det2* seedlings (defective in BR biosynthesis) (Tang et al. 2008). Mass spectrometry analysis of the gel spots identified BR signalling kinases (BSKs) with enhanced acidic forms detectable after BR treatment, suggesting phosphorylation. BSKs are among 12 members of the subfamily XII of receptor-like cytoplasmic kinases (RLCK) (Shiu et al. 2004) that feature an N-terminal kinase domain and a C-terminal tetratricopeptide repeat (TPR) domain. Interestingly, TPR domains are present in proteins from steroid receptor complexes, and may mediate protein–protein interactions (Galigniana et al. 2010). The acidic shift of BSK1 in response to BR was reduced in the *bri1-5* background, hinting at BRI1-mediated regulation of BSKs (Tang et al. 2008). Considering that confocal microscopy analysis of transgenic lines expressing BSK1-YFP indicated PM localization, these data point to BSKs as BRI1 or BAK1 substrates. Indeed, kinase assays confirmed phosphorylation of BSK1 by BRI1 in vitro, and mass spectrometry analysis combined with mutagenesis identified BSK1 Ser230 as a major BRI1 phosphorylation site (Tang et al. 2008). Although BSKs are RLCKs, BSK1 and BSK2 contain putative myristoylation sites that may be responsible for their apparent PM localization (Tang et al. 2008). A direct interaction between BRI1 and BSK1 was indicated by bimolecular fluorescence complementation (BiFC) analysis of transiently expressed YFP fusion proteins in *N. benthamiana* (Tang et al. 2008). Importantly, a similar interaction between BSK1 and BAK1 was not detectable. Furthermore, co-immunoprecipitation analysis using specific anti-BSK1 antibodies were used to confirm this association in *Arabidopsis*, which was reduced upon BR treatment. Since approximately half the amount of BSK1 could be immunoprecipitated following BR treatment, it is conceivable that BSK1 is released from BRI1 interactions following phosphorylation (Tang et al. 2008). Knockout lines for BSK1 were not available, thus phenotypic analysis was continued on the paralog BSK3, which also interacts with and is phosphorylated by BRI1 (Tang et al. 2008). *Bsk3* mutants display only reduced BR sensitivity, suggesting redundant or coinciding roles for BSK1 and its paralogs in BR signalling. In contrast, over-expression of BSK1, BSK3 or BSK5 suppressed *bri1-5* or *det2* dwarf phenotypes, as measured by differential expression of the downstream target *DWF4* gene. Importantly, BSK3 over-expression could partially overcome the strong dwarf phenotype of the null mutant *bri1-116*, but not that of a mutant of the downstream kinase BIN2, *bin2-1*. This points to BSK3 functioning as a positive regulator and substrate of BRI1 downstream of BRI1 but upstream of BIN2 (Tang et al. 2008).

2.2 Stem Cell Maintenance: CLAVATA Complexes

The *Arabidopsis* shoot apical meristem (SAM) is a complex arrangement of undifferentiated cells, dividing cells and differentiating cells that will ultimately develop into the aerial parts of the plant (leaves, stem, vasculature, flowers) (Bowman and Eshed 2000; Carles and Fletcher 2003; Dodsworth 2009). The stem cells are arranged in three layers at the apex of the SAM. Due to their preferred planes of division, they produce clonally distinct cell layers: stem cells of the two outer layers mainly divide anticlinally (perpendicular to the surface), giving rise to the epidermal (L1) and subepidermal layers (L2). This is in contrast to the stem cells below that divide anti- and periclinally (parallel to the surface), generating the interior tissue of the stem and lateral organs (L3). The SAM is divided into the central zone (CZ), comprising undifferentiated cells just above the organizing centre (OC), the peripheral zone of rapidly dividing cells, and the rib zone (RZ) (Fig. 1b inset). In adjacent areas of the meristem, RLKs and transcription factors mediate opposing effects on differentiation, requiring precise regulation of stem cell differentiation. In *Arabidopsis*, the stem cell pool is controlled by the CLAVATA pathway, where the peptide ligand CLAVATA 3 (CLV3) is secreted from stem cells in the CZ. CLV3 restricts expression of WUSCHEL (WUS), a homeodomain transcription factor that promotes stem cell fate, limiting its expression to a few cells of the OC (Clark et al. 1995; Brand et al. 2000; Schoof et al. 2000). WUS is also required for *CLV3* expression, and the balance of WUS and CLV3 levels is required for normal morphology (Schoof et al. 2000). For example, over-expression of *CLV3* and loss-of-function *wus* plants exhibit SAM termination, while *clv3* has an enlarged SAM with an enlarged WUS expression domain (Clark et al. 1995; Laux et al. 1996; Schoof et al. 2000).

CLV3 is a member of the CLV3/embryo-surrounding-region (CLE) family of arabinosylated glycopeptides, with a conserved C-terminal 14-amino-acid CLE domain (Kondo et al. 2006; Ohyama et al. 2009). Several RLK complexes are involved in CLV3 perception in the SAM of Arabidopsis (Fig. 1; Table 1). The LRR-RLK CLV1 is expressed primarily in the SAM centre, where it binds CLV3 to regulate stem cell specification and inhibit cell division in the SAM (Ogawa et al. 2008). Transient expression in *N. benthamiana* and fluorescence resonance energy transfer (FRET) analysis as well as co-immunoprecipitation in *Arabidopsis* and *N. benthamiana* showed that CLV1 forms homodimers at the plasma membrane (Bleckmann et al. 2010; Guo et al. 2010). More recently, live imaging of fluorescently tagged CLV1 in Arabidopsis meristems revealed that CLV1 relocalizes from the plasma-membrane to lytic vacuoles in the presence of CLV3 (Nimchuk et al. 2011).

CLAVATA 2 (CLV2) is an LRR-receptor-like protein (RLP) that is required for this pathway, although CLV2 is more widely expressed and actually functions in several organ development pathways (Jeong et al. 1999; Fiers et al. 2005; Wang et al. 2010). Genetic analysis revealed another regulator, the trans-membrane kinase CORYNE/SOL2 (CRN), that functions with CLV2 in the SAM regulatory pathway (Miwa et al. 2008; Muller et al. 2008). Importantly, CRN and CLV2 initially interact in the ER and this interaction is required for both to reach the PM (Bleckmann et al. 2010). Although CRN was initially considered as an adapter for correct CLV2 localization, recent work has shown that CRN kinase activity may be necessary for CLV signalling (Betsuyaku et al. 2010). Indeed, kinase-dead CRN could not complement CRN-deficient *sol2-1* mutants for its *clv1*-like carpel phenotype (Betsuyaku et al. 2010). Confocal microscopy, luciferase complementation imaging (LCI), as well as FRET analysis of heterologously expressed proteins were used to study the signalling complex(es) comprising CLV1, CLV2, and CRN. LCI and FRET provided evidence for the formation of CRN–CLV2 interactions (Bleckmann et al. 2010; Zhu et al. 2010), while CLV1 homodimerization was demonstrated by FRET analysis (Bleckmann et al. 2010). CLV1 and CRN homodimers are thought to be already assembled in the ER (Bleckmann et al. 2010) and later combine to form tetramers with separate CLV1–CLV2 and CLV2–CRN complexes (Muller et al. 2008; Meng and Feldman 2010; Zhu et al. 2010). Later, CLV2–CRN interactions were also shown in transgenic Arabidopsis, whereas no CLV2 homodimerization and only weak CLV1–CRN interactions were detected (Guo et al. 2010). Previous work showed that *clv2* and *sol* mutants are resistant to many CLE peptides (Miwa et al. 2009), and CLV2-CRN heterodimers bind CLV3 (Bleckmann et al. 2010). Remarkably, recent work employing radiolabelled CLE peptides has revealed that CLV1 and CLV2 share a similar affinity for diverse CLE peptides, independent of CRN (Guo et al. 2010).

Barely any meristem 1 (BAM1), BAM2 and BAM3 are CLV1-related RLKs that function antagonistically to CLV1 in stem cell maintenance (DeYoung et al. 2006). However, *CLV1* expression can rescue *bam* mutants, and meristem-expressed *BAM* can rescue *clv1* mutants (DeYoung et al. 2006), suggesting that these receptors are integral to CLV signalling and do not constitute an alternative pathway for meristem development. The main difference between BAMs and CLV1 is in their expression pattern: BAMs are more widely expressed than CLV1, are detectable throughout the plant (DeYoung et al. 2006), and concordantly play a broader role in floral organ development and leaf vascular patterning (Clark et al. 1993; DeYoung et al. 2006; Hord et al. 2006). Using co-immunoprecipitation analysis of transgenic *Arabidopsis* as well as transient expression in *N. benthamiana*, CLV1 was found to heteromerize with BAM1 or BAM2 (Guo et al. 2010). Given that several CLE domains could bind to CLV1 and CLV2, but BAMs have an even wider range of CLE binding capability (Miwa et al. 2009; Guo et al. 2010). The role of BAMs may be to protect the SAM from diffusion of CLE peptides from outside the meristem by sequestering CLE in the PZ (DeYoung and Clark 2008). This preserves the balance between *WUS* expression and CLV1 signalling required for stem cell maintenance (DeYoung and Clark 2008). Finally, over-expression of *BAM1/2* or *CLV1* can compensate for *clv2* phenotypes (Guo et al. 2010), suggesting that distinct CLV1-BAM and CLV2–CRN complexes can function independently.

Exogenously applied synthetic CLV3 peptides mimic CLV3 over-expression phenotypes, and can be employed to study this pathway more easily (Fiers et al. 2005; Kondo et al. 2006; Sawa et al. 2006; Kinoshita et al. 2007). Recent work employing synthetic CLV3 peptides (mCLV3) showed that RECEPTOR-LIKE PROTEIN KINASE 2/TOADSTOOL2 (RPK2/TOAD2) is an additional regulator in stem-cell maintenance (Kinoshita et al. 2010). *Rpk2* mutants mirror *clv* phenotypes, and double mutants with *clv1* have an additive phenotype (Kinoshita et al. 2010). RPK2-CLV complexes may constitute yet another avenue for CLV3 signal transduction. However, no CLV3 binding has been shown for RPK2, although *rpk2*/*toad2* mutants are partially insensitive to treatment with CLE peptide in the roots. This phenotype is also found in *crn* and *clv2* mutants, which bind CLV3. Furthermore, in *N. benthamiana* CLV1 is associated with CLV2 and RPK2 in a CRN-dependent manner (Betsuyaku et al. 2010).

Within the SAM, the PP2Cs POLTERGEIST (POL) and POL-like 1 (PLL1) also regulate the asymmetric nature of the stem cell division by promoting *WUS* expression in the basal daughter of dividing L3 stem cells (Song et al. 2006, 2008). Genetic evidence suggests that CLV1, CLV2, and CRN negatively regulate POL and PLL1 (Song et al. 2006; Müller et al. 2008). The molecular mechanism for POL/PLL1 is not known, but recent data have shown that POL/PLL1 are acylated, plasma membrane-localized proteins, and this is likely to be the site of their action (Gagne and Clark 2010). POL/PLL1 were also activated by binding the phospholipid phosphatidylinositol-4-phosphate, suggesting a potential role in trafficking of other CLV signalling components, though this has not yet been shown (Gagne and Clark 2010). It is tempting to speculate that POL/PLL1 may be directly associated with their regulatory receptors, but there is no evidence to support this idea. Negative regulation of POL/PLL1 is required to specify differential *WUS* expression in apical and basal daughter cells of periclinally dividing L3 cells (Song et al. 2006, 2008). One model suggests that due to apical expression of CLV3 in dividing cells, this acts as a directional signal causing polarization of POL/PLL1 expression or activity (Gagne and Clark 2010). Thus, the apical daughter cell would have low/absent POL/PLL1, while the basal daughter cell would have active POL/PLL1, leading to differential *WUS* expression and altered cell fate. The asymmetric distribution predicted by this model has been observed by confocal microscopy of fluorescently tagged POL/PLL1, localized to the plasma membrane of transgenic *Arabidopsis* roots (Gagne and Clark 2010).

CLV1 was previously found to form a large 450 kDa complex, including the phosphatase KAPP and a Rho GTPase-related protein (Trotochaud et al. 1999). Despite this evidence, the molecular basis for the role of these proteins in CLV signalling remains sketchy (Williams et al. 1997). Recent work used a phage display library of random peptides to identify in vitro binding partners of the heterologously expressed CRN cytoplasmic domain (Zhao et al. 2011). Several peptides matching the KAPP sequence were identified in this manner, further hinting at a role for KAPP in CLV signalling (Zhao et al. 2011). It is important to consider that many studies have been focused on the interactions between individual receptor pairs, and not simultaneous expression of all receptors. Recent work sought to address this deficit (Betsuyaku et al. 2010; Guo et al. 2010), by co-expressing the known receptors for CLV3 signalling together in *N. benthamiana* to observe their influence on each other. In summary, CLV1 is associated with CLV2 and RPK2 via CRN; CLV1 exists as homodimers, but mostly forms CLV1-BAM heterodimers and CLV2 exists largely as CLV2-CRN heterodimers, each heterodimer capable of autonomous signal perception (Betsuyaku et al. 2010; Guo et al. 2010).

In addition, sorting nexin 1 (SNX1) and calmodulin interact with CLV1 in vitro (Vanoosthuyse et al. 2003). However, similar to KAPP, these proteins were capable of interacting with several RLKs, including the stigma-specific S receptor kinase (SRK) (Fig. 1; Table 1), and their biological role in the regulation of the SAM has yet to be characterized.

Finally, phosphorylation is an important consequence and aspect of RLK complex formation. For CLV signalling, the study of receptor post-translational modification is in its infancy. However, a slight mobility shift in CLV1 protein migration during electrophoresis could be detected when proteins were extracted from *N. benthamiana* leaves where CLV1 was co-expressed with CLV3-YFP (Betsuyaku et al. 2010). This suggests auto- or trans-phosphorylation of the receptor upon ligand binding however, this has not yet been convincingly demonstrated and requires more in-depth characterization.

3 RLK Complexes in *Brassica* Self-incompatibility

Flowering plants have evolved complex stigma-pollen signalling to allow specific recognition of compatible pollen. Self-incompatibility (SI), which prevents fertilization of self-pollen, is, for example, present in the *Brassica* genus, and results in self-pollen rejection by prevention of pollen hydration and pollen tube penetration through the stigma (Hiscock and McInnis 2003; Takayama and Isogai 2005). Genetic and molecular studies have led to the identification of several components of this system. Control of the SI response is encoded by the polymorphic *S*-locus, which contains the gene for the stigma-specific S receptor kinase (SRK) and its cognate ligand S-locus Cys-rich/S-locus protein 11 (SCR/SP11), expressed in the pollen coat and anthers (Schopfer et al. 1999; Takasaki et al. 2000; Silva et al. 2001; Takayama et al. 2001). An SRK will only recognize SCR from the same haplotype. Following self-pollen attachment, SCR will engage with SRK to cause SRK phosphorylation and subsequent downstream signalling (Cabrillac et al. 2001; Takayama et al. 2001).

Screens for regulators of SRK revealed several interacting proteins (Bower et al. 1996; Gu et al. 1998; Vanoosthuyse et al. 2003; Kakita et al. 2007b), some of which turned out to be involved in SI signalling (discussed further below) (Fig. 1; Table 1). Genetic analysis of *Brassica rapa* var. Yellow Sarson revealed a recessive mutation in the *modifier* (*m*) gene, which suppressed stigmatic SI responses (Murase et al. 2004). *M* encodes the M-locus protein kinase (MLPK), which belongs to the RLCK family (subfamily VII) (Murase et al. 2004), and is expressed as two alternative transcripts (Kakita et al. 2007b). BiFC analysis in protoplasts indicated that MLPK is associated with the PM, where it directly interacts with un-stimulated SRK (Kakita et al. 2007b). MLPK is essential for SI responses (Murase et al. 2004), and in vitro kinase assays indicate that MLPK is a substrate for SRK kinase (Kakita et al. 2007a). Thus, MLPK may be a positive regulator of SI signalling mediated by SRK.

An Y2H assay employing the SRK kinase domain led to the identification of several additional SRK-interacting proteins, including thioredoxin-H-like 1 (THL1) and THL2 (Bower et al. 1996). In the absence of the SRK-activating ligand, the receptor is kept in an inhibited state by THL1, where auto-phosphorylation is prevented (Cabrillac et al. 2001). Pollen coat proteins relieve this haplotype-specific inhibition following self-incompatible interactions (Cabrillac et al. 2001). Importantly, a single cell can simultaneously accept cross-pollen and reject self-pollen. This duality can be explained by the unique sub-cellular localization of the receptor, recently revealed by specific antibodies (Ivanov and Gaude 2009b). SRK has an unusual sub-cellular distribution, with principally endosomal localization and lesser PM representation (Ivanov and Gaude 2009b). In endosomes, SRK is co-localized with THL1, suggesting that active SRK signalling occurs at the PM (Ivanov and Gaude 2009b). However, it is also possible that signalling continues from endosomes; something that has not been yet experimentally excluded. Over-expressed SRK can auto-activate (Giranton et al. 2000; Cabrillac et al. 2001), suggesting the existence of "SI domains" at the PM containing low amounts of SRK in order to prevent a non-specific activation and to restrict the SI response to restricted areas, where the activating ligand is present (Ivanov and Gaude 2009a).

The above-mentioned Y2H screen also identified the ARM-repeat containing 1 protein (ARC1), which interacts with SRK in a phosphorylation–dependent manner via the ARC1 C-terminus (Gu et al. 1998). Evidence for the role of ARC1 as a positive regulator of SI responses came with the observation that RNA interference of *Brassica napus ARC1* resulted in partial loss of SI (Stone et al. 1999). ARC1 possesses E3 ligase activity, though there is no evidence for SRK ubiquitination by ARC1 (Stone et al. 2003).

A model proposes that following SRK activation by SCR/SP11, THL1 is released, allowing SRK auto-phosphorylation and recruitment of ARC1. This may lead to ubiquitination and degradation of a negative regulator of SI and rejection of self-pollen. ARC1 is distributed within the cytoplasm and nucleus upon transient expression in tobacco cells; while it re-localizes to ER-associated proteasomes when activated SRK is co-expressed (Stone et al. 2003; Samuel et al. 2008). Furthermore, co-expression of ARC1 with MLPK leads to perinuclear localization for both proteins (Samuel et al. 2008). This dynamic localization of ARC1, as well as the ability of MLPK to phosphorylate ARC1 (Samuel et al. 2008), may point to a role for ARC1 as a target for MLPK activity, and correlates the positive roles of ARC1 and MLPK in the SI response (Stone et al. 1999; Murase et al. 2004).

An Y2H screen for ARC1-interacting proteins resulted in the identification of the exocyst complex subunit Exo70A1 (Samuel et al. 2009), which was recently found to be involved in secretory processes related to cell plate assembly during cytokinesis in *Arabidopsis* (Fendrych et al. 2010). Stigma-specific silencing of *Exo70A1* in the self-compatible *B. napus* cultivar Westar results in impaired self-pollination and reduced seed yield (Samuel et al. 2009), providing clear evidence for the role of Exo70A1 in stigmatic pollen recognition. Similarly, *Arabidopsis exo70A1* mutants suffer from poor pollen hydration and pollen tube growth (Samuel et al. 2009). In transgenic *Arabidopsis*, RFP-tagged Exo70A1 can be detected in punctate vesicles that co-localize with Golgi in immature stigmas, or at the PM in mature stigmas (Samuel et al. 2009). Notably, Exo70A1 is ubiquitinated by ARC1 in vitro (Samuel et al. 2009). ARC1-mediated degradation of Exo70A1 may explain the observed disappearance of Exo70A1 signal following self-pollination (Samuel et al. 2009). ARC1 was detected in the cytosol when expressed alone, but in the presence of SRK1, Exo70A1 and ARC1 were present in punctate structures co-localizing with the proteasomal subunit RPT2A and the late endosomal/prevacuolar marker Syp21 (Samuel et al. 2009). Hypothetically, if Exo70A1 is required for polarized secretion of vesicles containing factors required for compatible pollen recognition and resulting in pollen tube growth, then Exo70A1 could be degraded by ARC1 during an incompatible interaction to prevent such delivery. Further work remains to be done to identify these hypothetical factors delivered by Exo70A1, and to gather direct evidence for Exo70A1 degradation by ARC1.

Additional SRK interactors have been identified in vitro, but have not been fully characterized functionally. SRK was shown to interact with KAPP and was subject to KAPP-mediated de-phosphorylation in vitro (Vanoosthuyse et al. 2003). Furthermore, an Y2H screen identified sorting nexin 1 (SNX1) and calmodulin (CaM) as additional in vitro SRK interactors as shown for additional RLKs (Fig. 1; Table 1) (Vanoosthuyse et al. 2003). SNXs are involved in endosomal trafficking; thus, SNX1 could play a role in re-localization of SRK following activation. The biological functions of KAPP, SNX1 and calmodulin in SI responses have not been addressed yet.

4 RLK Complexes Involved in Innate Immunity

Plants detect the presence of invading microbes via the presence of conserved pathogen-associated molecular patterns (PAMPs) that are signature molecules found specifically in microbes and that play an important role in aspects of their life (Medzhitov and Janeway 1997). PAMP perception by surface-localized transmembrane pattern recognition receptors (PRRs) leads to PAMP-triggered immunity (PTI) restricting pathogen growth. Although several PAMPs recognized by plants are known, the number of plant PRRs is still sparse. Yet, several PRRs are LRR-RLKs and recent studies have demonstrated the role of associated proteins for PTI signalling.

4.1 Arabidopsis *EFR and FLS2 Complexes*

In *Arabidopsis*, the LRR-RLKs flagellin sensing 2 (FLS2) and EF-Tu receptor (EFR) bind the bacterial PAMPs flagellin (or the surrogate flg22 peptide) and elongation factor Tu (EF-Tu, or the surrogate elf18/elf26 peptides), respectively (Gomez-Gomez and Boller 2000; Kunze et al. 2004; Chinchilla et al. 2006; Zipfel et al. 2006). Surprisingly, FLS2 was found to form a ligand-dependent complex with the regulatory LRR-RLK BAK1 (Chinchilla et al. 2007; Heese et al. 2007) (Fig. 1; Table 1). This association occurs within seconds of flg22 binding, and leads to rapid phosphorylation of FLS2 and BAK1 (Schulze et al. 2010). Consistently, *bak1-4* mutants display reduced responsiveness in early and late flg22-induced responses (Chinchilla et al. 2007; Shan et al. 2008). Interestingly, loss of BAK1 also leads to reduced responsiveness to additional PAMPs, such as elf18, lipopolysaccharides (LPS), peptidoglycans (PGN), harpin Z (HrpZ), cold-shock protein 22 (CSP22), INF1, as well as to the endogenous danger-associated molecular pattern (DAMP) AtPep1 (Chinchilla et al. 2007; Heese et al. 2007; Shan et al. 2008; Krol et al. 2010), hinting that BAK1 may form a ligand-dependent complex with their corresponding PRRs. The receptors for AtPep1 are the LRR-RLKs PEP1 receptor 1 (PEPR1) and PEPR2 (Huffaker and Ryan 2007; Krol et al. 2010; Yamaguchi et al. 2010), and have been shown to interact with BAK1 in Y2H (Postel et al. 2010). Furthermore, elf26 treatment of *Arabidopsis* cell cultures leads to rapid phosphorylation of BAK1 and of a co-immunoprecipitated protein that migrates at the similar size as the glycosylated form of EFR (Schulze et al. 2010). Notably, the clear ligand-dependent heteromerization between EFR and BAK1 was recently demonstrated (Roux et al. 2011).

Importantly, the effect of *BAK1* loss-of-function on elf18 responses is less striking than for flg22 responses, with only early signalling responses being reduced in the null *bak1-4* mutant (Chinchilla et al. 2007; Shan et al. 2008). In addition, the *bak1-4* mutant is not fully insensitive to flg22 (Chinchilla et al. 2007; Heese et al. 2007). These observations indicate that EFR may preferentially interact with other RLKs than BAK1, and that additional complex components are required for full signalling downstream of FLS2 and EFR. The SERK family contains 5 closely related members in Arabidopsis, which are involved in diverse signalling pathways and can be functionally redundant (Albrecht et al. 2008). SERKs can combine in distinct oligomeric complexes specific to cellular pathways. Importantly, BKK1/SERK4 and BAK1/SERK3 are both required for cell death control and senescence and *bak1-4 bkk1-1* double mutants display a seedling lesion mimic phenotype (He et al. 2007; Kemmerling et al. 2007; Jeong et al. 2010).

We have recently shown using mass spectrometry and co-immunoprecipitation analyses that EFR is associated with SERK1, SERK2, BAK1 and BKK1 in planta in a ligand-inducible manner in *Arabidopsis* and *N. benthamiana* (Roux et al. 2011). Furthermore, FLS2 is also able to interact with SERK1, SERK2, BAK1 and BKK1 in *N. benthamiana* and *Arabidopsis*, but clearly shows a preferential association with BAK1 (Roux et al. 2011), consistent with the response of *bak1-4* mutants to flg22 and elf18 (Chinchilla et al. 2007; Heese et al. 2007). Using a novel *bak1* allele (*bak1-5*) that does not lead to cell death when combined with *bkk1* (Schwessinger et al. 2011), phenotypic analysis of double mutants revealed that BAK1 and BKK1 co-operate during FLS2-, EFR- and PEPR1/2-dependent responses and to mediate disease resistance against pathogenic bacteria and oomycetes (Roux et al. 2011).

bak1-5 plants are compromised for BAK1-dependent innate immunity, while brassinosteroid signalling and cell death control remain intact (Schwessinger et al. 2011). Inhibition of EFR signalling activity in *bak1-5* is kinase-dependent, indicating that phosphorylation-dependent differential regulation is in place for diverse signalling pathways (Schwessinger et al. 2011). Further in-depth analysis of the phosphorylation events between FLS2/EFR and the regulatory RLKs BAK1/BKK1 will be an important step towards a deeper understanding of the molecular mechanisms underlying receptor regulation and specificity of downstream signalling.

Recently, the membrane-associated RLCKs Botrytis-induced kinase 1 (BIK1) and related PBS1-like (PBL) proteins were identified as positive regulators of flg22 and elf18 responses (Lu et al. 2010; Zhang et al. 2010a). BIK1 forms a constitutive complex with FLS2. Flg22 treatment leads to BIK1 and PBL1 phosphorylation within minutes and to the partial dissociation of the FLS2-BIK1 complex. Notably, BIK1 also form a complex with EFR (Zhang et al. 2010a) and elf18 treatment leads to BIK1 phosphorylation (Lu et al. 2010). The FLS2-BIK1 association does not require BAK1 (Lu et al. 2010; Zhang et al. 2010a), and conversely the FLS2-BAK1 association does not require BIK1 (Lu et al. 2010; Zhang et al. 2010a). However, kinase-active BAK1 and FLS2 are required for flg22-dependent BIK1 phosphorylation (Lu et al. 2010; Zhang et al. 2010a). Whether BAK1 can directly interact with and phosphorylate BIK1 is still controversial (Lu et al. 2010; Zhang et al. 2010a). Interestingly, BIK1 and its paralogs are targeted by the *Pseudomonas syringae* effector AvrPphB (Zhang et al. 2010a), demonstrating the importance of these proteins for PTI. Future work should reveal how the dynamics of the FLS2/EFR-BAK1/BKK1 complexes and the associated phosphorylation events regulate BIK1 and potentially other substrates to trigger downstream signalling.

In addition to its interaction with RLKs and RLCKs, several other FLS2-associated proteins have been identified (Fig. 1; Table 1). In an Y2H screen, KAPP was found to interact with the FLS2 kinase domain, and KAPP over-expression resulted in an *fls2*-like phenotype, with severely compromised flg22 binding and responsiveness (Gómez-Gómez et al. 2001). These data suggest that KAPP down-regulates flg22 signalling by potentially inhibiting ligand binding, as KAPP over-expressing plants displayed reduced binding of labeled flg22 (Gómez-Gómez et al. 2001). However, the amount of FLS2 protein in the KAPP over-expressing plants was never assessed and no further studies into KAPP-mediated FLS2 de-phosphorylation have been published.

Stomatal cytokinesis defective 1 (SCD1) was recently discovered as an in vivo interaction partner for FLS2 by mass spectrometry analysis (Korasick et al. 2010). SCD1 is the only *Arabidopsis* protein possessing a DENN (differentially expressed in normal and neoplastic cells) domain, a conserved tripartite motif found in several eukaryotic proteins that regulates Rab GTPase activity. SCD1 was previously shown to be involved in growth and development; the discovery of the *scd1-1* allele, a point mutation in the DENN domain, allowed the first demonstration of its role in innate immunity. *Scd1-1* mutants display SA-dependent enhanced resistance to infection by syringe-infiltrated *Pseudomonas syringae* pv. *tomato* DC3000 (*Pto*

DC3000), correlated with hyper-accumulation of hydrogen peroxide. However, the same mutants are less sensitive to PAMPs, as manifested by reduced seedling growth inhibition and ROS production in response to flg22 or elf18 (Korasick et al. 2010). These somehow contradictory phenotypes may be due to misregulated signalling and not truly reflect the biological function of SCD1. The authors did not investigate FLS2 endocytosis in this mutant, although *SCD1* is co-expressed with coatomers, dynamins and adaptins (ATTED-II, http://atted.jp) suggesting a potential role in FLS2 trafficking.

PAMP-induced signalling may be facilitated by close association of proteins involved in the cascade, for example by association within detergent-resistant membranes (DRMs). A recent study used quantitative proteomics to investigate flg22-induced changes in *Arabidopsis* DRM-associated proteins (Keinath et al. 2010). Sixty-four proteins were enriched in DRMs upon flg22 elicitation, including FLS2; though unexpectedly BAK1 was not found (Keinath et al. 2010). Flg22-induced DRM-associated proteins identified include the LRR-RLKs FERONIA (FER) and HERCULES (HERK1); remorins; H^+-ATPases (AHA1, AHA2, AHA3, AHA4); Ca^{2+}-ATPases (ACA8 and ACA10); and the vacuolar ATPase subunit C de-etiolated 3 (DET3) (Keinath et al. 2010). Furthermore, AHA1, AHA2 and FER were previously found to be phosphorylated in response to flg22 treatment (Benschop et al. 2007; Nühse et al. 2007). An independent study of DRMs to identify core sterol-associated proteins also identified FLS2 in DRMs, but not strictly sterol-associated, likely due to its role in dynamic signalling (Kierszniowska et al. 2009). Mutants of FER, DET3 and AHA1 were compromised for flg22-induced stomatal closure and concordantly had enhanced susceptibility to weakly virulent strains of *Pto* DC3000 (Keinath et al. 2010). However, FER is also involved in root hair development (Duan et al. 2010), as well as cell elongation, where it functions with the related RLKs HERK1 and THESEUS1 (THE1) (Guo et al. 2009). Furthermore, FER also plays a role in pollen tube reception (Escobar-Restrepo et al. 2007), and in fungal penetration (Kessler et al. 2010). Given the promiscuous nature of this protein in several signalling pathways, FER could be a general signalling adaptor, and a specific FLS2-FER interaction has yet to be proven.

4.2 The Xa21 Complex in Rice

Xa21, initially characterized as a resistance gene derived from the wild rice species *Oryza longistaminata*, confers resistance to the bacterium *Xanthomonas oryzae* pv. *oryzae* (*Xoo*) (Khush et al. 1990; Song et al. 1995). Xa21 is an LRR-RLK with 23 extracellular LRRs and has a similar receptor architecture to EFR. Recently, it was shown that Xa21 is in fact a PRR, as it recognizes and binds the conserved small sulfated type-I secreted peptide ax21 (Lee et al. 2009).

An Y2H screen, using a truncation of the Xa21 intracellular domain as bait, led to the identification of several Xa21-binding (XB) proteins (discussed below) (Fig. 1; Table 1). First, the E3 ligase XB3, via its ankyrin repeat domain, was

gene transcription. Neither of these hypotheses has yet been confirmed, and recent cell biology studies of GFP-tagged Xa21 transgenic lines indicate that the majority of Xa21 is localized at the PM and nuclear localization was not observed, even upon *Xoo* inoculation (Chen et al. 2010a), suggesting that the second hypothesis is unlikely. Nonetheless, rice lines over-expressing *XB10* displayed suppression of defense marker genes, as well as enhanced basal and Xa21-mediated resistance upon *Xoo* infection (Peng et al. 2008).

XB24 is a unique ATPase from a previously unclassified subclass (Chen et al. 2010b). In yeast, XB24-Xa21 interactions were dependent on Xa21 kinase activity, but not on the ATPase activity of XB24 (Chen et al. 2010b). Furthermore, XB24 enhances Xa21 auto-phosphorylation in vitro and in planta in an ATPase-dependent manner (Chen et al. 2010b). A strong in planta interaction between Xa21 and endogenous XB24 could also be detected in transgenic rice upon immunoprecipitation. Interestingly, the amount of XB24 interacting with Xa21 decreased in the presence of *Xoo* PXO99 but not the *Xoo* PXO99 *ΔraxST* strain impaired in ax21 sulphation, suggesting dissociation of the interaction upon binding of active ax21 (Chen et al. 2010b). Silencing of *XB24* did not alter basal resistance to *Xoo* in Kitaake rice; however, it led to enhanced Xa21-mediated resistance, suggesting a negative role of XB24 in Xa21-dependent resistance (Chen et al. 2010b). Consistently, over-expression of *XB24* compromised disease resistance of Xa21 plants to *Xoo* PXO99, and this was dependent on XB24 ATPase activity (Chen et al. 2010b). Interestingly, the same Xa21 lines over-expressing *XB24* also showed an ax21-dependent decrease in Xa21 protein levels upon inoculation with *Xoo* PXO99 (Chen et al. 2010b).

ER quality control pathways are required for correct protein folding and biogenesis. EFR receptor biogenesis is reliant on several chaperones, including stromal-derived factor 2 (SDF2), which works in a complex with the Hsp70 luminal binding protein (BiP) (Li et al. 2009; Nekrasov et al. 2009). In rice, OsBiP3 was identified as an interactor of Xa21, and OsBiP3 over-expression lead to reduced Xa21 accumulation (Park et al. 2010). Additionally, silencing of *OsSDF2* in rice decreased Xa21-mediated resistance to *Xoo* (Park et al. 2010).

5 General Conclusions

Signalling from receptor tyrosine kinases (RTKs) in mammalian systems follows a series of steps that are largely conserved, even between highly divergent receptors. First, ligand activates the receptor, causing receptor dimerization or oligomerization. Importantly, receptor oligomerization affords accuracy and flexibility to the signalling cascade controlled by the ligand-binding receptor (Heldin 1995). Several molecular RTK dimerization mechanisms have been described (Lemmon and Schlessinger 2010); however, the same molecular detail is currently not available for plant receptors. Activated receptors then adopt a conformation that is optimal for phosphotransfer that relieves their auto-inhibition (Nolen et al. 2004). This initiates auto-phosphorylation of the kinase domain, often the activation loop, and the sequential phosphorylation of other areas of the intracellular domain, accompanied by enhanced receptor activity. This type of mechanism has been extensively studied in BRI1, but a lot of work remains towards understanding the molecular mechanisms of trans-phosphorylation in plant systems. Next, the phosphorylation sites are targets for attachment of proteins that possess affinity for phosphorylated amino acids. Examples include proteins containing Src-homology 2 (SH2)-domains or phosphotyrosine-binding (PTB) domains (such as SH2-domain-containing transforming protein (SHC) and growth factor receptor bound 2/Grb2) (Seet et al. 2006; Deribe et al. 2010). Although tyrosine phosphorylation is emerging as an important modification for plant RLKs (Sugiyama et al. 2008; Oh et al. 2009, 2010), only a few SH2-domain-containing proteins have been identified in plants (Williams and Zvelebil 2004; de la Fuente van Bentem and Hirt 2009) and a different mechanism is likely for plant receptors. Downstream components may be recruited directly to the receptor or via attachment to phosphorylated docking proteins. Docking proteins are usually membrane-tethered proteins, which may be associated with several different receptors (like Grb2-associated binder, Gab1) (Sármay et al. 2006) or specific to a particular receptor (FGF receptor substrate 2, specific to fibroblast growth factor receptor FGFR) (Gotoh 2009). The docking proteins are phosphorylated by the receptor to create further sites for additional signalling components to join the complex. Importantly, some signalling molecules are directly in contact with the receptor or docking proteins, while others may be more distally associated. In several cases, this distinction has not been possible for plant RK interactors, where often the only evidence of interaction is based on co-immunoprecipitation. Downstream signalling is often achieved via cytoplasmic kinases that relay the signal intracellularly, ultimately to achieve changes in gene expression. In plants, several cytoplasmic kinases are RLK substrates (Table 1), such as BSKs and BIK1, though their targets are largely unknown. Ubiquitination, usually of an activated receptor, is a common mechanism for restricting the influence of the RTK on the downstream signalling by promoting RTK degradation (Kirkin and Dikic 2007). This form of negative feedback regulation is in place for EGFR, which is ubiqutinated by CBL (Casitas B-lineage lymphoma proto-oncogen), an E3 ligase with affinity for diverse RTKs via its tyrosine kinase-binding domain (Swaminathan and Tsygankov 2006). Similarly in plants, the E3 ligases XB3 and ARC1 regulate Xa21 and SRK, respectively (Table 1). A direct mechanism of negative regulation is de-phosphorylation of the RTKs by receptor tyrosine phosphatases, such as Src-homology phosphatases (SHP1/2), that are recruited to EGFR (Lemmon and Schlessinger 2010). In rice, the PP2C XB15 has been shown to play this role in the negative regulation of Xa21. Importantly, despite the scarcity of SH2-domain-containing proteins, plants do possess proteins containing forkhead-associated (FHA) domains that bind to phosphothreonine residues. KAPP, harbouring such an FHA domain, appears to be a general negative regulator of diverse RLKs (Table 1) and has been linked with the regulation of CLV1, SRK, FLS2 and BRI1, though molecular details for this regulation are still limited.

In conclusion, despite all the knowledge gained thus far about plant receptor kinase complexes, most questions remain. Technological advancements, such as improved sensitivity and dynamic range in mass spectrometry, will surely facilitate new discoveries in years to come.

References

Albrecht C, Russinova E, Kemmerling B, Kwaaitaal M, de Vries SC (2008) Arabidopsis somatic embryogenesis receptor kinase proteins serve brassinosteroid-dependent and -independent signaling pathways. Plant Physiol 148:611–619

Asano K, Clayton J, Shalev A, Hinnebusch AG (2000) A multifactor complex of eukaryotic initiation factors, eIF1, eIF2, eIF3, eIF5, and initiator tRNAMet is an important translation initiation intermediate *in vivo*. Genes Dev 14:2534–2546

Benschop JJ, Mohammed S, O'Flaherty M, Heck AJR, Slijper M, Menke FLH (2007) Quantitative phosphoproteomics of early elicitor signaling in Arabidopsis. Mol Cell Proteomics 6:1198–1214

Betsuyaku S, Takahashi F, Kinoshita A, Miwa H, Shinozaki K, Fukuda H, Sawa S (2010) Mitogen-activated protein kinase regulated by the CLAVATA receptors contributes to the shoot apical meristem homeostasis. Plant Cell Physiol 52:14–29

Bleckmann A, Weidtkamp-Peters S, Seidel CAM, Simon R (2010) Stem cell signaling in Arabidopsis requires CRN to localize CLV2 to the plasma membrane. Plant Physiol 152:166–176

Bower MS, Matias DD, Fernandes-Carvalho E, Mazzurco M, Gu T, Rothstein SJ, Goring DR (1996) Two members of the thioredoxin-h family interact with the kinase domain of a Brassica S locus receptor kinase. Plant Cell 8:1641–1650

Bowman JL, Eshed Y (2000) Formation and maintenance of the shoot apical meristem. Trends Plant Sci 5:110–115

Brand U, Fletcher J, Hobe M, Meyerowitz E, Simon R (2000) Dependence of stem cell fate in *Arabidopsis* on a feedback loop regulated by CLV3 activity. Science 289:617–619

Cabrillac D, Cock JM, Dumas C, Gaude T (2001) The S-locus receptor kinase is inhibited by thioredoxins and activated by pollen coat proteins. Nature 410:220–223

Carles CC, Fletcher JC (2003) Shoot apical meristem maintenance: the art of a dynamic balance. Trends Plant Sci 8:394–401

Chen R-H, Miettinen PJ, Maruoka EM, Choy L, Derynck R (1995) A WD-domain protein that is associated with and phosphorylated by the type II TGF-beta receptor. Nature 377:548–552

Chen F, Gao M-J, Miao Y-S, Yuan Y-X, Wang M-Y, Li Q, Mao B-Z, Jiang L-W, He Z-H (2010a) Plasma membrane localization and potential endocytosis of constitutively expressed Xa21 proteins in transgenic rice. Mol Plant 3:917–926

Chen X, Chern M, Canlas PE, Ruan D, Jiang C, Ronald PC (2010b) An ATPase promotes autophosphorylation of the pattern recognition receptor XA21 and inhibits XA21-mediated immunity. Proc Natl Acad Sci USA 107:8029–8034

Chinchilla D, Bauer Z, Regenass M, Boller T, Felix G (2006) The Arabidopsis receptor kinase FLS2 binds flg22 and determines the specificity of flagellin perception. Plant Cell 18:465–476

Chinchilla D, Zipfel C, Robatzek S, Kemmerling B, Nürnberger T, Jones JD, Felix G, Boller T (2007) A flagellin-induced complex of the receptor FLS2 and BAK1 initiates plant defence. Nature 448:497–500

Choy L, Derynck R (1998) The type II Transforming growth factor (TGF)-beta receptor-interacting protein TRIP-1 acts as a modulator of the TGF-beta response. J Biol Chem 273:31455–31462

Clark SE, Running MP, Meyerowitz EM (1993) CLAVATA1, a regulator of meristem and flower development in Arabidopsis. Development 119:397–418

Clark AE, Running MP, Meyerowitz EM (1995) CLAVATA3 is a specific regulator of shoot and floral meristem development affecting the same processes as CLAVATA1. Development 121:2057–2067

Clouse SD, Langford M, McMorris TC (1996) A brassinosteroid-insensitive mutant in *Arabidopsis thaliana* exhibits multiple defects in growth and development. Plant Physiol 111:671–678

Cobb MH, Sang BC, Gonzalez R, Goldsmith E, Ellis L (1989) Autophosphorylation activates the soluble cytoplasmic domain of the insulin receptor in an intermolecular reaction. J Biol Chem 264:18701–18706

de la Fuente van Bentem S, Hirt H (2009) Protein tyrosine phosphorylation in plants: more abundant than expected? Trends Plant Sci 14:71–76

Deribe YL, Pawson T, Dikic I (2010) Post-translational modifications in signal integration. Nat Struct Mol Biol 17:666–672

DeYoung BJ, Clark SE (2008) BAM receptors regulate stem cell specification and organ development through complex interactions with CLAVATA signaling. Genetics 180:895–904

DeYoung BJ, Bickle KL, Schrage KJ, Muskett P, Patel K, Clark SE (2006) The CLAVATA1-related BAM1, BAM2 and BAM3 receptor kinase-like proteins are required for meristem function in Arabidopsis. Plant J 45:1–16

Ding Z, Wang H, Liang X, Morris ER, Gallazzi F, Pandit S, Skolnick J, Walker JC, Van Doren SR (2007) Phosphoprotein and phosphopeptide interactions with the FHA domain from Arabidopsis kinase-associated protein phosphatase. Biochemistry 46:2684–2696

Dodsworth S (2009) A diverse and intricate signalling network regulates stem cell fate in the shoot apical meristem. Dev Biol 336:1–9

Duan Q, Kita D, Li C, Cheung AY, Wu H-M (2010) FERONIA receptor-like kinase regulates RHO GTPase signaling of root hair development. Proc Natl Acad Sci USA 107:17821–17826

Ehsan H, Ray WK, Phinney B, Wang X, Huber SC, Clouse SD (2005) Interaction of Arabidopsis brassinosteroid-insensitive 1 receptor kinase with a homolog of mammalian TGF-beta receptor interacting protein. Plant J 43:251–261

Escobar-Restrepo JM, Huck N, Kessler S, Gagliardini V, Gheyselinck J, Yang WC, Grossniklaus U (2007) The FERONIA receptor-like kinase mediates male-female interactions during pollen tube reception. Science 317:656–660

Eulgem T, Somssich IE (2007) Networks of WRKY transcription factors in defense signaling. Curr Opin Plant Biol 10:366–371

Fendrych M, Synek L, Pecenkova T, Toupalova H, Cole R, Drdova E, Nebesarova J, Sedinova M, Hala M, Fowler JE, Zarsky V (2010) The Arabidopsis exocyst complex is involved in cytokinesis and cell plate maturation. Plant Cell 22:3053–3065

Fiers M, Golemiec E, Xu J, van der Geest L, Heidstra R, Stiekema W, Liu C-M (2005) The 14-amino acid CLV3, CLE19, and CLE40 peptides trigger consumption of the root meristem in Arabidopsis through a CLAVATA2-dependent pathway. Plant Cell 17:2542–2553

Gagne JM, Clark SE (2010) The Arabidopsis stem cell factor POLTERGEIST is membrane localized and phospholipid stimulated. Plant Cell 22:729–743

Galigniana MD, Echeverría PC, Erlejman AG, Piwien-Pilipuk G (2010) Role of molecular chaperones and TPR-domain proteins in the cytoplasmic transport of steroid receptors and their passage through the nuclear pore. Nucleus 1:290–308

Giranton J-LØ, Dumas C, Cock JM, Gaude T (2000) The integral membrane S-locus receptor kinase of Brassica has serine/threonine kinase activity in a membranous environment and spontaneously forms oligomers in planta. Proc Natl Acad Sci USA 97:3759–3764

Gómez-Gómez L, Boller T (2000) FLS2: an LRR receptor-like kinase involved in the perception of the bacterial elicitor flagellin in Arabidopsis. Mol Cell 5:1003–1011

Gómez-Gómez L, Bauer Z, Boller T (2001) Both the extracellular leucine-rich repeat domain and the kinase activity of FLS2 are required for flagellin binding and signaling in Arabidopsis. Plant Cell 13:1155–1163

Gotoh N (2009) Feedback inhibitors of the epidermal growth factor receptor signaling pathways. Int J Biochem Cell Biol 41:511–515

Gu T, Mazzurco M, Sulaman W, Matias DD, Goring DR (1998) Binding of an arm repeat protein to the kinase domain of the S-locus receptor kinase. Proc Natl Acad Sci USA 95:382–387

Guo H, Li L, Ye H, Yu X, Algreen A, Yin Y (2009) Three related receptor-like kinases are required for optimal cell elongation in *Arabidopsis thaliana*. Proc Natl Acad Sci USA 106:7648–7653

Guo Y, Han L, Hymes M, Denver R, Clark SE (2010) CLAVATA2 forms a distinct CLE-binding receptor complex regulating Arabidopsis stem cell specification. Plant J 63:889–900

He K, Gou X, Yuan T, Lin H, Asami T, Yoshida S, Russell SD, Li J (2007) BAK1 and BKK1 regulate brassinosteroid-dependent growth and brassinosteroid-independent cell-death pathways. Curr Biol 17:1109–1115

Hecht V, Vielle-Calzada J-P, Hartog MV, Schmidt EDL, Boutilier K, Grossniklaus U, de Vries SC (2001) The Arabidopsis somatic embryogenesis receptor kinase 1 gene is expressed in developing ovules and embryos and enhances embryogenic competence in culture. Plant Physiol 127:803–816

Heese A, Hann DR, Gimenez-Ibanez S, Jones AM, He K, Li J, Schroeder JI, Peck SC, Rathjen JP (2007) The receptor-like kinase SERK3/BAK1 is a central regulator of innate immunity in plants. Proc Natl Acad Sci USA 104:12217–12222

Heldin C-H (1995) Dimerization of cell surface receptors in signal transduction. Cell 80:213–223

Hiscock SJ, McInnis SM (2003) Pollen recognition and rejection during the sporophytic self-incompatibility response: Brassica and beyond. Trends Plant Sci 8:606–613

Hord CLH, Chen C, DeYoung BJ, Clark SE, Ma H (2006) The BAM1/BAM2 receptor-like kinases are important regulators of *Arabidopsis* early anther development. Plant Cell 18:1667–1680

Huffaker A, Ryan CA (2007) Endogenous peptide defense signals in Arabidopsis differentially amplify signaling for the innate immune response. Proc Natl Acad Sci USA 104:10732–10736

Ivanov B, Gaude T (2009a) Brassica self-incompatibility: a glimpse below the surface. Plant Signal Behav 4:996–998

Ivanov R, Gaude T (2009b) Endocytosis and endosomal regulation of the S-receptor kinase during the self-incompatibility response in *Brassica oleracea*. Plant Cell 21:2107–2117

Jeong S, Trotochaud AE, Clark SE (1999) The *Arabidopsis CLAVATA2* gene encodes a receptor-like protein required for the stability of the CLAVATA1 receptor-like kinase. Plant Cell 11:1925–1934

Jeong Y, Shang Y, Kim BH, Kim SY, Song JH, Lee JS, Lee MM, Li J, Nam KH (2010) BAK7 displays unequal genetic redundancy with BAK1 in brassinosteroid signaling and early senescence in Arabidopsis. Mol Cells 29:259–266

Jiang J, Clouse SD (2001) Expression of a plant gene with sequence similarity to animal TGF-beta receptor interacting protein is regulated by brassinosteroids and required for normal plant development. Plant J 26:35–45

Kakita M, Murase K, Iwano M, Matsumoto T, Watanabe M, Shiba H, Isogai A, Takayama S (2007a) Two distinct forms of M-locus protein kinase localize to the plasma membrane and interact directly with S-locus receptor kinase to transduce self-incompatibility signaling in *Brassica rapa*. Plant Cell 19:3961–3973

Kakita M, Shimosato H, Murase K, Isogai A, Takayama S (2007b) Direct interaction between S-locus receptor kinase and M-locus protein kinase involved in Brassica self-incompatibility signaling. Plant Biotechnol 24:185–190

Karlova R, Boeren S, Russinova E, Aker J, Vervoort J, de Vries S (2006) The *Arabidopsis* somatic embryogenesis receptor-like kinase 1 protein complex includes brassinosteroid-insensitive1. Plant Cell 18:626–638

Keinath NF, Kierszniowska S, Lorek J, Bourdais G, Kessler SA, Asano H, Grossniklaus U, Schulze W, Robatzek S, Panstruga R (2010) PAMP-induced changes in plasma membrane compartmentalization reveal novel components of plant immunity. J Biol Chem 285:39140–39149

Kemmerling B, Schwedt A, Rodriguez P, Mazzotta S, Frank M, Qamar SA, Mengiste T, Betsuyaku S, Parker JE, Müssig C, Thomma BP, Albrecht C, de Vries SC, Hirt H, Nürnberger T (2007) The BRI1-associated kinase 1, BAK1, has a brassinolide-independent role in plant cell-death control. Curr Biol 17:1116–1122

Kessler SA, Shimosato-Asano H, Keinath NF, Wuest SE, Ingram G, Panstruga R, Grossniklaus U (2010) Conserved molecular components for pollen tube reception and fungal invasion. Science 330:968–971

Khush GS, Bacalangco E, Ogawa T (1990) A new gene for resistance to bacterial blight from O. longistaminata. Rice Genet Newsl 7:121–122

Kierszniowska S, Seiwert B, Schulze WX (2009) Definition of *Arabidopsis* sterol-rich membrane microdomains by differential treatment with methyl-beta-cyclodextrin and quantitative proteomics. Mol Cell Proteomics 8:612–623

Kim T-W, Wang Z-Y (2010) Brassinosteroid Signal Transduction from Receptor Kinases to Transcription Factors. Annu Rev Plant Biol 61:681–704

Kinoshita T, Cano-Delgado A, Seto H, Hiranuma S, Fujioka S, Yoshida S, Chory J (2005) Binding of brassinosteroids to the extracellular domain of plant receptor kinase BRI1. Nature 433:167–171

Kinoshita A, Nakamura Y, Sasaki E, Kyozuka J, Fukuda H, Sawa S (2007) Gain-of-function phenotypes of chemically synthetic CLAVATA3/ESR-related (CLE) peptides in *Arabidopsis thaliana* and *Oryza sativa*. Plant Cell Physiol 48:1821–1825

Kinoshita A, Betsuyaku S, Osakabe Y, Mizuno S, Nagawa S, Stahl Y, Simon R, Yamaguchi-Shinozaki K, Fukuda H, Sawa S (2010) RPK2 is an essential receptor-like kinase that transmits the CLV3 signal in *Arabidopsis*. Development 137:3911–3920

Kirkin V, Dikic I (2007) Role of ubiquitin- and Ubl-binding proteins in cell signaling. Curr Opin Cell Biol 19:199–205

Kondo T, Sawa S, Kinoshita A, Mizuno S, Kakimoto T, Fukuda H, Sakagami Y (2006) A plant peptide encoded by CLV3 identified by in situ MALDI-TOF MS analysis. Science 313:845–848

Korasick DA, McMichael C, Walker KA, Anderson JC, Bednarek SY, Heese A (2010) Novel functions of stomatal cytokinesis-defective 1 (SCD1) in innate immune responses against bacteria. J Biol Chem 285:23342–23350

Krol E, Mentzel T, Chinchilla D, Boller T, Felix G, Kemmerling B, Postel S, Arents M, Jeworutzki E, Al-Rasheid KAS, Becker D, Hedrich R (2010) Perception of the *Arabidopsis* danger signal peptide 1 involves the pattern recognition receptor AtPEPR1 and its close homologue AtPEPR2. J Biol Chem 285:13471–13479

Kunze G, Zipfel C, Robatzek S, Niehaus K, Boller T, Felix G (2004) The N terminus of bacterial elongation factor Tu elicits innate immunity in Arabidopsis plants. Plant Cell 16:3496–3507

Laux T, Mayer K, Berger J, Jurgens G (1996) The WUSCHEL gene is required for shoot and floral meristem integrity in *Arabidopsis*. Development 122:87–96

Lee SW, Han SW, Sririyanum M, Park CJ, Seo YS, Ronald PC (2009) A type I-secreted, sulfated peptide triggers XA21-mediated innate immunity. Science 326:850–853

Lemmon MA, Schlessinger J (2010) Cell signaling by receptor tyrosine kinases. Cell 141:1117–1134

Li J, Chory J (1997) A putative leucine-rich repeat receptor kinase involved in brassinosteroid signal transduction. Cell 90:929–938

Li J, Wen J, Lease KA, Doke JT, Tax FE, Walker JC (2002) BAK1, an Arabidopsis LRR receptor-like protein kinase, interacts with BRI1 and modulates brassinosteroid signaling. Cell 110:213–222

Li J, Zhao-Hui C, Batoux M, Nekrasov V, Roux M, Chinchilla D, Zipfel C, Jones JDG (2009) Specific ER quality control components required for biogenesis of the plant innate immune receptor EFR. Proc Natl Acad Sci USA 106:15973–15978

Lu D, Wu S, Gao X, Zhang Y, Shan L, He P (2010) A receptor-like cytoplasmic kinase, BIK1, associates with a flagellin receptor complex to initiate plant innate immunity. Proc Natl Acad Sci USA 107:496–501

Lutz M, Knaus P (2002) Integration of the TGF-beta pathway into the cellular signalling network. Cell Signal 14:977–988

Massagué J (1998) TGF-beta signal transduction. Annu Rev Biochem 67:753–791

Medzhitov R, Janeway CA Jr (1997) Innate immunity: the virtues of a nonclonal system of recognition. Cell 91:295–298

Meng L, Feldman L (2010) CLE14/CLE20 peptides may interact with CLAVATA2/CORYNE receptor-like kinases to irreversibly inhibit cell division in the root meristem of Arabidopsis. Planta 232:1061–1074

Miwa H, Betsuyaku S, Iwamoto K, Kinoshita A, Fukuda H, Sawa S (2008) The receptor-like kinase SOL2 mediates CLE signaling in Arabidopsis. Plant Cell Physiol 49:1752–1757

Miwa H, Kinoshita A, Fukuda H, Sawa S (2009) Plant meristems: CLAVATA3/ESR -related signaling in the shoot apical meristem and the root apical meristem. J Plant Res 122:31–39

Müller R, Bleckmann A, Simon R (2008) The receptor kinase CORYNE of Arabidopsis transmits the stem cell-limiting signal CLAVATA3 independently of CLAVATA1. Plant Cell 20:934–946

Murase K, Shiba H, Iwano M, Che F-S, Watanabe M, Isogai A, Takayama S (2004) A membrane-anchored protein kinase involved in Brassica self-incompatibility signaling. Science 303:1516–1519

Nam KH, Li J (2002) BRI1/BAK1, a receptor kinase pair mediating brassinosteroid signaling. Cell 110:203–212

Nam KH, Li J (2004) The Arabidopsis transthyretin-like protein is a potential substrate of brassinosteroid-insensitive 1. Plant Cell 16:2406–2417

Nekrasov V, Li J, Batoux M, Roux M, Chu Z-H, Lacombe S, Rougon A, Bittel P, Kiss-Papp M, Chinchilla D, van Esse HP, Jorda L, Schwessinger B, Nicaise V, Thomma BPHJ, Molina A, Jones JDG, Zipfel C (2009) Control of the pattern-recognition receptor EFR by an ER protein complex in plant immunity. EMBO J 28:3428–3438

Nimchuk ZL, Tarr PT, Ohno C, Qu X, Meyerowitz EM (2011) Plant stem cell signaling involves ligand-dependent trafficking of the CLAVATA1 receptor kinase. Curr Biol 21:345–352

Nolen B, Taylor S, Ghosh G (2004) Regulation of protein kinases: controlling activity through activation segment conformation. Mol Cell 15:661–675

Nühse TS, Bottrill AR, Jones AM, Peck SC (2007) Quantitative phosphoproteomic analysis of plasma membrane proteins reveals regulatory mechanisms of plant innate immune responses. Plant J 51:931–940

Ogawa M, Shinohara H, Sakagami Y, Matsubayashi Y (2008) *Arabidopsis* CLV3 peptide directly binds CLV1 ectodomain. Science 319:294

Oh M-H, Wang X, Kota U, Goshe MB, Clouse SD, Huber SC (2009) Tyrosine phosphorylation of the BRI1 receptor kinase emerges as a component of brassinosteroid signaling in *Arabidopsis*. Proc Natl Acad Sci USA 106:658–663

Oh M-H, Wang X, Wu X, Zhao Y, Clouse SD, Huber SC (2010) Autophosphorylation of Tyr-610 in the receptor kinase BAK1 plays a role in brassinosteroid signaling and basal defense gene expression. Proc Natl Acad Sci USA 107:17827–17832

Ohyama K, Shinohara H, Ogawa-Ohnishi M, Matsubayashi Y (2009) A glycopeptide regulating stem cell fate in *Arabidopsis thaliana*. Nat Chem Biol 5(8):578–580. doi:10.1038/nchembio.182 DOI:dx.doi.org

Pandey SP, Somssich IE (2009) The role of WRKY transcription factors in plant immunity. Plant Physiol 150:1648–1655

Park C-J, Peng Y, Chen X, Dardick C, Ruan D, Bart R, Canlas PE, Ronald PC (2008) Rice XB15, a protein phosphatase 2 C, negatively regulates cell death and Xa21-mediated innate immunity. PLoS Biol 6:e231

Park C-J, Bart R, Chern M, Canlas PE, Bai W, Ronald PC (2010) Overexpression of the endoplasmic reticulum chaperone BiP3 regulates XA21-mediated innate immunity in rice. PLoS One 5:e9262

Peng Y, Bartley LE, Chen X, Dardick C, Chern M, Ruan R, Canlas PE, Ronald PC (2008) OsWRKY62 is a negative regulator of basal and Xa21-mediated defense against *Xanthomonas oryzae* pv. *oryzae* in rice. Mol Plant 1:446–458

Postel S, Küfner I, Beuter C, Mazzotta S, Schwedt A, Borlotti A, Halter T, Kemmerling B, Nürnberger T (2010) The multifunctional leucine-rich repeat receptor kinase BAK1 is implicated in *Arabidopsis* development and immunity. Eur J Cell Biol 89:169–174

Roux M, Schwessinger B, Albrecht C, Chinchilla D, Jones A, Holton N, Malinovsky FG, Tör M, de Vries S, Zipfel C (2011) The Arabidopsis leucine-rich repeat receptor-like kinases BAK1/SERK3 and BKK1/SERK4 are required for innate immunity to hemibiotrophic and biotrophic pathogens. Plant Cell 23:2440–2455

Samuel MA, Mudgil Y, Salt JN, Delmas F, Ramachandran S, Chilelli A, Goring DR (2008) Interactions between the S-domain receptor kinases and AtPUB-ARM E3 ubiquitin ligases suggest a conserved signaling pathway in Arabidopsis. Plant Physiol 147:2084–2095

Samuel MA, Chong YT, Haasen KE, Aldea-Brydges MG, Stone SL, Goring DR (2009) Cellular pathways regulating responses to compatible and self-incompatible pollen in brassica and Arabidopsis stigmas intersect at Exo70A1, a putative component of the exocyst complex. Plant Cell 21:2655–2671

Sármay G, Angyal A, KertÈsz K, Maus Mt, Medgyesi Dv (2006) The multiple function of Grb2 associated binder (Gab) adaptor/scaffolding protein in immune cell signaling. Immunol Lett 104:76–82

Sawa S, Kinoshita A, Nakanomyo I, Fukuda H (2006) CLV3/ESR-related (CLE) peptides as intercellular signaling molecules in plants. Chem Rec 6:303–310

Schlessinger J (1988) Signal transduction by allosteric receptor oligomerization. Trends Biochem Sci 13:443–447

Schlessinger J (2002) Ligand-induced, receptor-mediated dimerization and activation of EGF receptor. Cell 110:669–672

Schoof H, Lenhard M, Haecker A, Mayer K, Jurgens G, Laux T (2000) The stem cell population of *Arabidopsis* shoot meristems in maintained by a regulatory loop between the *CLAVATA* and *WUSCHEL* genes. Cell 100:635–644

Schopfer CR, Nasrallah ME, Nasrallah JB (1999) The male determinant of self-incompatibility in Brassica. Science 286:1697–1700

Schulze B, Mentzel T, Jehle AK, Mueller K, Beeler S, Boller T, Felix G, Chinchilla D (2010) Rapid heteromerization and phosphorylation of ligand-activated plant transmembrane receptors and their associated kinase BAK1. J Biol Chem 285:9444–9451

Schwessinger B, Roux M, Sklenar J, Kadota Y, Jones AM and Zipfel C (2011) Phosphorylation-dependent differential regulation of plant growth, cell death and innate immunity by the regulatory receptor-like kinase BAK1. PLoS Genet 7:e1002046

Seet BT, Dikic I, Zhou M-M, Pawson T (2006) Reading protein modifications with interaction domains. Nat Rev Mol Cell Biol 7:473–483

Shah K, Russinova E, Gadella TWJ, Willemse J, de Vries SC (2002) The Arabidopsis kinase-associated protein phosphatase controls internalization of the somatic embryogenesis receptor kinase 1. Genes Dev 16:1707–1720

Shan L, He P, Li J, Heese A, Peck SC, Nürnberger T, Martin GB, Sheen J (2008) Bacterial effectors target the common signaling partner BAK1 to disrupt multiple MAMP receptor-signaling complexes and impede plant immunity. Cell Host Microbe 4:17–27

Shi Y, Massagué J (2003) Mechanisms of TGF-beta signaling from cell membrane to the nucleus. Cell 113:685–700

Shiu SH, Bleecker AB (2001) Receptor-like kinases from Arabidopsis form a monophyletic gene family related to animal receptor kinases. Proc Natl Acad Sci USA 98:10763–10768

Shiu SH, Karlowski WM, Pan R, Tzeng YH, Mayer KF, Li WH (2004) Comparative analysis of the receptor-like kinase family in *Arabidopsis* and rice. Plant Cell 16:1220–1234

Silva N, Stone S, Christie L, Sulaman W, Nazarian K, Burnett L, Arnoldo M, Rothstein S, Goring D (2001) Expression of the S receptor kinase in self-compatible *Brassica napus* cv. Westar leads to the allele-specific rejection of self-incompatible *Brassica napus* pollen. Mol Genet Genomics 265:552–559

Song WY, Wang GL, Chen LL, Kim HS, Pi LY, Holsten T, Gardner J, Wang B, Zhai WX, Zhu LH, Fauquet C, Ronald P (1995) A receptor kinase-like protein encoded by the rice disease resistance gene, *Xa21*. Science 270:1804–1806

Song S-K, Lee MM, Clark SE (2006) POL and PLL1 phosphatases are CLAVATA1 signaling intermediates required for *Arabidopsis* shoot and floral stem cells. Development 133:4691–4698

Song S-K, Hofhuis H, Lee MM, Clark SE (2008) Key divisions in the early *Arabidopsis* embryo require POL and PLL1 phosphatases to establish the root stem cell organizer and vascular axis. Dev Cell 15:98–109

Stone SL, Arnoldo M, Goring DR (1999) A breakdown of Brassica self-incompatibility in ARC1 antisense transgenic plants. Science 286:1729–1731

Stone SL, Anderson EM, Mullen RT, Goring DR (2003) ARC1 is an E3 ubiquitin ligase and promotes the ubiquitination of proteins during the rejection of self-incompatible Brassica pollen. Plant Cell 15:885–898

Sugiyama N, Nakagami H, Mochida K, Daudi A, Tomita M, Shirasu K, Ishihama Y (2008) Large-scale phosphorylation mapping reveals the extent of tyrosine phosphorylation in Arabidopsis. Mol Syst Biol 4:193

Swaminathan G, Tsygankov AY (2006) The Cbl family proteins: ring leaders in regulation of cell signaling. J Cell Physiol 209:21–43

Takasaki T, Hatakeyama K, Suzuki G, Watanabe M, Isogai A, Hinata K (2000) The S receptor kinase determines self-incompatibility in Brassica stigma. Nature 403:913–916

Takayama S, Isogai A (2005) Self-incompatibility in plants. Annu Rev Plant Biol 56:467–489

Takayama S, Shimosato H, Shiba H, Funato M, Che F-S, Watanabe M, Iwano M, Isogai A (2001) Direct ligand-receptor complex interaction controls Brassica self-incompatibility. Nature 413:534–538

Tang W, Kim T-W, Oses-Prieto JA, Sun Y, Deng Z, Zhu S, Wang R, Burlingame AL, Wang Z-Y (2008) BSKs mediate signal transduction from the receptor kinase BRI1 in Arabidopsis. Science 321(5888):557–560

Trotochaud AE, Hao T, Wu G, Yang Z, Clark SE (1999) The CLAVATA1 receptor-like kinase requires CLAVATA3 for its assembly into a signaling complex that includes KAPP and a Rho-related protein. Plant Cell 11(3):393–406

Ullrich A, Schlessinger J (1990) Signal transduction by receptors with tyrosine kinase activity. Cell 61(2):203–212

Vanoosthuyse V, Tichtinsky G, Dumas C, Gaude T, Cock JM (2003) Interaction of calmodulin, a sorting nexin and kinase-associated protein phosphatase with the *Brassica oleracea* S locus receptor kinase. Plant Physiol 133:919–929

Wang X, Chory J (2006) Brassinosteroids regulate dissociation of BKI1, a negative regulator of BRI1 signaling, from the plasma membrane. Science 313:1118–1122

Wang X, Liu Y (2007) Regulation of innate immune response by MAP kinase phosphatase-1. Cell Signal 19:1372–1382

Wang Z-Y, Seto H, Fujioka S, Yoshida S, Chory J (2001) BRI1 is a critical component of a plasma-membrane receptor for plant steroids. Nature 410:380–383

Wang X, Li X, Meisenhelder J, Hunter T, Yoshida S, Asami T, Chory J (2005) Autoregulation and homodimerization are involved in the activation of the plant steroid receptor BRI1. Dev Cell 8:855–865

Ligands of RLKs and RLPs Involved in Defense and Symbiosis

Katharina Mueller and Georg Felix

Abstract Higher plants can detect a variety of molecular patterns that are indicative for the presence of mutualistic and antagonistic symbionts. These molecular signals comprise a variety of chemical structures from simple sterols and peptides to complex carbohydrates and modified proteins. Only for few of these signals the corresponding plant receptors have been identified so far. These known receptors, collectively termed pattern recognition receptors (PRRs), belong to the family of receptor-like kinases (RLKs) or to the related family of receptor-like proteins (RLPs).

1 Introduction

Higher plants interact with a variety of different types of mutualistic and antagonistic symbionts including species of bacterial and fungal microorganisms, insects and other animals, as well as other plants. Recognition of these organisms is a prerequisite for proper adapting of host plant metabolism and development, either to a state of defense or to a state of mutualistic exchange. The molecular signals that serve as triggers for these divergent programs in the host plants are the focus of this chapter. So far, the corresponding plant receptors have been identified only for few of these signals. These receptors are members of the RLK family or the family of receptor-like proteins (RLPs), membrane bound proteins with extracellular receptor domains like RLKs but lacking cytoplasmic kinase domains.

Here, we discuss some of the known ligand/receptor pairs involved in biotic interactions as examples of many more such pairs still to be identified. For some of the identified signals, notably the ones from mutualistic symbionts, genetic evidence implicates recognition via specific RLKs but conclusive demonstration of

K. Mueller • G. Felix (✉)
ZMBP Pflanzenbiochemie, University of Tübingen, Tübingen, Germany
e-mail: georg.felix@zmbp.uni-tuebingen.de

F. Tax and B. Kemmerling (eds.), *Receptor-like Kinases in Plants*,
Signaling and Communication in Plants 13,
DOI 10.1007/978-3-642-23044-8_10,

direct interaction between these receptor candidates and their signals is still missing. Numerous additional signals of microbial origin have been identified by their specific biological activity on plant cells. While their receptors still have to be identified there is good reason to assume that these receptors will eventually be found among the many orphan members of the RLK and RLP families.

1.1 Molecular Patterns that Act as Danger Signals

Plants have evolved ways to detect many microbial pathogens including bacteria, fungi and oomycetes and mount efficient defense responses. Molecular patterns associated with the presence of such pathogens serve as chemical signals for the detection systems of host plants. Adopting the nomenclature used in innate immunology of animals, microbial-derived molecules that act as elicitors of defense mechanism are referred to as pathogen-associated molecular patterns (PAMPs) or better microbe-associated molecular patterns (MAMPs) (Boller and Felix 2009; Medzhitov and Janeway 1997; Ausubel 2005). By analogy, endogenous elicitors that are associated with wounding and injury can be termed DAMPs (damage-associated molecular patterns) (Lotze et al. 2007). Receptors perceiving MAMPs and DAMPs, can collectively be termed pattern recognition receptors (PRRs). As detailed below, plant cells respond with an essentially identical set of responses when treated with various MAMPs and DAMPs. This indicates that plants interpret these various inputs as a common signal of "alert" and "danger" (Boller and Felix 2009). Indeed, one might extend this concept to the detection of bigger herbivores like insects. On one hand, insects feeding on plants cause mechanical wounding that can lead to the release of DAMPs. On the other hand, plants can also sense the presence of insects via molecular patterns characteristically occurring in the oral secretions of these herbivores (Wu and Baldwin 2010).

1.2 Functional Definition of MAMPs and DAMPs by Their Biological Activity

MAMPs/DAMPs can be defined functionally by the characteristic defense-related responses they trigger in plant cells. Regardless of whether these signals originate from bacteria, fungi, oomycetes or injured host tissue, they all seem to trigger a common, stereotypical set of initial responses (Boller and Felix 2009). These early responses include increased ion fluxes across the plasma membrane, activation of MAPK-cascades, production of reactive oxygen species (ROS), enhanced biosynthesis of ethylene, deposition of callose and altered expression of many genes. Some of these rapid responses can easily be monitored and are thus well suited as convenient bioassays to study MAMP/DAMP perception in plants (Fig. 1).

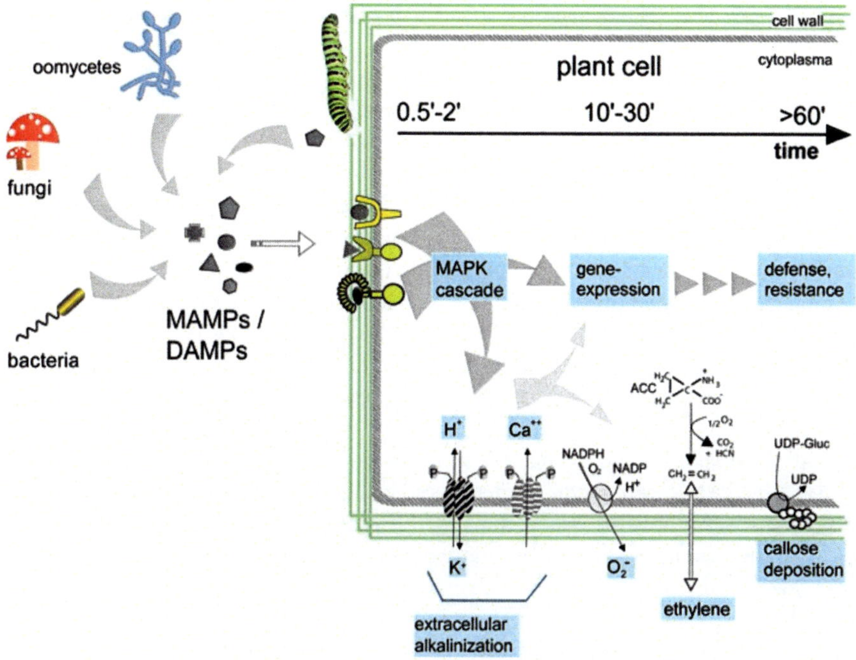

Fig. 1 Schematic view of MAMP/DAMP perception by plant cells

Under normal physiological conditions, cells polarize their plasma membrane and adjust their extracellular pH to slightly acidic conditions (~pH 5). Perception of MAMPs/DAMPs triggers opening of ion channels that allow influx of H^+ and Ca^{2+} to the cytoplasm, concomitant efflux of K^+, Cl^-, and NO_3^- and leads to membrane depolarization (Jeworutzki et al. 2010). As a consequence of these ion fluxes the extracellular pH starts to increase, a symptom that can be monitored easily, in particular with cells cultured in suspension. Extracellular alkalinization has been widely used as a sensitive assay for qualitative and quantitative characterization of MAMPs/DAMPs (reviewed in Boller 1995; Boller and Felix 2009).

MAMPs/DAMPs stimulate rapid influx of Ca^{2+} from the apoplast to the cytoplasm where Ca^{2+} serves as second messenger and activates calcium-dependent proteins such as Ca^{2+}-dependent protein kinases (CDPKs) (Boudsocq et al. 2010). In cells transgenic for the jellyfish protein aequorin, cytoplasmic Ca^{2+}-levels can be monitored elegantly via light emitted from the Ca^{2+} -dependent degradation of coelentheracin (Knight and Knight 1995).

Generation of reactive oxygen species (ROS), primarily superoxide anions ($O_2^{\cdot-}$), is another characteristic response that is widely used as a bioassay for studying M/DAMP (Apostol et al. 1989). In the presence of suitable chromogenic or luminogenic substrates, this response can easily be monitored in many types of plant tissues. ROS production, also termed oxidative burst, is a characteristic response of innate immunity in vertebrates as well (Kohchi et al. 2009). ROS are thought to directly damage and kill bacteria, e.g., in the process of phagocytosis by macrophages. In plants, direct antibiotic action has not been demonstrated so far. However, ROS production enhances crosslinking within the plant cell wall and renders these walls more resistant to degrading enzymes from pathogens (Apel and Hirt 2004).

Intracellular signaling triggered by perception of MAMPs/DAMPs involves activation of mitogen-activated protein kinase (MAPK) cascades. Rapid changes in the phosphorylation status and activity of these MAPKs have been successfully used to monitor and study MAMP perception (Asai et al. 2002; Mishra et al. 2006).

Enhanced ethylene biosyntheses is a well known response of plants exposed to different kinds of abiotic and biotic stresses (Morgan and Drew 1997). Induction of ethylene biosynthesis has been used as reliable and rapid assay to test for MAMP/DAMP perception in a broad variety of plant species (Albert et al. 2010a). As a bioassay it was successfully applied to purify and identify MAMPs such as the fungal glycopeptides and xylanase (Fuchs et al. 1989).

MAMP/DAMP perception triggers rapid and strong changes in expression of many different genes (Boudsocq et al. 2010; Basse et al. 1992). These changes can be monitored for individual genes or at the transcriptome level. Reporter constructs with MAMP-responsive promoters, such as the ones from the *PR-1*, *FRK1* and some *WRKY* genes, and GUS, luciferase or other reporter proteins have also been successfully used to monitor MAMP perception (Gust et al. 2007; Pandey and Somssich 2009; Asai et al. 2002).

In summary, MAMPs/DAMPs trigger a broad set of characteristic responses and symptoms, many of which can serve as rapid and convenient assays to test microbial substances for their activity as MAMPs/DAMPs in a qualitative and quantitative manner. However, this does not exclude further perception systems that involve different signaling pathways and distinct sets of responses. Fragments from the bacterial envelope such as lipopolysaccharides (LPS) and peptidoglycans (PGN) do not efficiently trigger early responses, yet they have been reported to activate increased expression of defense-related genes and disease resistance (Gust et al. 2007; Silipo et al. 2005). At present, little is known about such alternative pathways or about the types of receptors that might perceive these types of MAMPs in plants.

2 Polypeptide MAMPs

2.1 The Bacterial Flagellum as a Source of MAMP for Plants

Flagellin is the protein subunit that builds up the long filament of the bacterial flagellum. With up to 30,000 molecules per flagellum it forms an abundant and prominent target for recognition by the innate immune systems of plants and animals (Smith et al. 2003; Felix et al. 1999). The epitope of flagellin that acts as a MAMP in plants was identified as the most conserved part of this protein. Synthetic peptides like flg22 that represent 22 amino acids of this epitope are fully active in many different species of angiosperms and at least one species of gymnosperms (Albert et al. 2010a). In sensitive plant cells flg22 is active with a threshold of ~1 pM (Albert et al. 2010b; Jeworutzki et al. 2010). When applied to young *Arabidopsis* seedlings flg22 was found to cause growth inhibition. This effect was employed in a genetic screen for mutants growing in the presence of flg22 (Gomez-Gomez and Boller 2000; Gomez-Gomez et al. 1999). Several of the flg22-insensitive mutants were affected in a single gene termed *FLS2* (flagellin sensitive 2). *FLS2* encodes a leucine-rich repeat (LRR)-RLK that was later shown to be the bona fide receptor for flg22 (Gomez-Gomez and Boller 2000; Chinchilla et al. 2006).

Some bacterial species of *Ralstonia, Agrobacterium* and *Rhizobium* that live as pathogens or symbionts of plants have flagellins which are exceptionally divergent in their flg22-domains. Peptides representing these variant sequences are inactive as MAMPs, indicating that these bacteria were under selective pressure to avoid recognition by their plant hosts (Felix et al. 1999; Pfund et al. 2004).

In turn, plants might have evolved perception systems directed against epitopes of flagellin distinct from flg22. A second perception system for flagellin of certain bacterial strains has been reported for rice plants (Taguchi et al. 2003). It detects a domain distinct from flg22 via its state of glycosylation in a manner independent of FLS2.

2.2 *Abundant Cytoplasmic Proteins of Bacteria as MAMPs for Plants*

A second bacterial protein that acts as a MAMP in cells of Arabidopsis was identified as elongation factor Tu (EF-Tu) (Kunze et al. 2004). The minimal motif with full activity consists of the acetylated N-terminus of EF-Tu with the first 18 amino acid residues (elf18). Interestingly, N-terminal acetylation is known for only very few bacterial proteins. This particular modification of EF-Tu also serves as a recognition determinant for the perception in plants and peptides lacking acetylation were 20-fold less efficient as MAMPs (Kunze et al. 2004). Overall, elf18 induces the same set of responses as flg22 and experiments with radiolabeled elf18 identified specific, high-affinity binding to a protein with characteristics reminiscent of FLS2 (Kunze et al. 2004; Zipfel et al. 2006). Indeed, using a reverse genetic approach, mutations in one of the RLKs most closely related to FLS2 were found to abolish responses to elf18 (Zipfel et al. 2006). Heterologous expression in plant species that have no endogenous perception system for EF-Tu, such as *N. benthamiana* and tomato, could then demonstrate that this RLK functions as EF-Tu receptor (EFR) (Zipfel et al. 2006; Lacombe et al. 2010). An important finding of these studies was that presence of EFR, as occurring naturally in Arabidopsis plants or as expressed from a transgene in *N. benthamiana* and tomato, confers increased resistance to infection by pathogenic bacteria. On one hand, this shows the biotechnological potential for increased resistance in crop plants with an extended repertoire of MAMP recognition. On the other hand, these results also provide clear evidence that during the infection process a direct contact between EF-Tu and the EFR receptor does take place and is relevant for induction of plant defense. EF-Tu is a very abundant protein in the bacterial cytoplasm where it has an essential role in protein synthesis. Evidently, during the infection processes sufficient amounts of this protein get released from bacteria to act as ligand for the highly sensitive EFR receptor at the surface of plant cells. Bacterial cold shock protein (Csp) can serve as a further example of a cytoplasmic protein from bacteria that can be sensed as a MAMP in plants (Felix and Boller 2003). Tobacco, tomato and other solanaceous plants were found to have a highly sensitive perception system specific for the conserved RNA- and DNA-binding motif RNP-1 of bacterial Csp. This MAMP, represented by the csp15 peptide, is thus a candidate ligand for one of the many orphan RLKs in plants.

2.3 The Quorum Sensing Signal Ax21 as a MAMP for Rice

Xa21 from rice was identified as a resistance gene that confers immunity to diverse strains of the Gram-negative bacterium *Xanthomonas oryzae* pv. *oryzae (Xoo)* (Song et al. 1995). *Xa21* encodes a LRR-RLK related to EFR, suggesting that it might act as a receptor for a molecular pattern of the pathogen. Genetics on the side of the bacterial pathogen indicated that the *AvrXa21* gene of *Xoo*, encoding a 194-amino acid protein, determines recognition by Xa21 (Song et al. 1995). However, identification of the ligand that physically interacts with the Xa21 receptor remained difficult. Recently, this ligand was purified from supernatants of *Xoo* cultures and identified as the sulfated form of a 17 aa peptide, dubbed Ax21 for activator of XA21, derived from the N-terminal part of the AvrXa21 protein (Lee et al. 2009). Several bacterial genes were found to be involved in modification and export of this signal. In particular, tyrosine sulfation by a sulfotransferase is necessary for its recognition as MAMP in rice and, most likely, for its proposed function as a quorum sensing signal for the bacteria (Park et al. 2010). In summary, bacterial Ax21 and rice Xa21, formerly described as a bacterial "avirulence" protein and a corresponding resistance protein, can now be understood as a conserved molecular structure of bacteria, a MAMP, for which plants have evolved Xa21 as a specific PRR.

2.4 Secreted Proteins as MAMPs

Many pathogenic and saprophytic microorganisms release cell wall degrading enzymes and membrane disrupting substances to ensure their nutrition. Plants, in turn, can detect some of these secreted microbial enzymes and virulence factors as MAMPs.

Proteins that might act as signals for PRR comprise membrane interacting and pore forming factors such as harpinN and harpinZ from bacteria and the group of NLP-toxins that are produced by taxonomically unrelated microbial pathogens including bacteria, fungi and oomycetes (Wei et al. 1992; Engelhardt et al. 2009; Ottmann et al. 2009).

A well studied example for direct perception as a MAMP is the endo-β-1,4 xylanases from the ascomycete *Trichoderma viride* (Fuchs et al. 1989). In many varieties of tobacco and tomato this 22 kDa protein, named ethylene inducing xylanase (EIX), is a strong inducer of responses typically associated with cells responding to MAMPs (Belien et al. 2006; Felix et al. 1994). Protein denaturation by heat leads to concomitant loss of xylanase and MAMP activity. In contrast, inactivation of the enzyme activity by targeted mutation of the active center did not affect its activity as a MAMP, indicating that plants detect EIX via a PRR against a structural determinant of the protein (Enkerli et al. 1999; Furman-Matarasso et al. 1999). Using map-based cloning with a tomato variety that can detect EIX and a variety that cannot detect it, a locus encoding two closely related LRR-RLPs, *Le*EIX1 and *Le*EIX2, was identified (Ron and Avni 2004). Overexpression of *LeEIX1* or *LeEIX2* in plants of a non-responsive tobacco variety correlated with binding of the EIX-protein tagged with a fluorescence label. However, only plants expressing *LeEIX2* showed responses to EIX (Ron and Avni 2004). How plant RLPs like EIX2, lacking a cytoplasmic kinase domain, triggers intracellular signaling remains an open question. Nevertheless, as also emerging for RLKs, ligand-induced changes in the interaction with membrane proteins that act as co-receptors can be anticipated to play essential roles in this activation process.

In tomato, a number of additional LRR-RLPs have been identified that confer resistance against specific races of the fungal pathogens *Cladosporium fulvum* and *Verticillium albo-atrum* (Kruijt et al. 2005; Kawchuk et al. 2001). Clearly, these proteins are essential for the specific detection of products encoded by the corresponding effector genes of the pathogens. So far, however, none of these RLPs could be shown to directly interact with the fungal effector protein in a ligand-receptor like manner. Rather, at least some of these RLPs seem to sense these effectors indirectly by the modification they cause on components of the host itself. Thus, conceptually, sensing structures of "altered-self" in a guard-type manner bears resemblance to detection of signals released by injury or damage of the cells.

3 MAMPs Derived from Microbial Cell Walls or Cell Envelopes

Walls and cellular envelope of microorganisms consist of highly abundant and highly characteristic molecular structures that provide excellent targets for detection by host immune systems. Components that have been reported to trigger

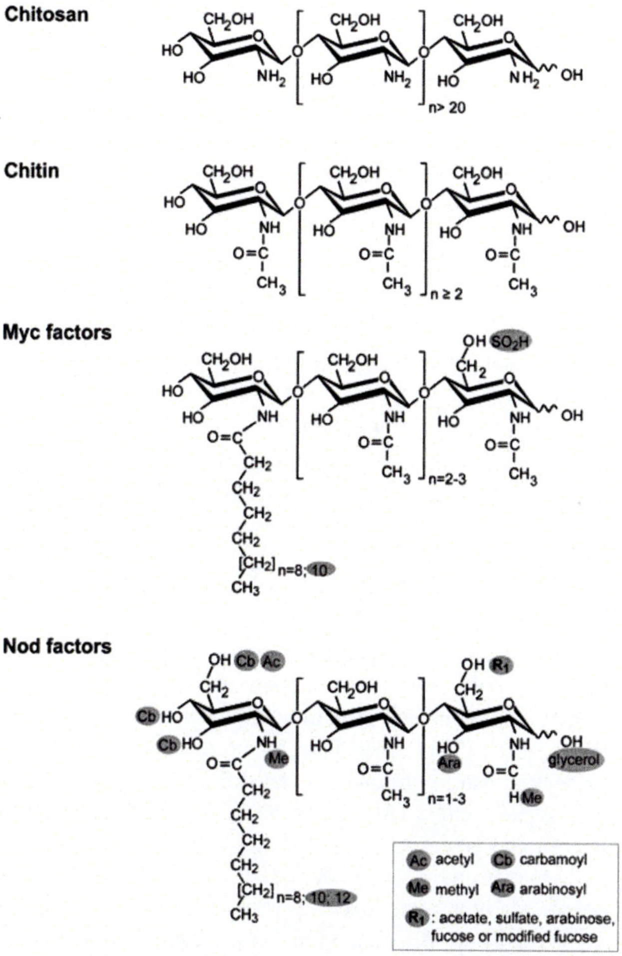

Fig. 2 Structure of (lipo)-chitooligosaccharides with biological activity as MAMPs/DAMPs or as signals for symbiosis with mycorrhiza and rhizobia

3.3 Chitin

The homopolymer chitin is built up by β-1,4 linked N-acetylglucosamine units and represents a structural component of fungal cell walls. It is also present in the exoskeleton of insects and crustaceans, but oligomers with more than 2 moieties of acetylglucosamine are not known to occur in plants. Chitin acts as a potent elicitor in many species of angiosperms, reviewed recently by (Hamel and Beaudoin 2010), and even gymnosperms (Albert et al. 2010a). For recognition by plants, the chain length of the chitin fragments is important. Chitin oligomers with ≥4 N-acetylglucosamine residues are fully active and induce responses at subnanomolar concentrations in tomato and soybean cells (Felix et al. 1993; Mithöfer et al.

1999). In contrast, other plant species including rice and wheat respond only to chitooligomers with a chain length of ≥7 monomers (Shibuya and Minami 2001). High-affinity chitin-binding sites were identified by binding studies and affinity crosslinking with radiolabeled chitooligomers in tomato, rice, soybean, wheat, barley, and carrot (Baureithel et al. 1994; Shibuya et al. 1993; Ito et al. 1997; Day et al. 2001; Okada et al. 2002). Up to now, two types of chitin-binding proteins that might act as chitin receptors have been indentified: one is CEBiP (chitin elicitor-binding protein) in rice (Kaku et al. 2006) and the second is the receptor-like kinase CERK1/LysM-RLK1 (Chitin-Elicitor Receptor Kinase 1) in Arabidopsis (Miya et al. 2007; Wan et al. 2008). Both proteins are characterized by lysine motif (LysM) domains but CEBiP, rather than being a receptor kinase, is a secreted protein attached to the membrane via a glycosylphosphatidylinositol (GPI) anchor. Silencing of CEBiP expression results in strongly reduced sensitivity of rice cells to chitin, indicating a key role in chitin perception (Kaku et al. 2006). For transmembrane activation of signaling, CEBiP probably depends on complex formation with other membrane proteins. Possible candidates for such co-receptor function are the *Os*CERK1 homologues in rice (Shimizu et al. 2010). In *A. thaliana*, the RLK CERK1 is essential for chitin perception, based on the analysis of mutants lacking *CERK1* (Miya et al. 2007). Pull-down assays with chitin beads show direct and specific binding of chitin to CERK1/LysM-RLK1 (Petutschnig et al. 2010; Iizasa et al. 2010) that can be competitively inhibited with defined, soluble chitin fragments with an IC_{50} of ~80 nM (Iizasa et al. 2010).

The LysM domain is an ancient protein domain originally identified in bacterial autolysin (Beliveau et al. 1991). Apart from enzymes involved in bacterial cell wall degradation, it occurs also in chitinases from different organisms. Occurrence of LysM motifs in the extracellular domains of RLKs and RLPs seems to be restricted to plants (Bateman and Bycroft 2000). The domain has a characteristic motif of ~40 amino acids that forms a βααβ secondary structure with the two α-helices stacking onto one side of a plate made up of a two-stranded antiparallel β-sheet (Bateman and Bycroft 2000; Bielnicki et al. 2006). Until now, it is not clear which structural features of the LysM domain determine the affinity for N-acetylglucosamine moieties occurring in chitin and the polysaccharide backbone of peptidoglycan from bacteria (Zhang et al. 2007). It also has to be elucidated how two or three of these domains in the LysM-type receptors might cooperate and lead to receptor activation.

3.4 *Lipo-Chitooligosaccharides as Microbial Signals in Mutualistic Symbiosis*

Plants can undergo different types of close interactions with beneficial and mutualistic symbionts. Thus, rather than distinction of "self" and "non-self," plants have to discriminate between friend and foe in order to respond with adequate developmental processes. Plants have evolved perception systems for "danger" and detection systems that allow interaction with mutualistic symbionts. As detailed in the symbiosis chapters of this book (Madsen and Stougaard, this issue), two important types of mutualistic interactions, those with mycorrhizal fungi and those with rhizobia, have been under extensive study. Here, we reemphasize the astounding finding that the microbial signals identified in both of these symbioses show structural similarities to the chitooligosacharides discussed as MAMP above (Fig. 2).

3.4.1 Nod Factors

Gram-negative bacteria of the genera *Azorhizobium*, *Bradyrhizobium*, *Mesorhizobium*, *Rhizobium*, and *Sinorhizobium*, collectively referred to as rhizobia, can undergo symbiosis with different species of legume plants. These bacteria are able to induce nodule organogenesis in the roots of their host plants. Rhizobial bacteria colonize the cells of this novel organ and, in turn, are employed by the plants for nitrogen fixation. Soluble factors secreted by the bacteria are sufficient to trigger nodule formation. These nodulation factors (nod factors) were identified as lipochitooligosaccharides consisting of a chitin-type backbone with 3 to 5 N-acetylglucosamine subunits and an acyl chain replacing the N-acetyl group in the sugar at the nonreducing end (Denarie et al. 1996) (Fig. 2). The production of nod factors in the bacteria is initiated after recognition of flavonoid signals released from the host plants (Spaink 2000). Expression of the bacterial *nod* genes leads to the biosynthesis and the secretion of the nod factors. The *nod* genes also determine the species-specific modification in the nod factors such as the length and the degree of saturation of the fatty acid, the number of sugar residues, as well as the substitutions at the reducing and at the nonreducing end (Denarie et al. 1996). These modifications are important determinants for recognition by the corresponding receptors of the plant hosts (Madsen and Stougaard, this issue).

3.4.2 The Myc Factor

The most common and widespread symbiosis occurs between land plants and glomeromycete fungi forming arbuscular mycorrhiza (AM). This root endosymbiosis improves uptake of water and mineral nutrients for the plants and provides products of photosynthesis to the fungal partner. Recent results show that *Glomus intraradices* secretes symbiotic signals that stimulate root growth and branching as early steps in the formation of AM (Maillet et al. 2011). These Myc factors were identified as a mixture of sulfated and nonsulfated simple lipochitooligosaccharides. Thus, structurally, they strongly resemble bacterial nod factors (Fig. 2).

3.5 Receptors for (Lipo)-Chitooligosaccharides

Due to their chitin backbone, nod factors have been found to exhibit high affinity for chitin binding sites present in plants like tomato and tobacco (Baier et al. 1999; Staehelin et al. 1994). In turn, unmodified chitin fragments also act as MAMPs in legume species. Early on, this hinted at the possibility that receptors related to the PRRs detecting chitin fragments might detect lipochitooligosaccharides important for symbiosis. Indeed, genetic approaches in several host species have led to the identification of LysM-type receptor-like kinase (LYKs) that are essential for nod factor perception (Madsen et al. 2003; Radutoiu et al. 2003; Limpens et al. 2003). However, despite much experimental efforts, these presumptive nod factor receptors (NFRs) have not yet been shown to physically interact with the nod factors. These negative results might indicate that the ligand or the receptor requires additional modifications for interaction to take place. Alternatively, additional host components, acting as co-receptors or adaptors might be required for functional ligand binding.

Analysis of mutants in Lotus unable to undergo both types of symbiosis, root nodule formation with rhizobial bacteria and AM with mycorrhizal fungi, led to the identification of a gene encoding a further receptor-like kinase with a small ectodomain consisting of 3 LRRs (Stracke et al. 2002; Endre et al. 2002). Analogous to the role of the receptor kinase BAK1, discussed in other chapters of this book, this SYMRK (symbiosis receptor-like kinase) might function as a common co-receptor for the receptors binding both types of lipochitooligosaccharides, the nod factors, and Myc factors, respectively.

As found for the receptors detecting unmodified chitooligosaccharides as MAMPs, RLKs, and RLPs with LysM ectodomains remain primary candidates for receptors of the lipooligosaccharides involved in symbiosis. Such receptors with LysM ectodomains may have evolved from a common receptor as supported by phylogenetic analysis of LysM domain containing proteins (Lohmann et al. 2010; Zhang et al. 2007). Mycorrhizal symbiosis occurred very early on in the evolution of terrestrial plants. This opens the intriguing question whether recognition of pathogenic microorganisms or mutualistic symbionts came first during evolution.

3.6 Damage-Associated Molecular Patterns

One possible mechanism for sensing damage is via perception of molecular patterns that are accessible to PRRs only after wounding of the cells. Oligogalacturonides (OGs), fragments of the plant cell wall component pectin, have been reported to elicit defense responses many years ago and can be considered as classical DAMPs of plants (Bruce and West 1982). Pectin fragments get released in the context of mechanical wounding or by the action of pectinolytic enzymes produced by many microbial pathogens (Ridley et al. 2001). Members of the wall associated receptor kinases (WAKs), RLKs with EGF-like ectodomains, have been implicated in perception of oligogalacturonides (Decreux and Messiaen 2005) and chapter on RLKs and cell wall in this book).

Similar to pectin fragments, breakdown products of cellulose and cutin have been reported to act as wound signals in some plants (Aziz et al. 2007; Schweizer et al. 1996). Conceptually, all components of cells that normally reside within the cytoplasm or organelles might serve as potential DAMP signals. When released from damaged cells such components could be sensed by neighboring cells as indicators of tissue integrity. This direct detection of damage seems to operate for DAMPs like heat shock protein (HSP) or high-mobility group box 1 protein (HMGB1) in the innate immune system of vertebrates (Jeannin et al. 2008). In plants, membrane disturbance by physical stress, detergents or by pore-forming proteins and toxins of microbes has been reported to induce responses similar to MAMPs/DAMPs (Lee et al. 2001; Ottmann et al. 2009; Klusener and Weiler 1999; Engelhardt et al. 2009). For some of these membrane-interacting factors, such as the harpinN, harpinZ, and the pore-forming NLP-toxins, host plants might have evolved specific PRRs for detection (Wei et al. 1992; Engelhardt et al. 2009; Ottmann et al. 2009). Alternatively, host plants might sense membrane disturbance via changes in the membrane potential or they might detect the release of intracellular components as DAMPs. ATP is a candidate for such a signal since exogenously applied ATP has been reported to trigger DAMP responses in plant cells (Roux and Steinebrunner 2007). A more indirect generation of a signal has been proposed for inceptin, a component of the chloroplastic ATP synthase, that is converted to a DAMP for cowpea plants after digestion by proteases of insect herbivores (Schmelz et al. 2007). Similarly, GmSubPep is part of a secreted subtilisin-like protease from soybean that has to be released from the intact protein in order to act as a DAMP (Pearce et al. 2010a).

3.7 Systemin and Other "Wound Hormones"

Several other peptidic signals have been identified from different plant species that do act as DAMPs in the context of wounding. The first of these signals, termed systemin, was isolated from tomato plants (Pearce et al. 1991). It triggers typical defense responses at subnanomolar concentrations in some species of the *Solanaceae* (*Lycopersicon peruvianum*, potato, pepper, and nightshade). Systemin, an 18-amino acid peptide is derived from a precursor protein (prosystemin) which is expressed exclusively in the cytosol of phloem parenchyma cells (Ryan and Pearce 2003). The peptide is thought to be released from injured cells and to initiate systemic defense response to herbivore attack. Grafting experiments revealed that it is not systemin itself that functions as a long-distance signal. Rather jasmonate or related compounds of the octadecanoid pathway seem to act as systemic signal (Li et al. 2002). Biochemical approaches using affinity-crosslinking of ligand and binding proteins identified the LRR-RLK SR160 as a systemin receptor (Scheer and Ryan 2002). Surprisingly, the polypeptide of 160 kDa turned out to be identical with CURL3, the tomato ortholog of BRI1. However, later results with *curl3* mutants showed full sensitivity to systemin indicating that the functional systemin receptor still has to be identified (Lanfermeijer et al. 2008; Holton et al. 2008).

Peptidic factors with biological activity similar to systemin have been reported for other species such as tobacco, soybean, petunia, and *Arabidopsis* (Pearce et al. 2010b; Pearce et al. 2007; Huffaker et al. 2006). Best studied among these peptides is the 23-amino acid peptide *At*Pep1 from *A. thaliana.* It is derived from the C-terminus of a presumably cytoplasmic precursor protein ProPep1 with 92-amino acids. ProPep1 belongs to a gene family more than 6 related proteins and is expressed at low levels in all tissues. The perception of the *At*Pep-peptides is mediated by two related LRR-RLKs, *At*PEPR1 and *At*PEPR2 and double mutants of *pepr1* and *pepr2* do not respond to any of the *At*Pep's (Yamaguchi et al. 2006, 2010; Krol et al. 2010). In analogy to systemin and its perception system a current model proposes a role for the *At*Pep/*At*PEPR system in the propagation and amplification of defense signals via the JA/ethylene pathway to mount a rapid and robust defense against microbial and nonmicrobial herbivores. Whether systemins and *At*Peps indeed have evolved as specific wound hormones or, alternatively, have additional, as yet unrecognized, roles in other processes of plant development remains an interesting topic for future work.

3.8 Numerous MAMPs/DAMPs and Orphan RLKs/RLPs

Apart from the examples discussed above, there are numerous other reports about perception of additional MAMPs and DAMPs by cells of various plant species. Table 1 lists some of these signals that represent molecules originating from different types of microbes and molecules with different chemical composition. Some of these signals are well defined and dose–response relationship for induction of physiological responses in plants suggest perception by specific, high-affinity PRRs. Importantly, this list of MAMPs and DAMPs is far from complete and many more of these signals can be expected to be identified in the future. However, not every plant detects the same array of MAMPs/DAMPs. Rather, the repertoire of molecular patterns recognized by different plant species is overlapping but not congruent. In Fig. 3, this is exemplified with a comparison of the recognition systems described for *Arabidopsis* and tomato. Interestingly, some of the perception systems, notably the ones for chitin fragments and flg22 of bacterial flagellin, occur very broadly in many species of the angiosperm and even gymnosperms (Albert et al. 2010a). Others, like the ones for heptaglucan, csp15 and elf18 seem to be limited to a few closely related species or families of plants. This diversity of recognition systems hints at rapid evolution of the corresponding pattern recognition receptors. The very large families of RLKs and RLPs occurring in all higher plants contain members that are highly conserved between all species but even more members that show clear divergence. These families of receptors thus provide

Table 1 Examples of MAMPs/DAMPs with activity in different plants

MAMPs from bacteria:	Active motif	Receptor	Active in
Flagellin, flg22	QRLSTGSRINSAKDDAAGLQIA	FLS2(LRR-RLK)	Many angiosperms and gymnosperms (Gomez-Gomez and Boller 2000)
EF-Tu, elf18	ac-SKEKFERTKPHVNVGTIG	EFR (LRR-RLK)	*Arabidopsis* and other *Brassicales* (Zipfel et al. 2006)
AvrXA21, Ax21	AENLSY^{sulf}NFVEGDYVRTP	XA21 (LRR-RLK)	Rice (Lee et al. 2009)
Xanthomonas harpin hpaG	NQGISEKQLDQLLTQFIFSMLLQD		Tobacco and other *Solanales* (Kim et al. 2004)
Harpins from *Erwinia* and *Pseudomonas*	hrpN and hrpZ proteins		Tobacco (Wei et al. 1992; He et al. 1993)
Cold shock protein, csp15	VKWFNAEKGFGFITP		Tobacco and other *Solanales* (Felix and Boller 2003)
Lipopolysaccharide, LPS	Lipid A and core saccharide		*Arabidopsis* (Silipo et al. 2005)
Peptidoglycan	??		*Arabidopsis* (Gust et al. 2007)
MAMPs from fungi:			
Chitin	β1,4-(N-acetylglucosamine)*n* $n \geq 8$	CERK1 (LysM-RLK)CeBIP (LysM-protein)	Many angiosperms and gymnosperms (reviewed by Hamel and Beaudoin, 2010)
Chitosan	β1,4-(glucosamine)*n* $n > 20$		Bean, rice and various other species (reviewed by Hamel and Beaudoin, 2010)
Wall fragments	Pentaglucan		Rice (Yamaguchi et al. 2000)
Ergosterol	Ergosterol		Tomato and other Solanales (Granado et al. 1995)
Sphingolipid	Cerebroside A, B		Rice (Koga et al. 1998)
Xylanase	TKLGE-motif in xylanase	EIX2 (LRR-RLP)	Tomato and other *Solanales* (Ron and Avni 2004)
Glycopeptides	(Mannose)10-(GLcNac)2-peptide		Tomato (Basse et al. 1992)
MAMPs from oomycetes:			
Wall fragments	Heptaglucan	GBP (glucanase)	Soybean (Umemoto et al. 1997)
Transglutaminase, pep13	VWNQPVRGFKVYE		Parsley, potato (Nürnberger et al. 1994)
Cryptogein and other elicitins, small secreted sterol binding proteins			Tomato and other *Solanales* (Mikes et al. 1997)
DAMPs and wound hormones:			
Oligogalacturonides	α-(1–4)-(D-galacturonic acid)*n* $n = 10$–15	WAKs (WAK-RLK)	Various plant species (Decreux and Messiaen 2005)
Cellodextrin	β-(1–4)-(D-glucose)*n* $n > 7$		Grapevine (Aziz et al. 2007)
Cutin monomers	10,16-dihydroxypalmitic acid		Tomato, potato (Schweizer et al. 1996)
ATPsynthase, inceptin	ICDINGVCVDA		Cowpea (Schmelz et al. 2007)
Systemin	AVQSKPPSKRDPPKMQTD		Tomato (Pearce et al. 1991)
Peptides with systemin-like function, HypSys	~16–20 aa peptides rich in hydroxyproline, glycosylated		Petunia, tomato, tobaccco (Pearce et al. 2007)
*At*Pep1	RGKEKVSSGRPGQHN	*At*PEPR1/2 (LRR-RLK)	*Arabidopsis* (Huffaker et al. 2006)
Subtilisin, GmSubPep	NTPPRRAKSRPH		Soybean (Pearce et al. 2010a)
RALF	YISY-motif in 49 aa protein		Various species (Pearce et al. 2010b)

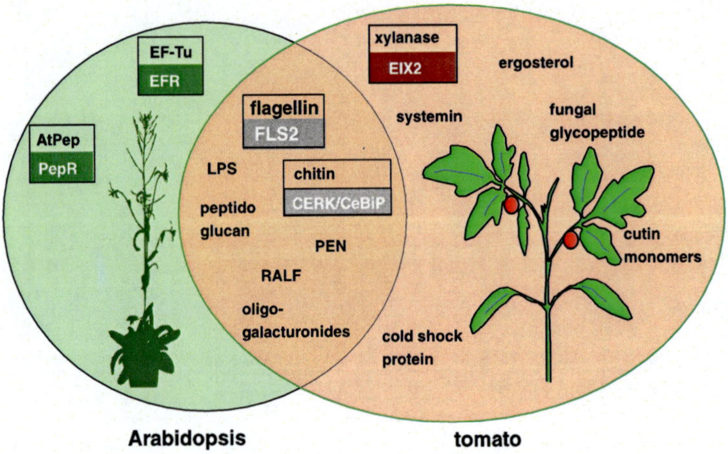

Fig. 3 Recognition systems for MAMPs/DAMPs in *Arabidopsis* and tomato

candidate receptors for numerous further ligands, including a multitude of MAMPs/DAMPs and other symbiosis signals. The identification of further ligand–receptor pairs will remain a challenge.

4 Conclusions

Plants have evolved a variety of sensory systems that enables them to use chemical cues from their environment for adaptive regulation of their metabolism and their development. Mechanistically, perception of signals indicating presence of pathogens or beneficial symbionts does not differ fundamentally from sensing endogenous signals like hormones and other regulators. Indeed, members of the same class of surface receptors, the RLKs and RLPs, serve roles as sensors for all of these signals. As detailed in other chapters of this book, the question of how diversification into different types of adaptive responses occurs is still open and at the focus of active research efforts.

References

Albert M, Jehle AK, Lipschis M, Mueller K, Zeng Y, Felix G (2010a) Regulation of cell behaviour by plant receptor kinases: pattern recognition receptors as prototypical models. Eur J Cell Biol 89:200–207

Albert M, Jehle AK, Mueller K, Eisele C, Lipschis M, Felix G (2010b) Arabidopsis thaliana pattern recognition receptors for bacterial elongation factor Tu and flagellin can be combined to form functional chimeric receptors. J Biol Chem 285:19035–19042

Apel K, Hirt H (2004) Reactive oxygen species: metabolism, oxidative stress, and signal transduction. Annu Rev Plant Biol 55:373–399

Apostol I, Heinstein PF, Low PS (1989) Rapid stimulation of an oxidative burst during elicitation of cultured plant cells. Role in defense and signal transduction. Plant Physiol 90:109–116

Asai T, Tena G, Plotnikova J, Willmann MR, Chiu WL, Gomez-Gomez L, Boller T, Ausubel FM, Sheen J (2002) MAP kinase signalling cascade in Arabidopsis innate immunity. Nature 415: 977–983

Ausubel FM (2005) Are innate immune signaling pathways in plants and animals conserved? Nat Immunol 6:973–979

Aziz A, Gauthier A, Bezier A, Poinssot B, Joubert JM, Pugin A, Heyraud A, Baillieul F (2007) Elicitor and resistance-inducing activities of beta-1,4 cellodextrins in grapevine, comparison with beta-1,3 glucans and alpha-1,4 oligogalacturonides. J Exp Bot 58:1463–1472

Baier R, Schiene K, Kohring B, Flaschel E, Niehaus K (1999) Alfalfa and tobacco cells react differently to chitin oligosaccharides and sinorhizobium meliloti nodulation factors. Planta 210: 157–164

Basse CW, Bock K, Boller T (1992) Elicitors and suppressors of the defense response in tomato cells. Purification and characterization of glycopeptide elicitors and glycan suppressors generated by enzymatic cleavage of yeast invertase. J Biol Chem 267:10258–10265

Bateman A, Bycroft M (2000) The structure of a LysM domain from E. coli membrane-bound lytic murein transglycosylase D (MltD). J Mol Biol 299:1113–1119

Baureithel K, Felix G, Boller T (1994) Specific, high affinity binding of chitin fragments to tomato cells and membranes. Competitive inhibition of binding by derivatives of chitooligosaccharides and a Nod factor of Rhizobium. J Biol Chem 269:17931–17938

Belien T, Van Campenhout S, Robben J, Volckaert G (2006) Microbial endoxylanases: effective weapons to breach the plant cell-wall barrier or, rather, triggers of plant defense systems? Mol Plant Microbe Interact 19:1072–1081

Beliveau C, Potvin C, Trudel J, Asselin A, Bellemare G (1991) Cloning, sequencing, and expression in Escherichia coli of a Streptococcus faecalis autolysin. J Bacteriol 173:5619–5623

Bielnicki J, Devedjiev Y, Derewenda U, Dauter Z, Joachimiak A, Derewenda ZS (2006) *B. subtilis* ykuD protein at 2.0 A resolution: insights into the structure and function of a novel, ubiquitous family of bacterial enzymes. Proteins 62:144–151

Boller T, Felix G (2009) A renaissance of elicitors: perception of microbe-associated molecular patterns and danger signals by pattern-recognition receptors. Annu Rev Plant Biol 60:379–406

Boudsocq M, Willmann MR, McCormack M, Lee H, Shan L, He P, Bush J, Cheng SH, Sheen J (2010) Differential innate immune signalling via Ca^{2+} sensor protein kinases. Nature 464: 418–422

Bruce RJ, West CA (1982) Elicitation of casbene synthetase activity in castor bean. The role of pectic fragments of the plant cell wall in elicitation by a fungal endopolygalacturonase. Plant Physiol 69:1181–1188

Cheong JJ, Hahn MG (1991) A specific, high-affinity binding site for the hepta-beta-glucoside elicitor exists in soybean membranes. Plant Cell 3:137–147

Chinchilla D, Bauer Z, Regenass M, Boller T, Felix G (2006) The Arabidopsis receptor kinase FLS2 binds flg22 and determines the specificity of flagellin perception. Plant Cell 18:465–476

Day RB, Okada M, Ito Y, Tsukada K, Zaghouani H, Shibuya N, Stacey G (2001) Binding site for chitin oligosaccharides in the soybean plasma membrane. Plant Physiol 126:1162–1173

Decreux A, Messiaen J (2005) Wall-associated kinase WAK1 interacts with cell wall pectins in a calcium-induced conformation. Plant Cell Physiol 46:268–278

Denarie J, Debelle F, Prome JC (1996) Rhizobium lipo-chitooligosaccharide nodulation factors: signaling molecules mediating recognition and morphogenesis. Annu Rev Biochem 65: 503–535

Endre G, Kereszt A, Kevei Z, Mihacea S, Kalo P, Kiss GB (2002) A receptor kinase gene regulating symbiotic nodule development. Nature 417:962–966

Engelhardt S, Lee J, Gabler Y, Kemmerling B, Haapalainen ML, Li CM, Wei Z, Keller H, Joosten M, Taira S, Nurnberger T (2009) Separable roles of the *Pseudomonas syringae* pv. phaseolicola accessory protein HrpZ1 in ion-conducting pore formation and activation of plant immunity. Plant J 57:706–717

Enkerli J, Felix G, Boller T (1999) The enzymatic activity of fungal xylanase is not necessary for its elicitor activity. Plant Physiol 121:391–397

Felix G, Boller T (2003) Molecular sensing of bacteria in plants. The highly conserved RNA-binding motif RNP-1 of bacterial cold shock proteins is recognized as an elicitor signal in tobacco. J Biol Chem 278:6201–6208

Felix G, Regenass M, Boller T (1993) Specific perception of subnanomolar concentrations of chitin fragments by tomato cells. Induction of extracellular alkalinization, changes in protein phosphorylation, and establishment of a refractory state. Plant J 4:307–316

Felix G, Regenass M, Spanu P, Boller T (1994) The protein phosphatase inhibitor calyculin A mimics elicitor action in plant cells and induces rapid hyperphosphorylation of specific proteins as revealed by pulse-labeling with [33 P]phosphate. Proc Natl Acad Sci USA 91: 952–956

Felix G, Duran JD, Volko S, Boller T (1999) Plants have a sensitive perception system for the most conserved domain of bacterial flagellin. Plant J 18:265–276

Fliegmann J, Mithofer A, Wanner G, Ebel J (2004) An ancient enzyme domain hidden in the putative beta-glucan elicitor receptor of soybean may play an active part in the perception of pathogen-associated molecular patterns during broad host resistance. J Biol Chem 279: 1132–1140

Fuchs Y, Saxena A, Gamble HR, Anderson JD (1989) Ethylene biosynthesis-inducing protein from cellulysin is an endoxylanase. Plant Physiol 89:138–143

Furman-Matarasso N, Cohen E, Du Q, Chejanovsky N, Hanania U, Avni A (1999) A point mutation in the ethylene-inducing xylanase elicitor inhibits the beta-1-4-endoxylanase activity but not the elicitation activity. Plant Physiol 121:345–351

Gomez-Gomez L, Boller T (2000) FLS2: an LRR receptor-like kinase involved in the perception of the bacterial elicitor flagellin in Arabidopsis. Mol Cell 5:1003–1011

Gomez-Gomez L, Felix G, Boller T (1999) A single locus determines sensitivity to bacterial flagellin in *Arabidopsis thaliana*. Plant J 18:277–284

Granado J, Felix G, Boller T (1995) Perception of fungal sterols in plants: subnanomolar concentrations of ergosterol elicit extracellular alkalinization in tomato cells. Plant Physiol 107: 485–490

Gust AA, Biswas R, Lenz HD, Rauhut T, Ranf S, Kemmerling B, Gotz F, Glawischnig E, Lee J, Felix G, Nurnberger T (2007) Bacteria-derived peptidoglycans constitute pathogen-associated molecular patterns triggering innate immunity in Arabidopsis. J Biol Chem 282:32338–32348

Hamel LP, Beaudoin N (2010) Chitooligosaccharide sensing and downstream signaling: contrasted outcomes in pathogenic and beneficial plant-microbe interactions. Planta 232: 787–806

He SY, Huang HC, Collmer A (1993) *Pseudomonas syringae* pv. syringae harpinPss: a protein that is secreted via the Hrp pathway and elicits the hypersensitive response in plants. Cell 73: 1255–1266

Holton N, Harrison K, Yokota T, Bishop GJ (2008) Tomato BRI1 and systemin wound signalling. Plant Signal Behav 3:54–55

Huffaker A, Pearce G, Ryan CA (2006) An endogenous peptide signal in Arabidopsis activates components of the innate immune response. Proc Natl Acad Sci USA 103:10098–10103

Iizasa E, Mitsutomi M, Nagano Y (2010) Direct binding of a plant LysM receptor-like kinase, LysM RLK1/CERK1, to chitin in vitro. J Biol Chem 285:2996–3004

Iriti M, Faoro F (2009) Chitosan as a MAMP, searching for a PRR. Plant Signal Behav 4:66–68

Ito Y, Kaku H, Shibuya N (1997) Identification of a high-affinity binding protein for N-acetylchitooligosaccharide elicitor in the plasma membrane of suspension-cultured rice cells by affinity labeling. Plant J 12:347–356

Jeannin P, Jaillon S, Delneste Y (2008) Pattern recognition receptors in the immune response against dying cells. Curr Opin Immunol 20:530–537

Jeworutzki E, Roelfsema MR, Anschutz U, Krol E, Elzenga JT, Felix G, Boller T, Hedrich R, Becker D (2010) Early signaling through the Arabidopsis pattern recognition receptors FLS2 and EFR involves Ca-associated opening of plasma membrane anion channels. Plant J 62: 367–378

Kaku H, Nishizawa Y, Ishii-Minami N, Akimoto-Tomiyama C, Dohmae N, Takio K, Minami E, Shibuya N (2006) Plant cells recognize chitin fragments for defense signaling through a plasma membrane receptor. Proc Natl Acad Sci USA 103:11086–11091

Kauss H, Jeblick W, Domard A (1989) The degrees of polymerization and N-acetylation of chitosan determine its ability to elicit callose formation in suspension cells and protoplasts of *Catharanthus roseus*. Planta 178:385–392

Kawchuk LM, Hachey J, Lynch DR, Kulcsar F, van Rooijen G, Waterer DR, Robertson A, Kokko E, Byers R, Howard RJ, Fischer R, Prufer D (2001) Tomato Ve disease resistance genes encode cell surface-like receptors. Proc Natl Acad Sci USA 98:6511–6515

Kim JG, Jeon E, Oh J, Moon JS, Hwang I (2004) Mutational analysis of Xanthomonas harpin HpaG identifies a key functional region that elicits the hypersensitive response in nonhost plants. J Bacteriol 186:6239–6247

Klusener B, Weiler EW (1999) Pore-forming properties of elicitors of plant defense reactions and cellulolytic enzymes. FEBS Lett 459:263–266

Knight H, Knight MR (1995) Chapter 14 Recombinant aequorin methods for intracellular calcium measurement in plants. In: David W, Galbraith HJB, Don PB (eds) Methods in cell biology, vol 49. Academic, San Diego, CA, pp 201–216

Koga J, Yamauchi T, Shimura M, Ogawa N, Oshima K, Umemura K, Kikuchi M, Ogasawara N (1998) Cerebrosides A and C, sphingolipid elicitors of hypersensitive cell death and phytoalexin accumulation in rice plants. J Biol Chem 273:31985–31991

Kohchi C, Inagawa H, Nishizawa T, Soma G (2009) ROS and innate immunity. Anticancer Res 29: 817–821

Krol E, Mentzel T, Chinchilla D, Boller T, Felix G, Kemmerling B, Postel S, Arents M, Jeworutzki E, Al-Rasheid KA, Becker D, Hedrich R (2010) Perception of the Arabidopsis danger signal peptide 1 involves the pattern recognition receptor AtPEPR1 and its close homologue AtPEPR2. J Biol Chem 285:13471–13479

Kruijt M, de Kock MJD, de Wit PJGM (2005) Receptor-like proteins involved in plant disease resistance. Mol Plant Pathol 6:85–97

Kunze G, Zipfel C, Robatzek S, Niehaus K, Boller T, Felix G (2004) The N terminus of bacterial elongation factor Tu elicits innate immunity in Arabidopsis plants. Plant Cell 16:3496–3507

Lacombe S, Rougon-Cardoso A, Sherwood E, Peeters N, Dahlbeck D, van Esse HP, Smoker M, Rallapalli G, Thomma BP, Staskawicz B, Jones JD, Zipfel C (2010) Interfamily transfer of a plant pattern-recognition receptor confers broad-spectrum bacterial resistance. Nat Biotechnol 28:365–369

Lanfermeijer FC, Staal M, Malinowski R, Stratmann JW, Elzenga JT (2008) Micro-electrode flux estimation confirms that the Solanum pimpinellifolium cu3 mutant still responds to systemin. Plant Physiol 146:129–139

Lee J, Klusener B, Tsiamis G, Stevens C, Neyt C, Tampakaki AP, Panopoulos NJ, Noller J, Weiler EW, Cornelis GR, Mansfield JW, Nürnberger T (2001) HrpZ(Psph) from the plant pathogen Pseudomonas syringae pv. phaseolicola binds to lipid bilayers and forms an ion-conducting pore in vitro. Proc Natl Acad Sci USA 98:289–294

Lee SW, Han SW, Sririyanum M, Park CJ, Seo YS, Ronald PC (2009) A type I-secreted, sulfated peptide triggers XA21-mediated innate immunity. Science 326:850–853

Li L, Li C, Lee GI, Howe GA (2002) Distinct roles for jasmonate synthesis and action in the systemic wound response of tomato. Proc Natl Acad Sci USA 99:6416–6421

Limpens E, Franken C, Smit P, Willemse J, Bisseling T, Geurts R (2003) LysM domain receptor kinases regulating rhizobial Nod factor-induced infection. Science 302:630–633

Lohmann GV, Shimoda Y, Nielsen MW, Jorgensen FG, Grossmann C, Sandal N, Sorensen K, Thirup S, Madsen LH, Tabata S, Sato S, Stougaard J, Radutoiu S (2010) Evolution and regulation of the *Lotus japonicus* LysM receptor gene family. Mol Plant Microbe Interact 23: 510–521

Lotze MT, Zeh HJ, Rubartelli A, Sparvero LJ, Amoscato AA, Washburn NR, Devera ME, Liang X, Tor M, Billiar T (2007) The grateful dead: damage-associated molecular pattern molecules and reduction/oxidation regulate immunity. Immunol Rev 220:60–81

Madsen EB, Madsen LH, Radutoiu S, Olbryt M, Rakwalska M, Szczyglowski K, Sato S, Kaneko T, Tabata S, Sandal N, Stougaard J (2003) A receptor kinase gene of the LysM type is involved in legume perception of rhizobial signals. Nature 425:637–640

Maillet F, Poinsot V, Andre O, Puech-Pages V, Haouy A, Gueunier M, Cromer L, Giraudet D, Formey D, Niebel A, Martinez EA, Driguez H, Becard G, Denarie J (2011) Fungal lipochitooligosaccharide symbiotic signals in arbuscular mycorrhiza. Nature 469:58–63

Medzhitov R, Janeway CA Jr (1997) Innate immunity: impact on the adaptive immune response. Curr Opin Immunol 9:4–9

Mikes V, Milat ML, Ponchet M, Ricci P, Blein JP (1997) The fungal elicitor cryptogein is a sterol carrier protein. FEBS Lett 416:190–192

Mishra NS, Tuteja R, Tuteja N (2006) Signaling through MAP kinase networks in plants. Arch Biochem Biophys 452:55–68

Mithöfer A, Ebel J, Bhagwat AA, Boller T, Neuhaus-Url G (1999) Transgenic aequorin monitors cytosolic calcium transients in soybean cells challenged with β-glucan or chitin elicitors. Planta 207:566–574

Miya A, Albert P, Shinya T, Desaki Y, Ichimura K, Shirasu K, Narusaka Y, Kawakami N, Kaku H, Shibuya N (2007) CERK1, a LysM receptor kinase, is essential for chitin elicitor signaling in Arabidopsis. Proc Natl Acad Sci USA 104:19613–19618

Morgan PW, Drew MC (1997) Ethylene and plant responses to stress. Physiol Plant 100:620–630

Nürnberger T, Nennstiel D, Jabs T, Sacks WR, Hahlbrock K, Scheel D (1994) High affinity binding of a fungal oligopeptide elicitor to parsley plasma membranes triggers multiple defense responses. Cell 78:449–460

Okada M, Matsumura M, Ito Y, Shibuya N (2002) High-affinity binding proteins for N-acetylchitooligosaccharide elicitor in the plasma membranes from wheat, barley and carrot cells: conserved presence and correlation with the responsiveness to the elicitor. Plant Cell Physiol 43:505–512

Ottmann C, Luberacki B, Kufner I, Koch W, Brunner F, Weyand M, Mattinen L, Pirhonen M, Anderluh G, Seitz HU, Nurnberger T, Oecking C (2009) A common toxin fold mediates microbial attack and plant defense. Proc Natl Acad Sci USA 106:10359–10364

Pandey SP, Somssich IE (2009) The role of WRKY transcription factors in plant immunity. Plant Physiol 150:1648–1655

Park CJ, Han SW, Chen X, Ronald PC (2010) Elucidation of XA21-mediated innate immunity. Cell Microbiol 12:1017–1025

Pearce G, Strydom D, Johnson S, Ryan CA (1991) A polypeptide from tomato leaves induces wound-inducible proteinase inhibitor proteins. Science 253:895–898

Pearce G, Siems WF, Bhattacharya R, Chen YC, Ryan CA (2007) Three hydroxyproline-rich glycopeptides derived from a single petunia polyprotein precursor activate defensin I, a pathogen defense response gene. J Biol Chem 282:17777–17784

Pearce G, Munske G, Yamaguchi Y, Ryan CA (2010a) Structure-activity studies of GmSubPep, a soybean peptide defense signal derived from an extracellular protease. Peptides 31: 2159–2164

Pearce G, Yamaguchi Y, Munske G, Ryan CA (2010b) Structure-activity studies of RALF, Rapid Alkalinization Factor, reveal an essential–YISY–motif. Peptides 31:1973–1977

Petutschnig EK, Jones AM, Serazetdinova L, Lipka U, Lipka V (2010) The LysM-RLK CERK1 is a major chitin binding protein in *Arabidopsis thaliana* and subject to chitin-induced phosphorylation. J Biol Chem 285:28902–28911

Pfund C, Tans-Kersten J, Dunning FM, Alonso JM, Ecker JR, Allen C, Bent AF (2004) Flagellin is not a major defense elicitor in Ralstonia solanacearum cells or extracts applied to *Arabidopsis thaliana*. Mol Plant Microbe Interact 17:696–706

Radutoiu S, Madsen LH, Madsen EB, Felle HH, Umehara Y, Gronlund M, Sato S, Nakamura Y, Tabata S, Sandal N, Stougaard J (2003) Plant recognition of symbiotic bacteria requires two LysM receptor-like kinases. Nature 425:585–592

Ridley BL, O'Neill MA, Mohnen D (2001) Pectins: structure, biosynthesis, and oligogalacturonide-related signaling. Phytochemistry 57:929–967

Ron M, Avni A (2004) The receptor for the fungal elicitor ethylene-inducing xylanase is a member of a resistance-like gene family in tomato. Plant Cell 16:1604–1615

Roux SJ, Steinebrunner I (2007) Extracellular ATP: an unexpected role as a signaler in plants. Trends Plant Sci 12:522–527

Ryan CA, Pearce G (2003) Systemins: a functionally defined family of peptide signals that regulate defensive genes in Solanaceae species. Proc Natl Acad Sci USA 100:14577–14580

Scheer JM, Ryan CA Jr (2002) The systemin receptor SR160 from Lycopersicon peruvianum is a member of the LRR receptor kinase family. Proc Natl Acad Sci USA 99:9585–9590

Schmelz EA, LeClere S, Carroll MJ, Alborn HT, Teal PE (2007) Cowpea chloroplastic ATP synthase is the source of multiple plant defense elicitors during insect herbivory. Plant Physiol 144:793–805

Schweizer P, Felix G, Buchala A, Muller C, Métraux JP (1996) Perception of free cutin monomers by plant cells. Plant J 10:331–341

Sharp JK, McNeil M, Albersheim P (1984) The primary structure of one elicitor-active and seven elicitor-inactive hexa(á -D-glucopyranosyl)-D-glucitols isolated from the mycelial walls of *Phytophthora megasperma* f. sp. glycinea. J Biol Chem 259:11321–11336

Shibuya N, Minami E (2001) Oligosaccharide signalling for defence responses in plant. Physiol Mol Plant Pathol 59:223–233

Shibuya N, Kaku H, Kuchitsu K, Maliarik MJ (1993) Identification of a novel high-affinity binding site for N -acetylchitooligosaccharide elicitor in the membrane fraction from suspension-cultured rice cells. FEBS Lett 329:75–78

Shimizu T, Nakano T, Takamizawa D, Desaki Y, Ishii-Minami N, Nishizawa Y, Minami E, Okada K, Yamane H, Kaku H, Shibuya N (2010) Two LysM receptor molecules, CEBiP and OsCERK1, cooperatively regulate chitin elicitor signaling in rice. Plant J 64:204–214

Silipo A, Molinaro A, Sturiale L, Dow JM, Erbs G, Lanzetta R, Newman MA, Parrilli M (2005) The elicitation of plant innate immunity by lipooligosaccharide of *Xanthomonas campestris*. J Biol Chem 280:33660–33668

Smith KD, Andersen-Nissen E, Hayashi F, Strobe K, Bergman MA, Barrett SL, Cookson BT, Aderem A (2003) Toll-like receptor 5 recognizes a conserved site on flagellin required for protofilament formation and bacterial motility. Nat Immunol 4:1247–1253

Song WY, Wang GL, Chen LL, Kim HS, Pi LY, Holsten T, Gardner J, Wang B, Zhai WX, Zhu LH, Fauquet C, Ronald P (1995) A receptor kinase-like protein encoded by the rice disease resistance gene, Xa21. Science 270:1804–1806

Spaink HP (2000) Root nodulation and infection factors produced by rhizobial bacteria. Annu Rev Microbiol 54:257–288

Staehelin C, Granado J, Mller J, Wiemken A, Mellor RB, Felix G, Regenass M, Broughton WJ, Boller T (1994) Perception of Rhizobium nodulation factors by tomato cells and inactivation by root chitinases. Proc Natl Acad Sci USA 91:2196–2200

Stracke S, Kistner C, Yoshida S, Mulder L, Sato S, Kaneko T, Tabata S, Sandal N, Stougaard J, Szczyglowski K, Parniske M (2002) A plant receptor-like kinase required for both bacterial and fungal symbiosis. Nature 417:959–962

Taguchi F, Shimizu R, Inagaki Y, Toyoda K, Shiraishi T, Ichinose Y (2003) Post-translational modification of flagellin determines the specificity of HR induction. Plant Cell Physiol 44: 342–349

Umemoto N, Kakitani M, Iwamatsu A, Yoshikawa M, Yamaoka N, Ishida I (1997) The structure and function of a soybean beta-glucan-elicitor- binding protein. Proc Natl Acad Sci USA 94: 1029–1034

Walker-Simmons M, Hadwiger L, Ryan CA (1983) Chitosans and pectic polysaccharides both induce the accumulation of the antifungal phytoalexin pisatin in pea pods and antinutrient proteinase inhibitors in tomato leaves. Biochem Biophys Res Commun 110:194–199

Wan J, Zhang XC, Neece D, Ramonell KM, Clough S, Kim SY, Stacey MG, Stacey G (2008) A LysM receptor-like kinase plays a critical role in chitin signaling and fungal resistance in Arabidopsis. Plant Cell 20:471–481

Wei ZM, Laby RJ, Zumoff CH, Bauer DW, He SY, Collmer A, Beer SV (1992) Harpin, elicitor of the hypersensitive response produced by the plant pathogen Erwinia amylovora. Science 257: 85–88

Wu J, Baldwin IT (2010) New insights into plant responses to the attack from insect herbivores. Annu Rev Genet 44:1–24

Yamaguchi T, Yamada A, Hong N, Ogawa T, Ishii T, Shibuya N (2000) Differences in the recognition of glucan elicitor signals between rice and soybean: beta-glucan fragments from the rice blast disease fungus Pyricularia oryzae that elicit phytoalexin biosynthesis in suspension- cultured rice cells. Plant Cell 12:817–826

Yamaguchi Y, Pearce G, Ryan CA (2006) The cell surface leucine-rich repeat receptor for AtPep1, an endogenous peptide elicitor in Arabidopsis, is functional in transgenic tobacco cells. Proc Natl Acad Sci USA 103:10104–10109

Yamaguchi Y, Huffaker A, Bryan AC, Tax FE, Ryan CA (2010) PEPR2 is a second receptor for the Pep1 and Pep2 peptides and contributes to defense responses in Arabidopsis. Plant Cell 22: 508–522

Zhang XC, Wu X, Findley S, Wan J, Libault M, Nguyen HT, Cannon SB, Stacey G (2007) Molecular evolution of lysin motif-type receptor-like kinases in plants. Plant Physiol 144: 623–636

Zipfel C, Kunze G, Chinchilla D, Caniard A, Jones JD, Boller T, Felix G (2006) Perception of the bacterial PAMP EF-Tu by the receptor EFR restricts Agrobacterium-mediated transformation. Cell 125:749–760

Receptor Ligands in Development

Melinka A. Butenko and Reidunn Birgitta Aalen

Abstract Although there are hundreds of genes encoding receptor-like kinases and putative secreted ligands, to date less that ten have been matched and been shown to control plant growth or development. Brassionsteroids (BRs) and peptide ligands are involved in signaling between cells in the close vicinity to each other, and not transported over long distances. BRs and sulfated peptide ligands (PSK and PSY) have growth-promoting activities, while cysteine-rich and proline-rich peptide ligands identified so far are involved in specific processes such as self-incompatibility, differentiation, meristem maintenance and cell separation. Here we review how ligands in development and their respective receptors have been identified, how they interact, as well as the functional redundancy found in ligand gene families.

1 Introduction

1.1 Cell-to-Cell Communication in Developmental Processes

Plant development, from embryogenesis to reproduction, requires a balance between cell proliferation and differentiation, and ordered orientation of cell elongation and cell divisions to provide directional growth. The classical plant hormones are small lipophilic compounds recognized as having such functions. For example, auxin, giberellin, and cytokinin levels are of crucial importance for the balance between cell proliferation, cell elongation, and differentiation during development of the primary root (Ubeda-Tomas and Bennett 2010). These phytohormones can be transported over some distances, and their absolute concentrations, and concentrations relative to that of other hormones are critical for the control of

M.A. Butenko • R.B. Aalen (✉)
Department of Molecular Biosciences, University of Oslo, Oslo, Norway
e-mail: m.a.butenko@imbv.uio.no; reidunn-aalen@imbv.uio.no

F. Tax and B. Kemmerling (eds.), *Receptor-like Kinases in Plants*,
Signaling and Communication in Plants 13,
DOI 10.1007/978-3-642-23044-8_11,

hormone-regulated genes. The concentrations are determined by the rates of synthesis and transport, and by the modifications and conjugations that may activate or deactivate the hormone. The traditional plant hormones are well suited for communication between different parts and organs of a plant. However, differentiation and specification of cell fate also requires communication between neighboring cells.

In animals, it has for a long time been recognized that cell fate and identity is mediated by ligand–receptor interactions between neighboring cells. Over the last decade the importance of such interactions in cell-to-cell communication in plants has gradually been discovered and recognized. Central to this form of transmitting information are small extracellular ligands that relay signals through membrane-bound receptor-like kinases (RLKs) to induce changes in gene expression (Butenko et al. 2009; De Smet et al. 2009; Fukuda et al. 2007; Matsubayashi 2003; Rychel et al. 2010).

1.2 *Identification of Ligands*

Small molecules that are ligands for receptor kinases have been identified by two approaches – biochemistry or genetics. The biochemical approach has started out from observations that certain plant extracts have had promoting or inhibiting effects on differentiation or cell proliferation in tissue cultures. Thereafter, the active substance has been isolated, purified, and the biological activity has been confirmed. This approach has most often been used in species other than Arabidopsis, species where less genetic information and tools are available. However, the novel bioactive compounds have thereafter been tested for their function in Arabidopsis, where the relevant genes have been identified. This underscores that the genes involved in generating these compounds have been conserved during plant evolution. Brassinosteroids (BRs), Phytosulfokine (PSK), Plant peptide containing sulfated tyrosine (PSY1), and Tracheary element differentiation inhibitory factor (TDIF) are examples of steroids and peptides identified by this biochemical approach (Grove et al. 1979; Ito et al. 2006; Matsubayashi and Sakagami 1996). An advantage of this approach is the direct identification of the biologically active secreted compound, which in the case of the peptide ligands, may arise by post-translational processing and modification of the precursor proteins.

Geneticists have started out with mutants with phenotypes of interest and subsequently cloned the mutated gene. This approach has been used successfully in identifying peptides such as CLAVATA 3 (CLV3), TAPETUM DETERMINANT1 (TPD1), INFLORESCENCE DEFICIENT IN ABSCISSION (IDA), and EPIDERMAL PATTERNING FACTORs (EPFs) (Butenko et al. 2003; Clark et al. 1995; Hara et al. 2009; Hunt and Gray 2009; Yang et al. 2003) and has the advantage of directly linking the gene in question to a given developmental process or morphological feature on which the genetic screen was based. The disadvantage is that the approach only gives direct access to the gene, and not to the final biologically active gene product.

Based on the genetic approach, putative receptors can be identified by the similarities of their mutant phenotypes, sometimes aided by a comparison of expression patterns to that of the hypothesized ligand. For biochemically identified ligands, genetic screens for mutants that are unresponsive to the normal effect observed for the hormone or peptide can be conducted and lead to the discovery of possible ligand–receptor pairs. However, the small number of genetically implied and biochemically proven ligand–receptor pairs to date, argues that the most challenging task in this field is to match ligands and receptors.

1.3 Ligand–Receptor Pairs and Their Role in Plant Development

With the exception of the BRs which are considered the plant analogs of steroid hormones in the animal kingdom (Kim and Wang 2010), the majority of confirmed ligands involved in development are all small peptides. In *Arabidopsis thaliana,* more than 1,000 genes encoding small, putative secreted peptides–potential ligands–have been identified using a bioinformatics approach (Lease and Walker 2006). Bioinformatics has also identified more than 400 genes encoding proteins with a receptor configuration, defined by the presence of a signal sequence, an amino-terminal domain with a transmembrane region, and a carboxyl-terminal kinase domain (Shiu and Bleecker 2001). Most of these RLKs can be classified into 21 different structural classes by their extracellular domains. Eight of the groups, altogether 216 RLKs, have extracellular domains with leucine-rich repeats (LRRs). These LRR-RLKs can further be subdivided in at least 13 classes based on the number and organization of the LRRs.

In spite of these large numbers, to date there is only biochemical evidence for seven distinct endogenous ligands in *Arabidopsis* interacting with specific receptors (Butenko et al. 2009). With the exception of the PEP1-6 peptide-PEPR1/2 receptor pair, suggested to participate in amplification of defense responses (Yamaguchi et al. 2006), these signaling modules are all identified as involved in developmental processes. The ligands of these modules and the peptide ligands of four other signaling systems that have been identified genetically are the subject of this review (Table 1).

2 Brassinosteroid (BR) Ligands: Controlling Cell Expansion and Proliferation

A minority of the RLK classes have so far been assigned ligands (see Table 1), and most of these ligands relay their signals through class X and XI LRR-RLKs with more than 20 LRRs in their extracellular domain (Shiu and Bleecker 2001). Interestingly, the BR receptor belongs to LRR-class X, although LRRs are mainly

Table 1 Peptide ligands with identified receptors involved in plant development

Ligand peptide(s)	Processed active peptide	Class	Receptor	RLK class	Biological function
PSK	$Y^{S}IY^{S}TQ$	Sulfated	PSKR1	LRR-X	Cellular proliferation/ differentiation
PSY1	$DYSGDPSANPKHDPGVO^{L\text{-}Ara\text{-}3}OS$	Sulfated	At1g72300	LRR-X	Cellular proliferation/ differentiation
TPD1	Unknown	cysteine-rich	EMS1	LRR-X	Cell fate determination in anthers
SCR/SCR-RELATED	Contains 8 conserved cysteines	cysteine-rich	SRK	S-domain	Self-incompatibility response
EPF/EPF-LIKE	Unknown	cysteine-rich	ERf	LRR-XIII	Stomata formation
CLV3	$RTVOSGO^{L\text{-}Ara\text{-}3}DPLHHH$	Proline-rich	CLV1/CLV2/CRN/ RPK2	LRR-XI/receptor-like protein/kinase protein/ LRR-XV	Shoot apical meristem maintenance
CLE40	Unknown	Proline-rich	ACR4	Crinkly 4-like	Root meristem maintenance
TDIF (CLE44)	HEVOSGPONPISN	Proline-rich	TDR/PXY	LRR-XI	Maintain procambial cells
IDA	Unknown	Proline-rich	HAE/HSL2	LRR-XI	Floral abscission

known to participate in protein-protein interactions and other LRR X receptors (PSKRs and EXS/EMS1) bind peptide or protein ligands.

2.1 Identification of BR

The first plant steroid compound with growth-promoting activity was identified in pollen extracts from oilseed rape (*Brassica napus*) in 1979 and therefore named brassinolide (BL) (Grove et al. 1979) (Fig. 1). BRs are hydroxylated derivatives of cholestane and their structural variations comprise substitution patterns on rings A and B as well as the C-17 side-chain. These compounds can be categorized as C_{27}, C_{28}, or C_{29} BRs, depending on the length of the side chain. At least 65 free BRs and five BR conjugates have been characterized from the plant kingdom (Bajguz 2007).

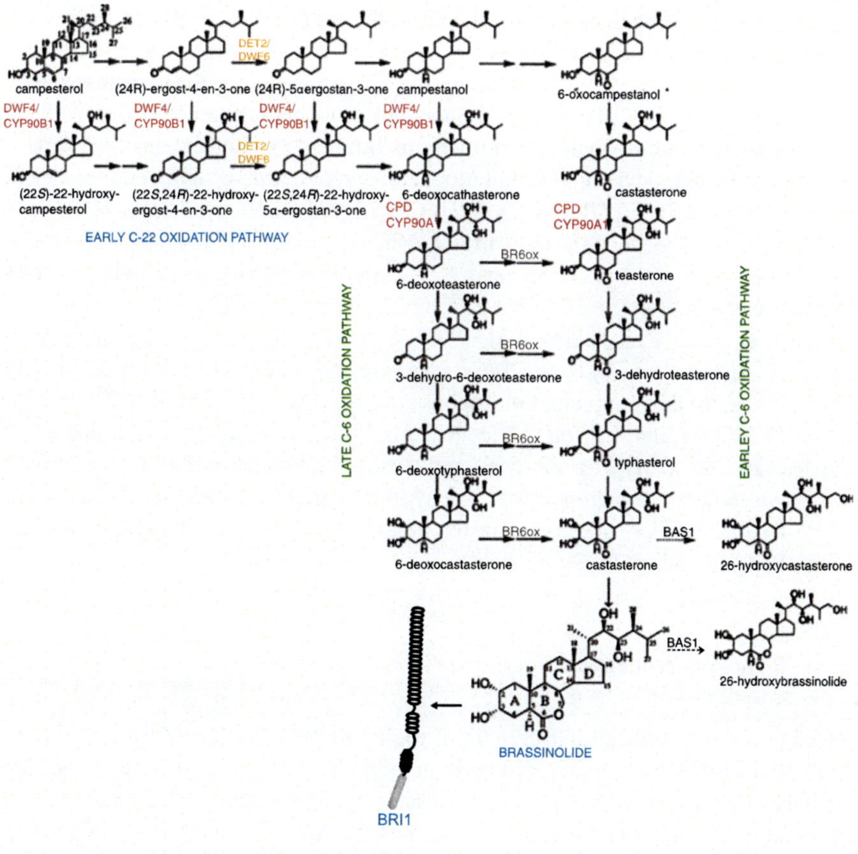

Fig. 1 Brassinosteroid biosynthesis pathway

2.2 BR Mutants and BR Production

After the identification of BR-deficient and BR-insensitive mutants, in the 1990s BRs were recognized as hormones (Clouse et al. 1996; Kauschmann et al. 1996; Li et al. 1996; Szekeres et al. 1996). Most Arabidopsis BR mutants are de-etiolated in the dark with short hypocotyls and open cotyledons. In the light, Arabidopsis BR mutants are characteristically dwarf, with rounded leaves, reduced apical dominance, delayed flowering and senescence, and reduced fertility. The short and rounded leaves are due to suppression of longitudinal cell expansion and proliferation (Nakaya et al. 2002), and the expression of genes associated with cell elongation, such as xyloglucan endotransglycosylases/hydrolases (XTHs), is reduced in BR mutants (Kauschmann et al. 1996).

BR-deficient mutants have defects in enzymes that catalyze steps in the conversion of the steroid campesterol to BL. This conversion occurs through a series of reductions, hydroxylations, epimerizations, and oxidations via two parallel routes, the early and late C-6 oxidation pathways, which are connected at multiple steps (Fujioka and Yokota 2003). The first steps reduce the C-5 double bond of campesterol, resulting in campestanol. One of these steps is catalyzed by the steroid 5α-reductase encoding gene *DE-ETIOLATED2* (*DET2*), also known as *DWARF6 (DWF6)* (Chory et al. 1991; Clouse et al. 1996; Li et al. 1996). Like many BR-deficient mutants, *det2* plants grown in the dark have many characteristics of light-grown plants, including hypocotyl growth inhibition, cotyledon expansion, primary leaf initiation, anthocyanin accumulation, and derepression of light-regulated gene expression. *CONSTITUTIVE AND PHOTOMORPHOGENESIS AND DWARFISM* (*CPD*) and *DWF4* encode the cytochrome P450 enzymes CYP90A1 and CYP90B1, respectively, which are C-22 and C-23 steroid hydroxylases that convert campestanol to teasterone or 6-deoxoteasterone, via the two C-6 oxidation pathways (Azpiroz et al. 1998; Szekeres et al. 1996). CYP90 catalyzes rate-limiting steps in BR synthesis, and the *CPD* and *DWF4* genes are subject to feedback regulation and are significantly downregulated in response to exogenous BL treatment (Choe et al. 1998). *AtBR6ox1* and *AtBR6ox2* encode steroid-6-oxidases linking the last steps of the late and the early C-6 pathway converting 6-deoxoteasterone to teasterone; 3-dehydro-6-deoxoteasterone to 3-dehydroteasterone; 6-deoxotyphasterol to typhasterol and finally 6-deoxocastasterone to castasterone, the immediate precursor of BL (Shimada et al. 2001, 2003).

2.3 The BR Receptor

BRs bind to the extracellular domain of the RLK BRASSINOSTEROID INSENSITIVE 1 (BRI1) to activate its kinase activity. This has been demonstrated first with co-immunoprecipitation, and thereafter by crosslinking of a biotin-tagged photoaffinity labeled biologically active biosynthetic precursor of BL (Kinoshita et al. 2005). BRI1 contains 25 LRRs and an island domain (ID) between 21st and

22nd LRRs. Binding studies using ^{3}H-labelled brassinolide and recombinant BRI1 fragments showed that the minimal binding domain for brassinosteroids consists of the 70-amino acid long ID together with the carboxy-terminal flanking LRR (ID-LRR22) (Kinoshita et al. 2005).

Interestingly, two BRI1-LIKE receptors (BRL1 and BRL3) also bind BL with high affinity, and expression of these using the BRI1 promoter can rescue *bri1* mutants. BRL2, however, neither binds BRs nor rescues *bri1*. BR activation of the receptor kinases leads to dephosphorylation of the transcription factors BZR1 and BZR2/BES1 resulting in transcriptional changes that promote growth. The majority of the known BR-regulated genes are associated with plant growth and development processes, such as cell wall modification, cytoskeleton formation, and hormone synthesis (Vert et al. 2005).

2.4 Transport and Perception of BRs

The BR biosynthesis genes are especially active in reproductive organs such as pollen, seeds, and fruits; in the shoot apex and in young, developing organs. Their expression is under tight developmental and organ-specific regulation (Shimada et al. 2003). There are indications that exogenous BR can move from the root to the shoot (Choe et al. 1998). However, tissue excision and reciprocal grafting experiments with BR mutants in pea and tomato suggest that endogenous BRs, unlike other phytohormones, cannot be transported over longer distances in the plant (Symons and Reid 2004; Symons et al. 2008). This implies that the homeostasis of bioactive BRs must be precisely controlled at tissue, or even at cellular levels, to ensure normal growth and development. Not only BR synthesis but also BR metabolism may regulate the amount of bioactive BR, as some modifications, such as 26-hydroxylation by BAS1 (PHYB ACTIVATION-TAGGED SUPRESSOR 1-DOMINANT), conjugation by the UDP-glycosyltransferase UGT73C5, or reduction catalyzed by BEN1 (BRI-ENHANCED1), lead to loss of activity (Poppenberger et al. 2005; Turk et al. 2003; Yuan et al. 2007). As the BR receptor resides in the plasma membranes of cells, BRs have to be exported out of the cells, where they are produced, to be perceived. Since there is a feedback regulation of BR synthesis genes, BR synthesis and perception can, in fact, take place in the very same cells. However, neighboring cells of the same tissue may perceive the BR signal. BR synthesis, export and perception could therefore be used for cell–cell communication and the outcome would be a homogenous response and behavior of similar cells in an organ.

3 Three Classes of Peptide Ligands

Surprisingly, it has been reported that the tomato BRI1 receptor also binds systemin, a wound-inducible 18 amino acid (aa) long peptide generated from a 200 aa precursor (Scheer and Ryan 2002; Szekeres 2003). Further studies suggest, however, that such binding does not transduce any signal into the cell (Malinowski et al. 2009). Interestingly, the known ligands of other LRR-RLKs in class X are peptides, mainly involved in general cell proliferation. The class XI ligand–receptor modules studied thus far seem to be more distinctively involved in organ-specific development or differentiation.

The peptide ligands involved in developmental processes can so far be assigned to three classes–the sulfated peptides, the cysteine-rich peptides, and the proline-rich peptides. These peptides are most often encoded by genes belonging to gene families, where functional redundancy is common. This can be not only an obstacle to the identification of novel ligand–receptor pairs, but also a guide, as similar ligands may interact with similar receptors (Butenko et al. 2009).

3.1 PSY1 and PSK: Sulfated Peptide Ligands Promoting Growth

In animals, tyrosine sulfation is a common posttranslational modification of proteins which are involved in protein–protein interactions, including ligand-receptor interactions, and this modification seems important for optimal proteolytic processing, and proteolytic activation of extracellular proteins (Moore 2003). In contrast to the dozens of sulfated peptides that have been identified in animals, only two, PSK and PSY1, have been identified in plants (Fig. 2).

PHYTOSULFOKINE (PSK)

```
PSK1 MKTKSEVLIFFFTLVLLLSMASSVILRE---DGFAP------PKPSPTTHEKASTK-G---DRDGV---ECKNSDSEEEC-LVKKTVA-AHTDYIYTQDLNLSP 86
PSK2 M---ANVSALL-TIALLL--CSTLMCT----ARPEPAISISITTAADPCNMEKKIE-GKLDDMHMVD-ENC-GAD-DEDC-LMRRTLV-AHTDYIYTQKKKH-P 87
PSK3 M---KQSLCLA-VLFLILSTSSSAIRRGKEDQEINPLV----SATSVEEDSVNKLM-G-------ME--YC-GEG-DEEC-LRRRMMTESHLDYIYTQHHK--H 81
PSK4 M---GKFTTIF-IMALLL--CSTLTYA----ARLTPTT----TTALSRENSVKEIE-G---DK--VEEESCNGIG-EEEC-LIRRSLV-LHTDYIYTQNHK--P 79
PSK5 M---VKFTTFLCIIALLL--CSTLTHAS---ARLNP------TSVYPEENSFKKLEQG---E---V---ICEGVG-EEECFLIRRTLV-AHTDYIYTQNHN--P 77
     *   :   :   : *:*   *:         *                   *        :    *     :*:* * :: :    * ******    :
```

PLANT PEPTIDE CONTAINING SULFATED TYROSINE (PSY)

```
At5g58650 MTFVV--R-LLVCL-L-LTLT-ITSSLARNPVSVSGGFENSGFQRS-LLMVNVEDYGDPSANPKHDPGVOOSATG--QRVVG-RG 75
At3g47295 MSFGT--R-LL--LFLILTLPLVTS-SSPNTLHVS-GIVKTGTTSRFLMMT-IEDYDDPSANTRHDPSVPTNAKADTTP------ 71
At2g29995 MGYSSSSRIGL-CLFLFFTFALLSSARI-SLSF-SENEMTVVPERS-L-MVSTNDYSDPTANGRHDP--PRG--G---R--GRRR 71
          *:     * :* :*:* :*: : :*    :    :* :    :  :: * *:  :** **:** :*** :*  : :    :  : :
```

Fig. 2 Alignment of PSK and PSY Precursor Homologs in *Arabidopsis*. Predicted N-terminal signal peptides are *underlined*. Identical amino acids (aa) are indicated by an *asterix*, and similar aa by a *colon*. The domains encoding the mature peptides are in *bold italics*. Sulfated tyrosine residues are *underlined*. Hydroxyprolines are indicated by the one-letter abbreviation: O. The mature PSY1 peptide (At5g58650) is glycosylated with three L-Arabinose residues on a hydroxyproline (*underlined*)

3.1.1 Identification of Sulfated Peptide Ligands

PSK and PSY1were both identified in cell-suspension culture medium based on their growth-promoting activity (Matsubayashi and Sakagami 1996). The 18 amino acid (aa) long tyrosine-sulfated PSY1 has two hydroxyprolines in the C-terminus, and is also glycosylated with three L-Arabinose residues (Amano et al. 2007). Overexpression of PSY1 results in larger cotyledons and longer leaves, mainly due to increases in cell size. Both the sulfate group and the L-Arabinose are needed for the activity as synthetic PSY1 without sulfation and glycosylation showed no activity. Like PSY1, the 5 aa long PSK with two sulfated tyrosines, promotes cellular proliferation in nanomolar concentrations (Amano et al. 2007; Yang et al. 2001). Exogenously applied PSK has also been reported to have an effect in diverse species for example as a trigger of tracheary element (TE) differentiation, somatic embryogenesis, adventitious root formation, and pollen germination (Chen et al. 2000; Kobayashi et al. 1999; Matsubayashi et al. 1999; Yamakawa et al. 1998).

3.1.2 PSK and PSY1 Genes and Ligand Synthesis

PSK is produced from an ~80 aa precursor proteins by posttranslational sulfation of tyrosine residues and proteolytic processing, while PSY1 originates from a 75 aa protein. In the N-terminus the unprocessed proteins have secretory signal peptides, indicating that they are transported out of the cell where they are produced. In Arabidopsis, there are five paralogous genes encoding PSK precursors and two genes encoding proteins containing peptides very similar to PSY1 (Yang et al. 2001). The only conserved amino acids within PSK precursors are the five amino acid PSK domain and several conserved residues immediately upstream of the PSK domain (including dibasic amino acid residues). The *At*PSK genes are expressed in many tissues, including meristems, while PSY1 is expressed at particularly high levels in the marginal region of leaves, in the shoot apical meristem, and in the elongation zone of roots and is highly up regulated by wounding (Amano et al. 2007; Yang et al. 2001).

No mutant phenotypes have been published for the PSK and PSY1 genes, suggesting functional redundancy. As overexpression of PSK does not induce any phenotypic changes, it is possible that PSK can trigger cell proliferation only under specific conditions or that the overexpressed PSK is not sufficiently processed (Matsubayashi and Sakagami 2006). The natural role of these sulfated peptides is therefore still not clear.

3.1.3 PSK and PSY1 Receptors

A receptor for PSK was first identified from carrot plasma membrane fractions after photo affinity labeling using a photoactivatable ^{125}I-labeled PSK and subsequent purification from microsomal fractions, using ligand-based affinity chromatography (Matsubayashi et al. 2002). The PSK-binding protein turned out to be an LRR-RLK (PSKR1), which when overexpressed in transgenic carrot cells produced accelerated growth and a significant increase in PSK-binding sites on their cell membranes. PSKR1 is strongly expressed in cultured carrot cells, and also found in leaves, the apical meristem, hypocotyl, and root. PSKR1 antisense calli can survive and proliferate from smaller calli than the wild type. Disruption or overexpression of the *Arabidopsis* ortholog of PSKR1 (AtPSKR1) alters cellular longevity and callus-forming potential (Matsubayashi et al. 2005, 2006).

Like PSK, PSY1 induces cell division of asparagus mesophyll cells, and it was therefore hypothesized that PSK and PSY1 belonged to the same pathway. In addition to *At*PSKR1, there are two *At*PSKR1-like genes in Arabidopsis, *At*PSKR2 and At1g72300. Root growth is promoted by exogenous PSK and PSY1 in wt seedling, but significantly less with PSK in *atpskr1–2* and PSY1 in *At1g72300* mutant background, suggesting that the At1g72300 LRR-RLK is the receptor of PSY1. In the absence of PSKR1, PSK has been shown to interact with *At*PSKR2 (Amano et al. 2007). Interestingly, similar to the BRs/BRI1 ligand–receptor interaction, PSK binds specifically the 35 aa ID of *At*PSKR1 (Shinohara et al. 2007).

Triple *At*PSKR1, *At*PSKR2, and Atlg72300 mutants are semi-dwarf plants with decreased cell number and cell size and insufficiency in tissue repair after wounding, suggesting that these three similar RLKs work together to relay the signal from the two sulfated ligands. Little is known about the regulation of PSK and PSY1 peptides. One can, however, suggest that the processing of the preprotein might be important. A tyrosylprotein sulfotransferase (AtTSPT) was recently identified in *Arabidopsis*, which specifically interacts with the sulfation motif of the PSY1 precursor peptide, and can catalyze tyrosine sulfation of both PSY1 and PSK precursor polypeptide in vitro (Komori et al. 2009). *At*TPST is expressed throughout the plant body, and the highest level of expression is in the root apical meristem. A loss-of-function mutant of *At*TPST displayed a marked dwarf phenotype accompanied by stunted roots, pale green leaves, reduction in higher order veins, early senescence, and a reduced number of flowers and siliques, suggesting that sulfation of PSK and PSY1 or additional peptides is important for function.

3.2 Cysteine-Rich Peptide Ligands

More than 800 *Arabidopsis* genes encoding small putatively secreted cysteine-rich peptides (CRPs) of various classes, including defensin-like proteins and lipid transfer proteins, have been identified by bioinformatics analysis (Silverstein et al. 2007). These all have an N-terminal signal peptide (NSP) and a divergent charged or polar mature peptide, but in each family of CPRs there is a specific pattern of conserved cysteine residues. Structural determination of these CRPs indicated that these cysteines are important for intramolecular bond-formation and generation of a knotted, stable, and compact structure. Thus far, three different types of CRPs have been identified as ligands of receptor-like kinases in plants (Fig. 3).

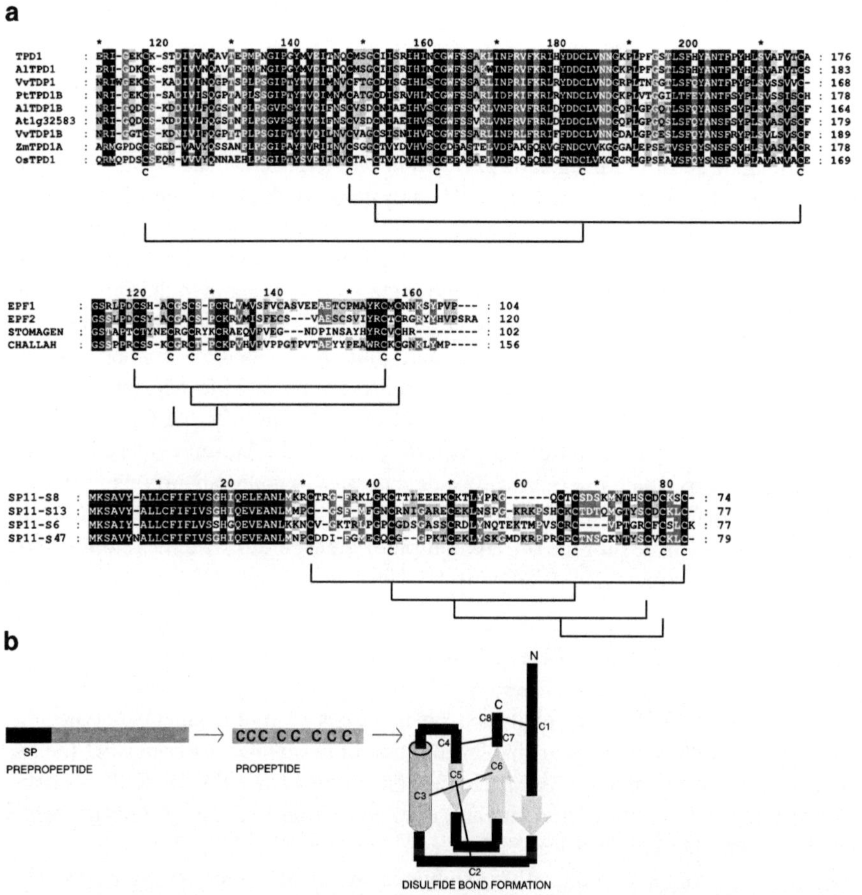

Fig. 3 Alignment and Disulfide Bond Formation of Cysteine-Rich Peptide Ligands. (**a**) The TAPETUM DETERMINANT 1 (TPD) of *Arabidopsis* and homologs found in otherplant species have six conserved cysteines (C) that can be predicted to form the disulfide bridges indicated under the alignments. EPIDERMAL PATTERNING FACTOR (EPF) and EPF-LIKE (EPFL) proteins have six to eight conserved Cs in the C-terminal part, which for STOMAGEN has been shown to form the indicated disulfide bridges important for ligand activity. The S-LOCUS CYSTEINE-RICH (SCR) proteins have eight conserved C residues that form disulfide bridges as indicated under the alignment. (**b**) The SCR peptide ligands are processed from a longer prepropeptide with a signal peptide(SP) and contain eight conserved cysteines (C). The correct disulfide bond formation (C1–C8,C2–C5, C3–C6, and C4–C7) is a prerequisite for receptor interaction and activity. *Arrows* indicate β strands and *cylinder* an α helice

3.2.1 TPD1: Necessary for Formation of Tapetum and Microspores

Identification of the TPD1 Ligand

The *tapetum determinant 1* (*tpd1*) mutant was identified in a genetic screen by its complete male sterility (Yang et al. 2003). The *TPD1* gene, which was tagged with a Ds element in the *tpd1* mutant, encodes a cysteine-rich, plant-specific protein of 176 aa. The mature *tpd1* anther contains no pollen grains. Achesporial cells in the anther, undergo several periclinal and anticlinal cell divisions to give rise to both sporogenous cells that develop further into microsporocytes, and to the surrounding endothecium, middle layer and tapetum (Sanders et al. 1999). In the *tpd1* anther, the endothecium is morphologically normal, but *tpd1* locules contain excess microsporocytes and lack the tapetum layer. Although meiotic nuclear divisions occur in these microsporocytes, they do not complete normal meiotic divisions, resulting in the degeneration of microsporocytes and the presence of cell debris. The middle layer, which in the wild-type degenerates, remains abnormal in the *tpd1* anther until the final stage of development when it becomes highly vacuolated. Thus, *TPD1* is required for the differentiation of the tapetum layer, development of functional microspores, and degeneration of the middle layer (Yang et al. 2003).

The TPD1 Protein and the TPD1 Receptor

The receptor for TPD1 was identified on the basis of its mutant phenotype: the phenotype of *tpd1* is indistinguishable from that of *excess microsporocytes1* (*ems1*, also known as *extra sporogenous cells*, *exs*) encoding a class X LRR-RLK. Genetic interactions between these mutants and overexpression lines suggested that these genes were in the same pathway. TPD1 was dependent on EMS1 for function, and the *ems1 tpd1* double-mutant anther exhibits an identical phenotype to the *ems1* and *tpd1* single mutants (Jia et al. 2008; Yang et al. 2005). Furthermore, ectopic expression of TPD1 using the cauliflower mosaic virus (*CaMV*) *35S* promoter results in abnormal anther development, reduced fertility, and changed silique morphology in a wild-type background, but *ems1* mutants are insensitive to the overexpression phenotypes. This is consistent with TPD1 being the ligand of EMS1 (Jia et al. 2008; Yang et al. 2005).

The direct interaction between TPD1 and EMS1 has been shown by several methods; in vitro by GST pull down, *in planta* by co-immunoprecipitation, and in yeast by two-hybrid interaction, which also allowed the identification of the TPD1-interacting region (TIR) (Jia et al. 2008). EMS1 has 29 LRRs and the TIR fragment contains four typical LRRs and one LRR with low similarity to the classic LRR.

Transport and Perception of TPD1

Homologs of the *TDP1* gene are found in a number of plant species, and one undescribed homolog is present in Arabidopsis. Alignment of the protein sequences reveals the presence of an NSP, a variable region and a region with six conserved cysteine residues (Fig. 3a). It is reasonable to assume that these cysteines are involved in disulfide bonds, creating a knotted protein ligand that interacts with the receptor.

TPD1 is expressed in microsporocytes only, while the *EMS1* gene is predominantly expressed in the tapetum (Jia et al. 2008; Yang et al. 2003). Thus, the current model suggests that TPD1 is secreted from microsporocytes or their precursors and then bind to EMS1 localized on the surface of tapetum precursor cells. The signals generated via EMS1 phosphorylation direct tapetum formation, while in the absence of this signaling, microsporocyte differentiation might be the default pathway. Alternatively, a normally developed tapetum layer might restrict further proliferation of microsporocytes.

3.2.2 S-locus Cysteine-Rich Protein (SCR) and Self-Incompatibility

Identification of the Self-Incompatibility System

Self-incompatibility (SI) systems are found in many flowering plants to prevent inbreeding. In *Brassica*, SI is controlled by a single multi-allelic locus named the sterility locus (S-locus). When pollen and pistil share the same allele, the SI response inhibits metabolic activation of the pollen grain and subsequent growth of the pollen tube, ensuring that pollen from closely related individuals is rejected (Takayama and Isogai 2005). The system is based on the interaction of S-locus cysteine-rich peptides (SCR), also called S-locus protein 11 (SP11), and the S-locus receptor kinase (SRK), which so far is the only non-LRR-RLK involved in development, for which biochemical interaction with a ligand has been shown (Table 1).

An S-locus glycoprotein (SLG), whose function is still not understood, and the receptor, the female part of the system, were identified first (Franklin-Tong 2002) by cloning a 76-kb genomic fragment in the *S9* haplotype of *B. rapa*. Years later, the small *SCR* gene was identified by screening cDNA libraries of immature anthers and was confirmed to encode the male determinant (Schopfer et al. 1999; Shiba et al. 2001; Takayama et al. 2000). Recombinant or chemically synthesized SCR/SP11 peptide applied to the stigma at concentrations as low as 50 fmol per stigma inhibit the hydration of compatible pollen. Strong S-haplotype-specific ligand–receptor interactions have been confirmed between SRKs and ^{125}I-labeled or Flag-tagged SCR/SP11 (Takayama et al. 2001).

The different SRC/SP11 alleles encode highly polymorphic secretory peptides of 74–81 aa; they contain, however, eight conserved cysteine residues that have been shown to form disulfide bridges (Fig. 3b) (Mishima et al. 2003). Correct disulfide bond formation (C1–C8, C2–C5, C3–C6, and C4–C7) is a prerequisite for SCR/SP11 activity. The haplotype variation is found in the C3–C4 hypervariable domain, which together with the C5–C6 region contributes to this ligand–receptor interaction.

Transport and Perception of SCR

The SRK receptor is expressed on the stigma surface, while the SCR/SP11 protein is specifically expressed in the tapetum layer and pollen grain of the anther and accumulates in the coat of mature pollen grains. Consistent with this, the SCR/SP11 genes encode NSPs. *A. thaliana* is self-compatible, although several closely related Arabidopsis species, such as *A. lyrata* and *A. halleri*, are not. Recently, an inversion was identified in the S-locus of *A. thaliana* ecotype Col-0 that encodes an SCR with only three cysteine residues. This haplotype or its derivatives are found in 95% of roughly 300 accessions, strongly suggesting a positive selection for this haplotype. Interestingly, restoration of the gene by a reinversion could restore self-incompatibility (Tsuchimatsu et al. 2010).

The Arabidopsis genome also includes the large *SCR-LIKE (SCRL)* gene family with diverse expression patterns, encoding 4.4–9.5 kD basic and hydrophilic peptides that have NSPs and the eight conserved cysteine residues (Vanoosthuyse et al. 2001). These proteins may represent potential cysteine-rich ligands of unknown receptors.

3.2.3 EPF and EPF-LIKE Ligands Controlling Stomatal Development

Identification of EPF and EPF-LIKE Ligands

The major sites of photosynthesis are the leaf mesophyll cells, which are dependent on CO_2 uptake from the environment and water loss by transpiration. Gas exchange takes place through the pores of the stomata found in the leaf epidermis. The two guard cells of the stomata can regulate pore size, but the density of stomata also influences gas exchange rates (Bergmann and Sack 2007). Besides environmental factors, stomatal density is controlled by genes involved in the generation of meristemoid cells arising from asymmetric cell divisions of epidermal cells. The meristemoid cell becomes the guard mother cell, which divide symmetrically to produce the two guard cells (Bergmann and Sack 2007).

Four peptides involved in controlling the number and positioning of the stomata have been identified through mutant screens. They are all members of a small gene family encoding EPIDERMAL PATTERNING FACTOR (EPF) and EPF-LIKE (EPFL) proteins of less than 120 aa (Rowe and Bergmann 2010). These proteins have an NSP and a 40–50 aa conserved C-terminal part with six to eight cysteine residues that can be suggested to form a knotted structure.

Mutation in *EPF2* results in an increased number of stomata which, nevertheless, are evenly spaced (Hara et al. 2009; Hunt and Gray 2009). Hence, EPF2 may be a negative regulator of the asymmetric cell divisions, and is expressed in cells that can divide asymmetrically (meristemoid mother cells, MMCs) and cells that have divided (meristemoids). However, EPF2 does not act directly on asymmetric cell division, but rather inhibits cells from acquiring the MMC fate (Hara et al. 2009).

This suggests that EPF2 concentrations above a certain threshold inhibit protodermal cells from becoming MMCs.

The positioning of the asymmetric cell divisions are however, controlled by EPF1, which is expressed in the meristemoids and their daughter cells, the guard cells (Geisler et al. 2000). *efp1* mutants show clustering of stomata (Hara et al. 2009; Hunt and Gray 2009). CHALLAH (CHAL¸ EPFL6) is also an inhibitor of stomatal density, as the *chal1* mutant increases the number of stomata 20-fold in the hypocotyls (Abrash and Bergmann 2010). This mutation has, however, no effect in leaves. STOMAGEN (EPLF9) is in contrast to EPF1, EPF2, and CHAL, a positive regulator of stomatal development, and its expression level correlates with stomatal density (Hunt et al. 2010; Kondo et al. 2010; Sugano et al. 2010).

Putative EPF/EPFL Receptors

Overexpression of EPF1 and EPF2 reduces stomatal density and this phenotype is dependent on the receptor protein TOO MANY MOUTHS (TMM) and the highly similar LRR-RLKs ERECTA, ERECTA-LIKE1 (ERL1) and ERL2, which belong to class XIII of the RLKs and have 22 LRRs (Hara et al. 2009; Hunt and Gray 2009). This dependency suggests that these cysteine-rich peptides are the ligands of these receptors. Both TMM and the ERECTA-family (ERf) receptors restrict stomata formation in leaves; the *tmm* mutant forms clusters of stomata in leaves (Bhave et al. 2009), while *er erl1 erl2* triple mutants form super-numerous clustered stomata (Shpak et al. 2005).

Surprisingly, the *tmm* mutant entirely lacks stomata in stems. STOMAGEN requires TMM for function, as STOMAGEN knockdown or overexpression has no effect on stomatal density in *tmm* mutants (Kondo et al. 2010; Sugano et al. 2010). *CHAL* was identified as a gene affecting stomata development in a genetic screen for mutations rescuing the stomataless phenotype of the hypocotyl and inflorescence stem in *tmm* mutants (Abrash and Bergmann 2010). When CHAL is overexpressed using the 35S promoter, stomatal numbers are reduced in leaves. This is even more evident in *tmm* background. In contrast, if, in addition the function of any two ERf members is lost, the overexpression severity is reduced, such that these triple mutants with CHAL overexpression may form stomata. These data indicate that any of the three ERECTA family members may serve as a potential receptor for CHAL and that this receptor–ligand interaction is repressed by TMM. Interestingly, *tmm chal er* triple mutants fail to produce stomata in hypocotyls, suggesting that ER may serve as a buffer to limit the activation of ERL1 and ERL2 (Abrash and Bergmann 2010).

Transport and Perception of EPF/EPFL Ligands

All the ERf receptors are expressed uniformly in protodermal cells in young leaves, and expression of ERL1 and ERL2 persist in the stomatal lineage (Shpak et al. 2005).

EPF2 is expressed in the stomatal lineage already in the meristemoid mother cell, while *EPF1* is expressed in both the meristemoid cell and in the guard cells. In contrast, *CHAL* has a tissue-specific expression pattern in cells surrounding the vascular elements of the stem, consistent with its specific role in the stem (Abrash and Bergmann 2010). *STOMAGEN* is expressed in the subepidermal tissues in the mesophyll cells (Hunt et al. 2010; Kondo et al. 2010; Sugano et al. 2010). Using a STOMAGEN fusion protein, it has been shown that STOMAGEN is secreted from the mesophyll cells to the apoplastic space (Sugano et al. 2010), as suggested from the NSP found in all EPFand EPF-LIKE proteins.

The current model suggests that EPF2 can activate ERf signaling both in the absence and in the presence of TMM, and is most important in restricting acquisition of MMC identity in protodermal cells. EPF1 signals through ERf receptors in a TMM-dependent manner to restrict meristemoid identity to the smaller product of the asymmetric division; this also ensures an even distribution of stomata. STOMAGEN from the mesophyll may compete with EPFs from the epidermis for the TMM protein to modulate the number of stomata. In stems and hypocotyls, TMM can titrate CHAL to regulate the amount of CHAL, which when overexpressed would activate ERf-signaling complexes, and inhibit development of the stomatal linage.

3.3 Proline-Rich Peptide Ligands of LRR-RLKs Class XI

Peptides involved in developmental processes that function as ligands of class XI LRR-RLKs have some common characteristics: they are encoded by small intronless genes encoding proteins of less than 100 aa, with an NSP for export out of the cell and a conserved 12–20 aa C-terminal motif with a proline-rich core. Two such gene families have been identified, the CLAVATA3/ ENDOSPERM SURROUNDING REGION (CLE) with more than 40 members in Arabidopsis, which share a 12–13 aa active CLE motif, and the IDA/IDA-LIKE family with six members and a 20 aa conserved EPIP motif (Fig. 4).

3.3.1 CLAVATA3: Control of Apical Meristem Size

Identification of the CLV3 Ligand

CLV3 is the best studied plant peptide ligand so far. It is crucial for regulation of the number of stem cells in the shoot apical meristem (SAM) (Fletcher et al. 2000; Schoof et al. 2000). Stem cells are found in a small niche in the meristem, the central zone (CZ), where they divide to replenish themselves and also divide into the peripheral zone (PZ) that will undergo several additional rounds of cell divisions as transit-amplifying cells before they differentiate and generate organs (Stahl and Simon 2010). Mutation in *CLV3* leads to an enlarged apical meristem

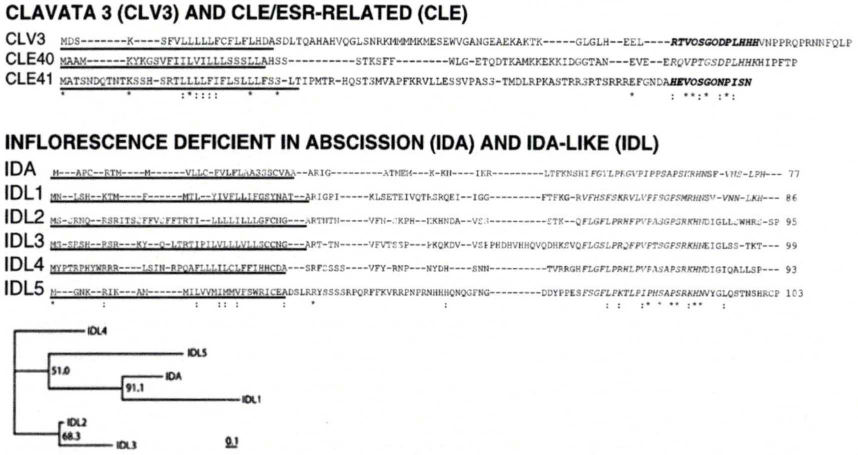

Fig. 4 Alignment of CLV/CLE and IDA/IDL Homologs in *Arabidopsis* Predicted N-terminal signal peptides are *underlined*. Identical amino acids (aa) are indicated by an *asterix*, and similar aa by a *colon*. The domain encoding the mature CLV3/CLEpeptides are in ***bold italics***. Hydroxyprolines are indicated by the one-letter abbreviation: O.The mature CLV3 peptide is glycosylated with three Arabinose residues on a hydroxyproline (*underlined*). The domains predicted to encode the mature IDA/IDL peptides are in *italics* as is the predicted mature peptide of CLE40

already in the mature embryo and throughout the life span of the plant (Clark et al. 1995). Thus, *clv3* mutants develop fasciated stems, produce more cells that will give rise to flower buds, and have flowers with an increased number of floral organs. Strong, ectopic expression of CLV3 using the 35S promoter results, on the contrary, in termination of the apical meristem, a phenotype also found in the *wuschel* (*wus*) mutant (Brand et al. 2000). The expression of the homeodomain encoding transcription factor *WUS* gene in the organizing center (OC) of the meristem is needed for maintenance of a population of dividing, undifferentiated stem cells. *CLV3* works in a negative feedback loop with *WUS* to maintain the correct balance of meristematic cells (Brand et al. 2000; Fletcher et al. 2000; Schoof et al. 2000). Enhanced expression of *CLV3* by its endogenous promoter showed however, that the shoot apical meristem can buffer alterations in CLV3 levels (Muller et al. 2006).

Identification of CLV3 Receptors

Mutation in the LRR-RLK *CLAVATA1* gene results in a phenotype similar to *clv3*, as is the case for *CLAVATA2*, which encodes an LRR receptor protein without a kinase domain (Kayes and Clark 1998). Based on these similarities, it was suggested that CLV3 functions as a ligand of a CLV1/CLV2 heterodimer (Trotochaud et al. 1999). However, only in 2008 was a direct physical interaction between CLV3 and CLV1 first demonstrated. CLV1 with the kinase domain replaced with a HaloTag was overexpressed in tobacco BY-2 cells and was found to bind a 12 aa radioactively labeled CLV3 peptide (Ogawa et al. 2008).

The fact that *clv1* null alleles confer weaker phenotypes than *clv3* suggested that CLV3 is able to signal through other RLKs. Suppressor mutation screens have identified CORYNE (CRN), also known as SUPPRESSOR OF OVEREXPRESSION OF LIGAND–LIKE PROTEIN1 2 (SOL2) as a mediator of CLV3 signaling (Miwa et al. 2008; Muller et al. 2008). CRN/SOL2 is unrelated to CLV1, and lacks an extracellular domain, but is capable of interacting with CLV2 (Bleckmann et al. 2010; Zhu et al. 2010). Receptor complexes containing CLV1, CLV2, and CRN have also been shown to form. In addition, it has been demonstrated that the CLV1 homologs BARELY ANY MERISTEM (BAM) 1 and BAM2 can bind to the CLV3-derived CLE peptide in vitro (Guo et al. 2010).

Interestingly, even *clv1 crn* double mutants have only slightly smaller inflorescence meristems and fewer carpels than single *clv3* mutants, suggesting the presence of an yet unknown additional receptor complex that allows CLV3 signaling (Muller et al. 2008). A screen for *clv3 peptide insensitive* (*cli*) mutants has recently identified yet another LRR-RLK, RECEPTOR-LIKE PROTEIN KINASE 2 (RPK2), also known as TOADSTOOL 2 (TOAD2), as a receptor candidate for relaying the CLV3 signal (Kinoshita et al. 2010). RPK2 is an LRR-RLK with 22 LRRs, forms homo-oligomers but does not associate with CLV1 or CLV2. These genetic and biochemical findings suggest that three major receptor complexes, RPK2 homomers, CLV1 homomers, and CLV2-CRN/SOL2 heteromers, are likely to mediate three signaling pathways, mainly in parallel but potentially with crosstalk to regulate homeostasis of the SAM.

Transport and Perception of CLV3

By expressing a CLV3-GFP fusion protein under the control of an endogenous *CLV3* promoter, it was demonstrated that the protein has a broader cellular distribution than its mRNA and spreads outward and downward close to the OC (Lenhard and Laux 2003). The fusion construct could almost fully complement the mutant phenotype, and export of CLV3 was shown to be a prerequisite for function (Lenhard and Laux 2003; Sharma et al. 2003). In fact, *CLV3* on the one hand and its receptors are not expressed in the same regions of the shoot apex, suggesting that CLV3 functions noncell-autonomously. *CLV3* is expressed in the CZ of the meristem, while *CLV1*, on the other hand, is expressed in a small group of cells just beneath the *CLV3* expression domain, the BAM receptors are expressed in the PZ of the apical meristem (Guo et al. 2010; Trotochaud et al. 1999). It has therefore been suggested that the presence of CLV1 and/or the BAM1 receptors in the PZ of the meristem act to sequester the CLV3 ligand to maintain a sustainable stem cell population.

3.3.2 CLE40: Control of Root Stem Cell Differentiation

Identification of the CLE40 Ligand

CLE40 is closely related to CLV3, it can substitute for CLV3 function in the shoot meristem. Overexpression of CLV3 and CLE40 results not only in the termination of the SAM, but also in the consumption of the root meristematic cells (Fiers et al. 2005; Hobe et al. 2003). *CLE40* is expressed in the differentiated columella cells, and is the only *CLE* gene thus far with a knock-out mutation resulting in a root phenotype (Hobe et al. 2003; Stahl et al. 2009). A very subtle root waving phenotype was first identified (Hobe et al. 2003), but more recently it was found that *cle40* mutant roots have multiple layers of columella stem cells together with an enlarged expression region of *WOX5*, encoding a WUS-like homeobox transcription factor (Stahl et al. 2009). Increased levels of CLE40 terminates stem cell production by promoting stem cell differentiation.

Identification of a CLE40 Receptor, Transport and Perception

Inhibition of root growth by exogenous addition of a CLE40 peptide is not found in the *crn* mutant suggesting that CRN, together with CLV2, can act as a receptor complex for CLE40. However, despite expression in the root apical meristem, *crn/sol2* and *clv2* mutants do not have any root phenotype, so this complex may not be the natural receptor of CLE40 (Miwa et al. 2008; Muller et al. 2008). Genetic evidence instead suggests that CLE40 is the ligand of the non-LRR RLK *ARABIDOPSIS* CRINKLY4 (ACR4) (Stahl et al. 2009), as this gene when mutated also displays an overproduction of columella stem cells (De Smet et al. 2008). In vitro, application of a synthetic CLE40 peptide rescues *cle40* mutants, but not *acr4* mutants. Interestingly, the maintenance of a balance between proliferation and differentiation of distal root stem cells is achieved by expression of *CLE40* in differentiated root cells, and not like *CLV3* in undifferentiated SAM cells (Stahl et al. 2009).

3.3.3 TDIF (CLE41 and CLE 44): Control of Tracheary Element Differentiation

Identification of the TDIF Ligand

The TRACHEARY ELEMENT DIFFERENTIATION INHIBITORY FACTOR (TDIF) has been identified as an inhibitor of xylem differentiation of mesophyll cells in *Zinnia elegans* (Asteraceae). It is a 12 aa peptide identical to the CLE motif in CLE41 and CLE44 (HEVP^{h}SGP^{h}NPISN, where P^{h} are hydroxyprolines) and very similar to CLE42 (Ito et al. 2006). The peptides containing the CLE motif of these three proteins are the only CLE-peptides with the ability to inhibit tracheary element differentiation, whereas CLV3 peptides promote differentiation. The difference in function between TDIF and CLV3 has been assigned to the characteristic terminal regions of the peptides, where the R/H at the N-terminus and the HH/SN on the C-terminus are crucial (Kondo et al. 2008). TDIF peptide was isolated from liquid medium in which submerged cultures of CLE44-overexpressing transgenic plants were grown, confirming that *CLE44* and probably *CLE41* encode the precursor of the TDIF peptide (Ohyama et al. 2008). Addition of synthetic TDIF to submerged cultures of Arabidopsis seedlings, or overexpression of *CLE41* and *CLE44*, specifically causes discontinuity of xylem vessel strands in rosette leaves, while the phloem and procambium strands are intact (Hirakawa et al. 2008). TDIF specifically inhibits differentiation from procambial cells into tracheary elements (TEs) and has, in addition, been shown to promote procambium proliferation in the vascular system.

Identification of the TDIF/CLE41 Receptor

The ability of TDIF to inhibit xylem vessel formation from stem-cell-like procambial cells was used to screen a population of T-DNA mutant LRR-RLK genes expressed in the procambium, for insensitivity to TDIF. This led to the identification of the putative TDIF Receptor (TDR) (Hirakawa et al. 2008). TDR encodes an LRR-RLK belonging to the subclass XI, and is closely related to CLV1. Direct interaction between TDIF and TDR has been demonstrated by co-immunoprecipitation of the ligand and the receptor expressed in tobacco BY-2 cells and by photoaffinity assays (Hirakawa et al. 2008). Interestingly, TDR is identical to PHLOEM INTERCALATED WITH XYLEM (PXY), known to be involved in the maintenance of cell polarity during vascular development (Fisher and Turner 2007). Furthermore, TDR–PXY mutants display genetic interactions with mutants in two closely related RLKs, PXL1, and PXL3 (Fisher and Turner 2007).

Genetic evidence compatible with the CLE peptides of CLE41 and the almost identical CLE 42 being ligands of TDR/PXY has also been provided. In the Arabidopsis inflorescence stem, vascular bundles, consisting of phloem and xylem separated by procambium, are generated by a coordinated process of orientated cell divisions and differentiation. Mutations in *PXY* affect this organization and result in irregular cell divisions (Fisher and Turner 2007). Overexpression of the respective genes using the *35S* promoter generate vascular bundles in the inflorescence stem with significantly more cells than wild-type bundles, and irregularly oriented cell divisions (Etchells and Turner 2010). This phenotype was PXY dependent as *pxy 35S::CLE41* and *pxy 35S::CLE42* plants displayed a *pxy* single mutant phenotype (Etchells and Turner 2010).

Transport and Perception of the CLE41 Ligand

PXY and *CLE41* are expressed in adjacent, nonoverlapping domains; CLE41 is produced in and secreted from phloem cells and perceived by PXY/TDR located in the plasma membrane of mainly procambial cells. This signal enhances division of procambial cells and suppresses the differentiation of procambial cells into xylem cells. Overexpression and ectopic expression of CLE41 has been used to understand how CLE41 is involved in establishing polarized cell divisions in the procabium (Hirakawa et al. 2010; Sieburth 2007).

3.3.4 IDA: Control of Floral Organ Abscission

Identification of the IDA Ligand

Mutation in the gene *INFORESCENCE DEFICIENT IN ABSCISSION* (*IDA*) results in an indefinite retention of floral organs, which are normally abscised in *Arabidopsis* shortly after pollination has taken place (Butenko et al. 2003). Abscission is a physiologically controlled process of cell separation taking place at sites with abscission zones (AZ), which are files of small densely cytoplasmic cells at the base of the organs to be shed. Separation of sepals, petals, and stamens results from degradation of the middle lamella connecting the cell walls of adjacent cell layers in the AZs at the base of these organs. Overexpression of *IDA*, which encodes a small peptide of 77 aa, results in the shedding of other organs, such as pedicels, cauline leaves and branches, normally not abscised in Arabidopsis (Stenvik et al. 2006). This suggests that IDA is an inducer of cell separation, rather than an inhibitor of a repair process that hypothetically could be triggered by cell wall loosening. Interestingly, vestigial AZs are found at branch points and at the base of pedicels and cauline leaves. Thus, ectopic expression of *IDA* is sufficient for cell separation to take place at these sites, indicating that the other components needed to promote cell separation are present in the vestigial AZs.

Identification of IDA Receptors

Double mutants of the closely related class XI LRR-RLKs HAESA (HAE) and HAESA-LIKE2 (HSL2) display a phenotype very similar to *ida*, a total deficiency in floral organ abscission despite the presence of morphologically distinct AZ cells (Cho et al. 2008; Stenvik et al. 2008). The two RLKs seem to have redundant functions as the single mutants display no abscission phenotype. In *hae hsl2* plants overexpressing IDA, none of the overexpression phenotypes were seen, which is consistent with *IDA* being the ligand of *HAE* and *HSL2*.

Transport and Perception of IDA

The function of IDA is dependent on export out of the cells where it is produced, as overexpression constructs without the signal peptide did not result in any overexpression phenotypes (Stenvik et al. 2008). The expression pattern of *HAE* and *HSL2* are virtually identical in the floral organ AZ and are restricted to the AZ cells of the petals, sepals and filaments (Cho et al. 2008). *IDA* has a broader expression sector in the AZ region, suggesting that IDA functions noncell-autonomously, and moves to the AZ cells where the receptors are found. Interestingly, *HAE* and *HSL2* are, however, also expressed in the vestigial AZs, which provide an explanation for the ability *IDA* has to induce cell separation at these sites. In addition to expression in the AZ of the floral organs, *IDA* and *HAE* were found expressed the silique dehiscence zone and endodermis, cortex, and epidermis cells, which separate during lateral root emergence, facilitated by expression of cell wall remodeling enzymes (Butenko et al. 2006; Peret et al. 2009). Thus, it is conceivable that *IDA* signaling is involved in not only floral organ abscission, but also other cell separation processes in Arabidopsis.

3.4 Functional Redundancy of Proline-Rich Peptides

3.4.1 The CLE and EPIP Peptides

Bioinformatic analysis has identified more than 30 *CLE* genes, and five *IDA-LIKE (IDL)* genes in Arabidopsis, and both classes of peptides are also found in other plant species. The consensus sequence of the CLE domain, SKRLVPSGPNPLHN, where the most highly conserved residues are underlined, has some resemblance to the PIP motif conserved in the IDA/IDL peptides (P[I,L]PPS[A,G]PS[R,K][R,K]HN), most notably the PSGP core (Stenvik et al. 2008).

The signaling activities of these two peptide groups are conferred by the CLE motif and an extended PIP motif (EPIP), respectively. Synthetic 14-amino-acid polypeptides corresponding to the CLE motif of CLV3, CLE19, and CLE40, applied to Arabidopsis roots mimics for instance, the overexpression phenotype of these *CLE* genes, and a synthetic 20 aa EPIP of IDA or IDL1 can induce floral organ abscission (Fiers et al. 2005; Stenvik et al. 2008). Such data have highlighted the question of whether these proteins are processed to release an active peptide. In vitro both CLV3 and IDA can be processed by an activity from cauliflower meristems (Ni and Clark 2006; Stenvik et al. 2008). More importantly, overexpression of CLV3 in calli or secreted by submerged seedlings into the liquid growth medium, has led to the identification of secreted peptides corresponding to 12-mers or 13-mers of the CLE motif, RTVP^{h}SGP^{h}DPLHH or RTVP^{h}SGP^{h}DPLHHH (Ito et al. 2006; Kondo et al. 2006; Ohyama et al. 2009). The hydroxyproline in position 7 was found to be arabinosylated, a modification that increased the biological activity and binding affinity of the peptide to the receptor many fold (Ohyama et al. 2009). Based on the similarity between CLE and EPIP it can be suggested that EPIP also contains hydroxyprolines important for function.

Both, for CLV3 and for IDA, deletion analysis has shown that only the NSP and the CLE/EPIP motif are necessary for function (Fiers et al. 2006; Stenvik et al. 2008). Domain swap experiments, where the conserved domain has been replaced between family members and expressed under the regulatory elements of *CLV3* and *IDA*, respectively, showed that some CLE and EPIP sequences could replace the activity of CLV3 or IDA, respectively (Ni and Clark 2006; Stenvik et al. 2008). This indicates that there exists some functional redundancy within each of these families, possibly mediated by the ability of different peptides of the same family to bind the same receptor(s) (Butenko et al. 2009; Guo et al. 2010). The common overexpression phenotype for IDA and the IDL peptides provides further evidence for genetic redundancy between these genes (Stenvik et al. 2006, 2008). Likewise, overexpression of *CLE* genes have been used to study functional similarity. Originally, CLE peptides were classified in A-type and B-type, where the A-type were involved in the suppression of stem cell development in the root and shoot; and the B-type (CLE41 and CLE44) in the suppression of the differentiation of xylem cells from stem-cell-like procambial cells and promotion of procambial cell proliferation (Ito et al. 2006; Whitford et al. 2008). More recently, sequence similarity and overexpression phenotypes have been used to classify the *CLE* genes into 13 groups (Oelkers et al. 2008). Overexpression typically results in (a) premature mortality and/or developmental timing delays in transgenic *Arabidopsis* plants, (b) arrest of growth from the shoot apical meristem (SAM), (c) inhibition or stimulation of root growth, or (d) dwarf plants.

Although overexpression and swap-experiments can indicate functional redundancy, biologically relevant redundancy requires an overlapping or adjacent expression pattern, and a sufficiently high expression level. Hence, the IDL peptides of *Arabidopsis* are not expected to be involved in floral organ abscission, but rather in cell separation processes at the sites where they are expressed, like root cap sloughing, seed abscission and dehiscence (Butenko et al. 2009; Stenvik et al. 2008). A recent comprehensive analysis of the expression patterns of the *CLE* genes (Jun et al. 2010) will be valuable in guiding crosses to generate double or triple mutants to identifying natural roles of more *CLE* genes. The expression patterns of the *CLE* and *IDL* genes should also be compared to those of potential receptors. Redundancy on the peptide side might have a counterpart on the receptors side, suggesting, for example, that IDL peptides may interact with HAESA-LIKE receptors (Butenko et al. 2009). In the case of the CLE peptides, there are many candidate receptors in the LRR-RLKs class XI of which CLV1, BAM1, BAM2, and PXY/TDR all have already been shown to bind a number of CLE peptides, although with slight different affinities (Guo et al. 2010).

4 Export, Concentration Gradients and Movement of Ligands

In contrast to traditional hormones, ligands of membrane-bound receptor kinases do not move over long distances. However, peptide ligands affecting growth and development normally arise from preproteins with N-terminal signal sequences, and are exported out of the cells where they are produced. It is therefore relevant to ask whether exported ligands signal to their own cell, only to the next-neighbor cells or can move over some distances in the apoplastic space and exert an effect on more distant cells, given that these cells express the matching receptor. For the BRs, it is assumed that the steroid ligand can be produced and perceived by the same cell, as the amount of ligand is controlled by a feedback mechanism (Symons et al. 2008) (Fig. 5a). Peptide signaling between adjacent cells is assumed to be the case for TPD1-EMS1 during tapetum and microspore differentiation (Fig. 5b). In the case of floral organ abscission, the expression pattern of *IDA* is broader than that of *HAE* and *HSL2*, which are found specifically in the AZ cells (Fig. 5c) (Cho et al. 2008; Stenvik et al. 2008). The outcome of the IDA/HAE-HSL2 signaling is specific, with cell wall degradation between adjacent cell layers in the floral organ AZ, but so far it is not clear whether ligand and receptors are expressed in the very same cells. Continuous exposure of the AZ region to IDA results, however, in cell enlargement also of cells surrounding the AZs, possibly due to excessive secretion of cell wall remodeling enzymes (Stenvik et al. 2006).

The relationships are more complex with the peptides and receptors involved in stomata development, involving communication between tissues and competition for receptors facilitated by partial functional redundancies. CHAL is expressed only in cells surrounding the vascular bundle in stems and appears to diffuse to the epidermis where TMM and the ERf receptors are expressed, and where it restricts stomatal differentiation (Fig. 5d). In leaves, the EPF1 and EPF2 peptides, TMM and the ERf receptors, are all expressed in the stomata lineage cells of the epidermis. There are, however, differences in the spatial and temporal expression patterns both for ligands and for receptors (Hara et al. 2009). This, and differences between the affinity of the ligands for the receptors, may be important for the distinct roles of these peptides. The negative regulators EPF1 and EPF2 may furthermore compete for TMM with the STOMAGEN peptide, which is promoting stomata formation and exported from the mesophyll cells to the apoplastic space (Rychel et al. 2010). Finally, it has been suggested that secretion of EPF1 from the meristemoidal cells will create a concentration gradient of ligands, which allows asymmetric division of the MMC only at a distance from the meristemoid (Rowe and Bergmann 2010) (Fig. 5d).

That polarized cell division can arise as a result of a concentration gradient of ligand in the apoplastic space. This has been suggested from experiments with CLE41; when this peptide is expressed not only in the phloem, but also in the xylem, that is on both sides of the procambium cells expressing the PXY receptor, the cell division plane in the procambium, which is generating new xylem cells, is altered (Fig. 5e) (Etchells and Turner 2010).

Concentration gradients may also be important when ligand and receptor are produced more distantly apart, as in the cases of CLV3 and CLV40 and their receptors in the shoot and root meristem, respectively (Fig. 5f) (Stahl and Simon 2010). In these systems, designed perhaps for a continuous maintenance of a balance between stem cell population and differentiated cells, the ligand is the limiting signal. Such systems call for tight control of ligand production, for instance through a feedback mechanism between the ligand-producing and the receptor-producing cell. Another possibility is the one suggested for CLV3 and CLE

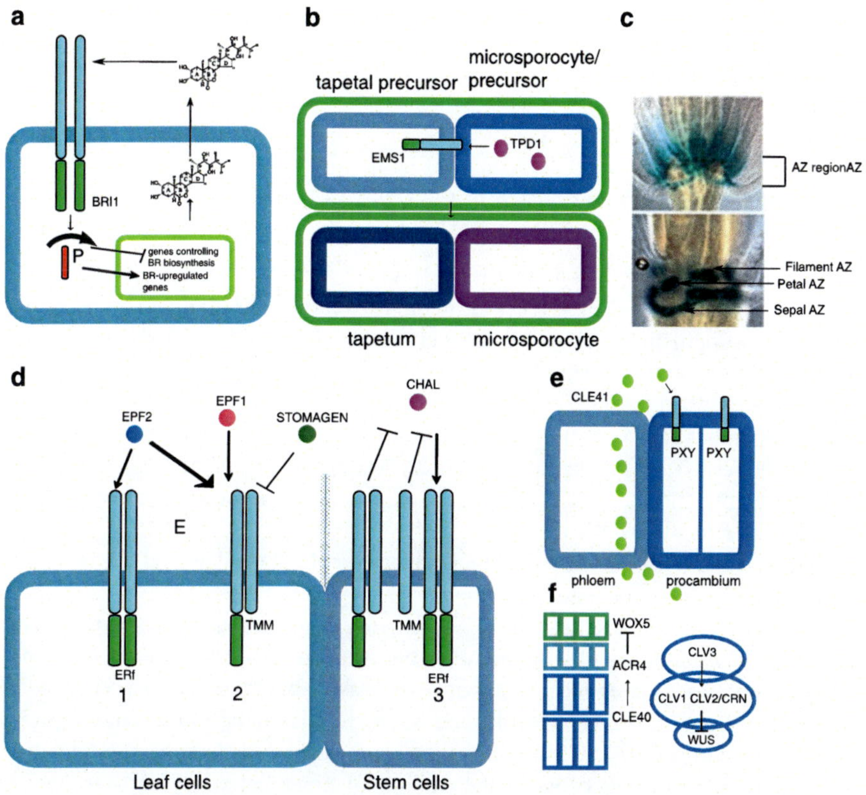

Fig. 5 Transport and Perception of Peptide Ligands (**a**) Self-signaling. The steroid ligand brassinosteroid (BR) is produced and perceived by the (**b**) Neighbor-cell signaling. TDP1 is secreted from microsporocytes or their precursors and binds to EMS1 RLK localized to tapetal precursors to induce a signaling cascade that ensures the specification of tapetal cell fate.same cell. A negative feedback loop ensures the regulation of BR biosynthesis. (**c**) Partly overlapping expression patterns. *IDA* has a broad expression covering the entire abscission zone (AZ) region prior to floral organ shedding. The RLKs relaying the IDA signal during floral abscission have a specific expression pattern localized to the filament, petal and sepal AZ cells. (**d**) Cascades and competition. In the leaf, EFP2 is secreted by stomatal lineage cells, which block the protoderm to meristemoid mother cell (MMC) transition by ERf-signaling both in the presence and absence of TMM (1). EFP1 is produced in the meristemoid generated after asymmetric division of MMC cells, and restricts the ability to generate guards cells to meristemoids in a TMM dependent manner (2). STOMAGEN is produced in the mesophyll and diffuses to the epidermis to promote stomatal development. It may inhibit TMM mediatedsignaling by binding TMM receptor complexes in competition with EPF peptides (3). In the stems, CHAL originates in cells surrounding the vascular bundle and may diffuse to the epidermis, where it restricts stomatal differentiation. TMM titrates CHAL thereby preventing CHAL from activating ERf-signalling complexes (3). (**e**) Concentration gradients. CLE41 is expressed in the phloem and signals to PXY in the procambium. A concentration gradient of the CLE41 peptide provides positional information to the dividing procambial cell and sets an appropriate division plane (indicated by *blue vertical line*). (**f**) Distance and feed-back loops. CLE40 in the columella cells signals through ACR4 in the overlaying columella stem cells to control the root stem cell population by repressing the activity of WOX5 in the quiescent center. Cells in the outer layer of the shoot meristem secrete CLV3 peptides, which bind to receptors in the L2 and L3 layers of the meristem and repress WUS in the subajacent organizing center, thereby adjusting the homeostasis of the stem cell population in the shoot

peptides in the shoot apex – the presence of several receptors in the regions around the production site of the ligand can titrate out superfluous ligand (DeYoung and Clark 2008).

5 Future Perspectives

Since the vast majority of putative ligands and receptors in plants have not been assigned a function yet, it is very difficult to estimate the number involved in distinct development processes. It is likely that we so far have discovered only the simplest examples of peptide ligand signaling, where mutation in single genes, such as *CLV3*, *IDA*, or *TDF1*, results in aberrant development. There are, however, increasing indications that the stability and fine-tuning of developmental processes may be governed by an intricate cross-talk between antagonistic peptide ligands with differential affinity to multiple related receptors, as seen, for instance, in stomata development and control of meristem size.

Over the last few years the assumed hegemonic role of the classical hormones as governors of plant growth and development has been challenged by the discovery of novel compounds, BRs and peptides, relaying communication between neighboring cells. As indicated by the involvement of the auxin-induced peptide ligand *IDA* in lateral root emergence, it is, however, reasonable that there is no *in planta* competition, but rather an interplay between traditional hormones and peptide hormones. Future understanding of development must therefore incorporate signaling networks and cross-talk between the traditional plant hormones and the newly developing field of ligand–receptor signaling.

References

Abrash EB, Bergmann DC (2010) Regional specification of stomatal production by the putative ligand CHALLAH. Development 137:447–455

Amano Y, Tsubouchi H, Shinohara H, Ogawa M, Matsubayashi Y (2007) Tyrosine-sulfated glycopeptide involved in cellular proliferation and expansion in Arabidopsis. Proc Natl Acad Sci USA 104:18333–18338

Azpiroz R, Wu Y, LoCascio JC, Feldmann KA (1998) An Arabidopsis brassinosteroid-dependent mutant is blocked in cell elongation. Plant Cell 10:219–230

Bajguz A (2007) Metabolism of brassinosteroids in plants. Plant Physiol Biochem 45:95–107

Bergmann DC, Sack FD (2007) Stomatal development. Ann Rev Plant Biol 58:163–181

Bhave N, Veley K, Nadeau J, Lucas J, Bhave S, Sack F (2009) *TOO MANY MOUTHS* promotes cell fate progression in stomatal development of Arabidopsis stems. Planta 229:357–367

Bleckmann A, Weidtkamp-Peters S, Seidel CAM, Simon R (2010) Stem cell signaling in Arabidopsis requires CRN to localize CLV2 to the plasma membrane. Plant Physiol 152:166–176

Brand U, Fletcher JC, Hobe M, Meyerowitz EM, Simon R (2000) Dependence of stem cell fate in Arabidopsis on a feedback loop regulated by CLV3 activity. Science 289:617–619

Butenko MA, Patterson SE, Grini PE, Stenvik GE, Amundsen SS, Mandal A, Aalen RB (2003) *INFLORESCENCE DEFICIENT IN ABSCISSION* controls floral organ abscission in Arabidopsis and identifies a novel family of putative ligands in plants. Plant Cell 15:2296–2307

Butenko MA, Stenvik G-E, Alm V, Sæther B, Patterson SE, Aalen RB (2006) Ethylene dependent and -independent pathways controlling floral abscission are revelaed to converge using promoter::reporter gene constructs in the *ida* abscission mutant. J Exp Bot 57:3627–3637

Butenko MA, Vie AK, Brembu T, Aalen RB, Bones AM (2009) Plant peptides in signalling: looking for new partners. Trends Plant Sci 14:255–263

Chen YF, Matsubayashi Y, Sakagami Y (2000) Peptide growth factor phytosulfokine-alpha contributes to the pollen population effect. Planta 211:752–755

Cho SK, Larue CT, Chevalier D, Wang H, Jinn TL, Zhang S, Walker JC (2008) Regulation of floral organ abscission in *Arabidopsis thaliana*. Proc Natl Acad Sci USA 105:15629–15634

Choe S, Dilkes BP, Fujioka S, Takatsuto S, Sakurai A, Feldmann KA (1998) The *DWF4* gene of Arabidopsis encodes a Cytochrome P450 that mediates multiple 22a-hydroxylation steps in brassinosteroid biosynthesis. Plant Cell 10:231–244

Chory J, Nagpal P, Peto CA (1991) Phenotypic and genetic analysis of *det2*, a new mutant that affects light-regulated seedling development in Arabidopsis. Plant Cell 3:445–459

Clark SE, Running MP, Meyerowitz EM (1995) *CLAVATA3* is a specific regulator of shoot and floral meristem development affecting the same processes as *CLAVATA1*. Development 121:2057–2067

Clouse SD, Langford M, McMorris TC (1996) A brassinosteroid-insensitive mutant in *Arabidopsis thaliana* exhibits multiple defects in growth and development. Plant Physiol 111:671–678

De Smet I, Vassileva V, De Rybel B, Levesque MP, Grunewald W, Van Damme D, Van Noorden G, Naudts M, Van Isterdael G, De Clercq R, Wang JY, Meuli N, Vanneste S, Friml J, Hilson P, Jurgens G, Ingram GC, Inze D, Benfey PN, Beeckman T (2008) Receptor-like kinase ACR4 restricts formative cell divisions in the Arabidopsis root. Science 322:594–597

De Smet I, Voss U, Jurgens G, Beeckman T (2009) Receptor-like kinases shape the plant. Nat Cell Biol 11:1166–1173

DeYoung BJ, Clark SE (2008) BAM receptors regulate stem cell specification and organ development through complex interactions with CLAVATA signaling. Genetics 180:895–904

Etchells JP, Turner SR (2010) The PXY-CLE41 receptor ligand pair defines a multifunctional pathway that controls the rate and orientation of vascular cell division. Development 137:767–774

Fiers M, Golemiec E, Xu J, van der Geest L, Heidstra R, Stiekema W, Liu C-M (2005) The 14-Amino Acid CLV3, CLE19, and CLE40 peptides trigger consumption of the root meristem in Arabidopsis through a *CLAVATA2*-dependent pathway. Plant Cell 17:2542–2553

Fiers M, Golemiec E, van der Schors R, van der Geest L, Li KW, Stiekema WJ, Liu C-M (2006) The CLAVATA3/ESR motif of CLAVATA3 is functionally independent from the nonconserved flanking sequences. Plant Physiol 141:1284–1292

Fisher K, Turner S (2007) PXY, a receptor-like kinase essential for maintaining polarity during plant vascular-tissue development. Curr Biol 17:1061–1066

Fletcher JC, Brand U, Hobe M, Meyerowitz EM, Simon R (2000) Stem cell fate in *Arabidopsis thaliana* depends on a feedback loop regulated by CLV3 activity. Mol Biol Cell 11:1497

Franklin-Tong N (2002) Receptor-ligand interaction demonstrated in *Brassica* self-incompatibility. Trends Genet 18:113–115

Fujioka S, Yokota T (2003) Biosynthesis and metabolism of brassinosteroids. Annu Rev Plant Biol 54:137–164

Fukuda H, Hirakawa Y, Sawa S (2007) Peptide signaling in vascular development. Curr Opin Plant Biol 10:477–482

Geisler M, Nadeau J, Sack FD (2000) Oriented asymmetric divisions that generate the stomatal spacing pattern in Arabidopsis are disrupted by the *too many mouths* mutation. Plant Cell 12:2075–2086

Grove MD, Spencer GF, Rohwedder WK, Mandava N, Worley JF, Warthen JD, Steffens GL, Flippenanderson JL, Cook JC (1979) Brassinolide, a plant growth-promoting steroid isolated from *Brassica napus* pollen. Nature 281:216–217

Guo Y, Han L, Hymes M, Denver R, Clark SE (2010) CLAVATA2 forms a distinct CLE-binding receptor complex regulating Arabidopsis stem cell specification. Plant J 63:889–900

Hara K, Yokoo T, Kajita R, Onishi T, Yahata S, Peterson KM, Torii KU, Kakimoto T (2009) Epidermal cell density is autoregulated via a secretory peptide, EPIDERMAL PATTERNING FACTOR 2 in Arabidopsis leaves. Plant Cell Physiol 50:1019–1031

Hirakawa Y, Shinohara H, Kondo Y, Inoue A, Nakanomyo I, Ogawa M, Sawa S, Ohashi-Ito K, Matsubayashi Y, Fukuda H (2008) Non-cell-autonomous control of vascular stem cell fate by a CLE peptide/receptor system. Proc Natl Acad Sci USA 105:15208–15213

Hirakawa Y, Kondo Y, Fukuda H (2010) Regulation of vascular development by CLE peptide-receptor systems. J Integr Plant Biol 52:8–16

Hobe M, Muller R, Grunewald M, Brand U, Simon R (2003) Loss of CLE40, a protein functionally equivalent to the stem cell restricting signal CLV3, enhances root waving in Arabidopsis. Dev Genes Evol 213:371–381

Hunt L, Gray JE (2009) The signaling peptide EPF2 controls asymmetric cell divisions during stomatal development. Curr Biol 19:864–869

Hunt L, Bailey KJ, Gray JE (2010) The signalling peptide EPFL9 is a positive regulator of stomatal development. New Phytol 186:609–614

Ito Y, Nakanomyo I, Motose H, Iwamoto K, Sawa S, Dohmae N, Fukuda H (2006) Dodeca-CLE peptides as suppressors of plant stem cell differentiation. Science 313:842–845

Jia G, Liu X, Owen HA, Zhao D (2008) Signaling of cell fate determination by the TPD1 small protein and EMS1 receptor kinase. Proc Natl Acad Sci USA 105:2220–2225

Jun J, Fiume E, Roeder A, Meng L, Sharma VK, Osmont KS, Baker C, Ha CM, Meyerowitz EM, Feldman LJ, Fletcher J (2010) Comprehensive analysis of CLE polypeptide signaling gene expression and over-expression activity in Arabidopsis. Plant Physiol 154:1721–1736

Kauschmann A, Jessop A, Koncz C, Szekeres M, Willmitzer L, Altmann T (1996) Genetic evidence for an essential role of brassinosteroids in plant development. Plant J 9:701–713

Kayes JM, Clark SE (1998) *CLAVATA2*, a regulator of meristem and organ development in Arabidopsis. Development 125:3843–3851

Kim TW, Wang ZY (2010) Brassinosteroid signal transduction from receptor kinases to transcription factors. Annu Rev Plant Biol 61:681–704

Kinoshita T, Cano-Delgado A, Seto H, Hiranuma S, Fujioka S, Yoshida S, Chory J (2005) Binding of brassinosteroids to the extracellular domain of plant receptor kinase BRI1. Nature 433:167–171

Kinoshita A, Betsuyaku S, Osakabe Y, Mizuno S, Nagawa S, Stahl Y, Simon R, Yamaguchi-Shinozaki K, Fukuda H, Sawa S (2010) RPK2 is an essential receptor-like kinase that transmits the CLV3 signal in Arabidopsis. Development 137:3911–3920

Kobayashi T, Eun CH, Hanai H, Matsubayashi Y, Sakagami Y, Kamada H (1999) Phytosulphokine-alpha, a peptidyl plant growth factor, stimulates somatic embryogenesis in carrot. J Exp Bot 50:1123–1128

Komori R, Amano Y, Ogawa-Ohnishi M, Matsubayashi Y (2009) Identification of tyrosylprotein sulfotransferase in Arabidopsis. Proc Natl Acad Sci USA 106:15067–15072

Kondo T, Sawa S, Kinoshita A, Mizuno S, Kakimoto T, Fukuda H, Sakagami Y (2006) A plant peptide encoded by CLV3 identified by in situ MALDI-TOF MS analysis. Science 313:845–848

Kondo T, Nakamura T, Yokomine K, Sakagami Y (2008) Dual assay for MCLV3 activity reveals structure-activity relationship of CLE peptides. Biochem Biophys Res Commun 377:312–316

Kondo T, Kajita R, Miyazaki A, Hokoyama M, Nakamura-Miura T, Mizuno S, Masuda Y, Irie K, Tanaka Y, Takada S, Kakimoto T, Sakagami Y (2010) Stomatal density is controlled by a mesophyll-derived signaling molecule. Plant Cell Physiol 51:1–8

Lease KA, Walker JC (2006) The Arabidopsis unannotated secreted peptide database, a resource for plant peptidomics. Plant Physiol 142:831–838
Lenhard M, Laux T (2003) Stem cell homeostasis in the Arabidopsis shoot meristem is regulated by intercellular movement of CLAVATA3 and its sequestration by CLAVATA1. Development 130:3163–3173
Li J, Nagpal P, Vitart V, McMorris TC, Chory J (1996) A role for brassinosteroids in light-dependent development of Arabidopsis. Science 272:398–401
Malinowski R, Higgins R, Luo Y, Piper L, Nazir A, Bajwa VS, Clouse SD, Thompson PR, Stratmann JW (2009) The tomato brassinosteroid receptor BRI1 increases binding of systemin to tobacco plasma membranes, but is not involved in systemin signaling. Plant Mol Biol 70:603–616
Matsubayashi Y (2003) Ligand-receptor pairs in plant peptide signaling. J Cell Sci 116:3863–3870
Matsubayashi Y, Sakagami Y (1996) Phytosulfokine, sulfated peptides that induce the proliferation of single mesophyll cells of *Asparagus officinalis L.* Proc Natl Acad Sci USA 93:7623–7627
Matsubayashi Y, Sakagami Y (2006) Peptide hormones in plants. Annu Rev Plant Biol 57:649–674
Matsubayashi Y, Takagi L, Omura N, Morita A, Sakagami Y (1999) The endogenous sulfated pentapeptide phytosulfokine-alpha stimulates tracheary element differentiation of isolated mesophyll cells of Zinnia. Plant Physiol 120:1043–1048
Matsubayashi Y, Ogawa M, Morita A, Sakagami Y (2002) An LRR receptor kinase involved in perception of a peptide plant hormone, phytosulfokine. Science 296:1470–1472
Matsubayashi Y, Ogawa M, Sakagami Y (2005) Analysis of *pskr1-1*, an Arabidopsis knockout mutant of PSK receptor gene. Plant Cell Physiol 46:S177–S177
Matsubayashi Y, Shinohara H, Ogawa M (2006) Identification and functional characterization of phytosulfokine receptor using a ligand-based approach. Chem Rec 6:356–364
Mishima M, Takayama S, Sasaki K, Jee J, Kojima C, Isogai A, Shirakawa M (2003) Structure of the male determinant factor for Brassica self-incompatibility. J Biol Chem 278:36389–36395
Miwa H, Betsuyaku S, Iwamoto K, Kinoshita A, Fukuda H, Sawa S (2008) The receptor-like kinase SOL2 mediates CLE signaling in Arabidopsis. Plant Cell Physiol 49:1752–1757
Moore KL (2003) The biology and enzymology of protein tyrosine O-sulfation. J Biol Chem 278:24243–24246
Müller R, Borghi L, Kwiatkowska D, Laufs P, Simon R (2006) Dynamic and compensatory responses of Arabidopsis shoot and floral meristems to CLV3 signaling. Plant Cell 18:1188–1198
Müller R, Bleckmann A, Simon R (2008) The receptor kinase CORYNE of Arabidopsis transmits the stem cell-limiting signal CLAVATA3 independently of CLAVATA1. Plant Cell 20:934–946
Nakaya M, Tsukaya H, Murakami N, Kato M (2002) Brassinosteroids control the proliferation of leaf cells of *Arabidopsis thaliana*. Plant Cell Physiol 43:239–244
Ni J, Clark SE (2006) Evidence for functional conservation, sufficiency, and proteolytic processing of the CLAVATA3 CLE domain. Plant Physiol 140:726–733
Oelkers K, Goffard N, Weiller G, Gresshoff P, Mathesius U, Frickey T (2008) Bioinformatic analysis of the CLE signaling peptide family. BMC Plant Biol 8:1
Ogawa M, Shinohara H, Sakagami Y, Matsubayashi Y (2008) Arabidopsis CLV3 peptide directly binds CLV1 ectodomain. Science 319:294
Ohyama K, Ogawa M, Matsubayashi Y (2008) Identification of a biologically active, small, secreted peptide in Arabidopsis by in silico gene screening, followed by LC-MS-based structure analysis. Plant J 55:152–160
Ohyama K, Shinohara H, Ogawa-Ohnishi M, Matsubayashi Y (2009) A glycopeptide regulating stem cell fate in *Arabidopsis thaliana*. Nat Chem Biol 5:578–580
Peret B, Larrieu A, Bennett MJ (2009) Lateral root emergence: a difficult birth. J Exp Bot 60:3637–3643

Poppenberger B, Fujioka S, Soeno K, George GL, Vaistij FnE, Hiranuma S, Seto H, Takatsuto S, Adam G, Yoshida S, Bowles D (2005) The UGT73C5 of *Arabidopsis thaliana* glucosylates brassinosteroids. Proc Natl Acad Sci USA 102:15253–15258

Rowe MH, Bergmann DC (2010) Complex signals for simple cells: the expanding ranks of signals and receptors guiding stomatal development. Curr Opin Plant Biol 13:548–555

Rychel AL, Peterson KM, Torii KU (2010) Plant twitter: ligands under 140 amino acids enforcing stomatal patterning. J Plant Res 123:275–280

Sanders PM, Bui AQ, Weterings K, McIntire KN, Hsu YC, Lee PY, Truong MT, Beals TP, Goldberg RB (1999) Anther developmental defects in *Arabidopsis thaliana* male-sterile mutants. Sex Plant Reprod 11:297–322

Scheer JM, Ryan CA (2002) The systemin receptor SR160 from *Lycopersicon peruvianum* is a member of the LRR receptor kinase family. Proc Natl Acad Sci USA 99:9585–9590

Schoof H, Lenhard M, Haecker A, Mayer KFX, Jurgens G, Laux T (2000) The stem cell population of Arabidopsis shoot meristems is maintained by a regulatory loop between the *CLAVATA* and *WUSCHEL* genes. Cell 100:635–644

Schopfer CR, Nasrallah ME, Nasrallah JB (1999) The male determinant of self-incompatibility in Brassica. Science 286:1697–1700

Sharma VK, Carles C, Fletcher JC (2003) Maintenance of stem cell populations in plants. Proc Natl Acad Sci USA 100:11823–11829

Shiba H, Takayama S, Iwano M, Shimosato H, Funato M, Nakagawa T, Che FS, Suzuki G, Watanabe M, Hinata K, Isogai A (2001) A pollen coat protein, SP11/SCR, determines the pollen S-specificity in the self-incompatibility of Brassica species. Plant Physiol 125:2095–2103

Shimada Y, Fujioka S, Miyauchi N, Kushiro M, Takatsuto S, Nomura T, Yokota T, Kamiya Y, Bishop GJ, Yoshida S (2001) Brassinosteroid-6-oxidases from Arabidopsis and tomato catalyze multiple C-6 oxidations in brassinosteroid biosynthesis. Plant Physiol 126:770–779

Shimada Y, Goda H, Nakamura A, Takatsuto S, Fujioka S, Yoshida S (2003) Organ-specific expression of brassinosteroid-biosynthetic genes and distribution of endogenous brassinosteroids in Arabidopsis. Plant Physiol 131:287–297

Shinohara H, Ogawa M, Sakagami Y, Matsubayashi Y (2007) Identification of ligand binding site of phytosulfokine receptor by on-column photoaffinity labeling. J Biol Chem 282:124–131

Shiu SH, Bleecker AB (2001) Receptor-like kinases from Arabidopsis form a monophyletic gene family related to animal receptor kinases. Proc Natl Acad Sci USA 98:10763–10768

Shpak ED, McAbee JM, Pillitteri LJ, Torii KU (2005) Stomatal patterning and differentiation by synergistic interactions of receptor kinases. Science 309:290–293

Sieburth LE (2007) Plant development: PXY and polar cell division in the procambium. Curr Biol 17:R594–R596

Silverstein KAT, Moskal WA, Wu HC, Underwood BA, Graham MA, Town CD, VandenBosch KA (2007) Small cysteine-rich peptides resembling antimicrobial peptides have been underpredicted in plants. Plant J 51:262–280

Stahl Y, Simon R (2010) Plant primary meristems: shared functions and regulatory mechanisms. Curr Opin Plant Biol 13:53–58

Stahl Y, Wink RH, Ingram GC, Simon R (2009) A signaling module controlling the stem cell niche in Arabidopsis root meristems. Curr Biol 19:909–914

Stenvik G-E, Butenko MA, Urbanowicz BR, Rose JK, Aalen RB (2006) Overexpression of INFLORESCENCE DEFICIENT IN ABSCISSION activates cell separation in vestigial abscission zones in Arabidopsis. Plant Cell 18:1467–1476

Stenvik G-E, Tandstad NM, Guo Y, Shi C-L, Kristiansen W, Holmgren A, Clark SE, Aalen RB, Butenko MA (2008) The EPIP peptide of INFLORESCENCE DEFICIENT IN ABSCISSION is sufficient to induce abscission in Arabidopsis through the receptor-like kinases HAESA and HAESA-LIKE2. Plant Cell 20:1805–1817

Sugano SS, Shimada T, Imai Y, Okawa K, Tamai A, Mori M, Hara-Nishimura I (2010) Stomagen positively regulates stomatal density in Arabidopsis. Nature 463:241–U130

Symons GM, Reid JB (2004) Brassinosteroids do not undergo long-distance transport in pea. implications for the regulation of endogenous brassinosteroid levels. Plant Physiol 135:2196–2206

Symons GM, Ross JJ, Jager CE, Reid JB (2008) Brassinosteroid transport. J Exp Bot 59:17–24

Szekeres M (2003) Brassinosteroid and systemin: two hormones perceived by the same receptor. Trends Plant Sci 8:102–104

Szekeres M, Németh K, Koncz-Kálmán Z, Mathur J, Kauschmann A, Altmann T, Rédei GP, Nagy F, Schell J, Koncz C (1996) Brassinosteroids rescue the deficiency of CYP90, a Cytochrome P450, controlling cell elongation and de-etiolation in Arabidopsis. Cell 85:171–182

Takayama S, Isogai A (2005) Self-incompatibility in plants. Annu Rev Plant Biol 56:467–489

Takayama S, Shiba H, Iwano M, Shimosato H, Che FS, Kai N, Watanabe M, Suzuki G, Hinata K, Isogai A (2000) The pollen determinant of self-incompatibility in *Brassica campestris*. Proc Natl Acad Sci USA 97:1920–1925

Takayama S, Shimosato H, Shiba H, Funato M, Che FS, Watanabe M, Iwano M, Isogai A (2001) Direct ligand-receptor complex interaction controls Brassica self-incompatibility. Nature 413:534–538

Trotochaud AE, Hao T, Wu G, Yang ZB, Clark SE (1999) The CLAVATA1 receptor-like kinase requires CLAVATA3 for its assembly into a signaling complex that includes KAPP and a Rho-related protein. Plant Cell 11:393–405

Tsuchimatsu T, Suwabe K, Shimizu-Inatsugi R, Isokawa S, Pavlidis P, Stadler T, Suzuki G, Takayama S, Watanabe M, Shimizu KK (2010) Evolution of self-compatibility in Arabidopsis by a mutation in the male specificity gene. Nature 464:1342–1346

Turk EM, Fujioka S, Seto H, Shimada Y, Takatsuto S, Yoshida S, Denzel MA, Torres QI, Neff MM (2003) CYP72B1 inactivates brassinosteroid hormones: an intersection between photomorphogenesis and plant steroid signal transduction. Plant Physiol 133:1643–1653

Ubeda-Tomas S, Bennett MJ (2010) Plant development: size matters, and it's all down to hormones. Curr Biol 20:R511–R513

Vanoosthuyse V, Miege C, Dumas C, Cock JM (2001) Two large *Arabidopsis thaliana* gene families are homologous to the Brassica gene superfamily that encodes pollen coat proteins and the male component of the self-incompatibility response. Plant Mol Biol 46:17–34

Vert Gg, Nemhauser JL, Geldner N, Hong F, Chory J (2005) Molecular mechanisms of steroid hormone signaling in plants. Annu Rev Cell Dev Biol 21:177–201

Whitford R, Fernandez A, De Groodt R, Ortega E, Hilson P (2008) Plant CLE peptides from two distinct functional classes synergistically induce division of vascular cells. Proc Natl Acad Sci USA 105:18625–18630

Yamaguchi Y, Pearce G, Ryan CA (2006) The cell surface leucine-rich repeat receptor for AtPep1, an endogenous peptide elicitor in Arabidopsis, is functional in transgenic tobacco cells. Proc Natl Acad Sci USA 103:10104–10109

Yamakawa S, Sakuta C, Matsubayashi Y, Sakagami Y, Kamada H, Satoh S (1998) The promotive effects of a peptidyl plant growth factor, phytosulfokine-alpha, on the formation of adventitious roots and expression of a gene for a root-specific cystatin in cucumber hypocotyls. J Plant Res 111:453–458

Yang H, Matsubayashi Y, Nakamura K, Sakagami Y (2001) Diversity of Arabidopsis genes encoding precursors for phytosulfokine, a peptide growth factor. Plant Physiol 127:842–851

Yang S-L, Xie L-F, Mao H-Z, Puah CS, Yang W-C, Jiang L, Sundaresan V, Ye D (2003) *TAPETUM DETERMINANT1* is required for cell specialization in the *Arabidopsis* anther. Plant Cell 15:2792–2804

Yang S-L, Jiang L, Puah CS, Xie L-F, Zhang X-Q, Chen L-Q, Yang W-C, Ye D (2005) Overexpression of *TAPETUM DETERMINANT1* alters the cell fates in the Arabidopsis carpel and tapetum via genetic interaction with *EXCESS MICROSPOROCYTES1/EXTRA SPOROGENOUS CELLS*. Plant Physiol 139:186–191

Yuan T, Fujioka S, Takatsuto S, Matsumoto S, Gou X, He K, Russell SD, Li J (2007) BEN1, a gene encoding a dihydroflavonol 4-reductase (DFR)-like protein, regulates the levels of brassinosteroids in *Arabidopsis thaliana*. Plant J 51:220–233

Zhu Y, Wang Y, Li R, Song X, Wang Q, Huang S, Jin JB, Liu C-M, Lin J (2010) Analysis of interactions among the CLAVATA3 receptors reveals a direct interaction between CLAVATA2 and CORYNE in Arabidopsis. Plant J 61:223–233

Phosphorylation and RLK Signaling

Steven D. Clouse, Michael B. Goshe, and Steven C. Huber

Abstract Plant genomes encode hundreds of receptor-like kinases (RLKs) with an organization of functional domains similar to those of animal receptor kinases. Ligand-dependent phosphorylation has now been demonstrated for several plant RLKs and identification of specific phosphorylation sites followed by their functional characterization has advanced our understanding of RLK signaling mechanisms regulating growth, morphogenesis, and disease resistance. Advances in mass spectrometry and phosphopeptide enrichment technology have been applied to plant phosphoproteomics, revealing hundreds of novel in vivo RLK phosphorylation sites and allowing comparative analysis of phosphorylation site sequence motifs. This chapter examines recent studies on both targeted RLK phosphorylation site analysis and global phosphoproteomic studies that have generated data useful for understanding mechanisms of RLK phosphorylation and its role in plant signal transduction.

1 Introduction

Reversible phosphorylation of Ser, Thr, and Tyr residues is the most-studied and best understood posttranslational modification of protein structure and leads to critical modulation of an enormous number of eukaryotic signal transduction

---

S.D. Clouse (✉)
Department of Horticultural Science, North Carolina State University, 228 Kilgore Hall, 7609, Raleigh, NC 27695-7609, USA
e-mail: steve_clouse@ncsu.edu

M.B. Goshe
Department of Molecular and Structural Biochemistry, North Carolina State University, Raleigh, NC 27695, USA

S.C. Huber
USDA/ARS, University of Illinois, Urbana, IL 61801, USA

F. Tax and B. Kemmerling (eds.), *Receptor-like Kinases in Plants*,
Signaling and Communication in Plants 13,
DOI 10.1007/978-3-642-23044-8_12,

pathways. The end result of signaling pathways involving receptor kinases is most often the change in expression of hundreds of genes resulting from a cascade of protein interactions and phosphorylation events that alter the activity of specific transcription factors. The role of phosphorylation in mammalian receptor kinase function, including families of receptor tyrosine kinase (RTK) and transforming growth factor β (TGF-β) receptor kinases, has been extensively characterized and a general paradigm for their mechanism of action has been established (Lemmon and Schlessinger 2010; Rahimi and Leof 2007). Ligand binding by the extracellular domain leads to oligomerization with other receptor kinase partners and a consequent activation of the cytoplasmic kinase domain of the receptor, most often by initial auto- or transphosphorylation of conserved residues essential for kinase function, such as the activation loop of kinase subdomains VII and VIII, and by phosphorylation of sites in the juxtamembrane (JM) and carboxyl terminal (CT) segments of the cytoplasmic domain that may release inhibition of kinase activity by conformational changes (Hubbard 2004; Pawson 2004).

Phosphorylation of JM and CT residues also generates docking sites for specific downstream substrates of the receptor kinase, which may include additional receptor kinases, soluble cytoplasmic kinases, phosphatases, 14-3-3 proteins, metabolic enzymes, transcription factors, translation initiation factors, and scaffolding proteins. Phosphorylation of these substrates on specific Ser, Thr, or Tyr residues by the receptor kinase may alter protein stability, subcellular localization, interactions with other proteins, and/or a critical protein function such as kinase activity or DNA binding of a transcription factor, thus generating a phosphorelay in which initial perception of ligand by the receptor kinase extracellular domain results in transduction of the signal to the nucleus, where resulting alterations in gene expression yield a ligand-dependent cellular response.

Arabidopsis receptor-like kinases (RLKs) belong to a monophyletic gene family of 610 members that represents over 2% of predicted protein-encoding genes in this plant (Shiu and Bleecker 2001). More than 400 Arabidopsis RLKs have an organization of functional domains similar to that of animal receptor kinases, including a putative extracellular ligand-binding domain, a single-pass transmembrane sequence and an intracellular kinase domain (Cock et al. 2002). Several of these plant RLKs have proven functional roles in the regulation of plant growth, morphogenesis, disease resistance, and responses to environmental stress signals (Becraft 2002; de Lorenzo et al. 2009; Tor et al. 2009; Torii 2004), but the functions of many members of this large family of membrane proteins remain unknown. Conservation of plant RLK structural domain organization with mammalian receptor kinases suggests a parallel in functional mechanisms as well, and recent work has indeed demonstrated that ligand-dependent phosphorylation of plant RLKs occurs on specific Ser, Thr, and Tyr residues in vivo (Oh et al. 2009b; Wang et al. 2005a, 2008).

To characterize plant RLK function thoroughly, it is essential to understand the role of ligand-dependent receptor oligomerization and cytoplasmic domain phosphorylation, including identification of specific phosphorylation sites and examination of their functional significance in protein/protein interactions and modification of protein function. Furthermore, identification of phosphorylation sites in RLK kinase domain cytoplasmic substrates and interacting partners that propagate the specific signal downstream is also required for a complete picture of RLK action.

The very large number of RLKs found in Arabidopsis present immense possibilities for heterodimerization and thus the diversification and amplification of signaling pathways controlling plant growth and development and responses to stress. Building a network of RLK protein interactions and clarifying the role phosphorylation plays in generating these networks are keys to understanding the role of this family in regulating plant form and function.

The identification and functional characterization of in vivo RLK phosphorylation sites remains a technical challenge that requires multiple experimental approaches including molecular genetics, mutational analysis, kinase biochemistry, membrane proteomics, and phosphopeptide analysis by mass spectrometry. Recent advances in phosphoproteomics as well as targeted functional analyses of individual RLK phosphorylation events have generated a large number of RLK phosphorylation sites that form the basis for modeling RLK action and future experimentation on RLK function. This chapter surveys recent studies at the levels of both phosphoproteomics and individual RLK functional analysis.

2 Strategies for Phosphorylation Site Identification of RLKs Using Mass Spectrometry

Phosphorylation of the solvent-accessible hydroxyl of Ser, Thr, and Tyr residues at site-specific motifs converts this polar functional group into a larger, more polar anionic group that modulates protein activity. This modification occurs transiently and usually at levels of low stoichiometry. The identification of pSer and pThr-containing peptides employing collision-induced dissociation (CID) to generate b- and y-ions during MS/MS analysis is more difficult than unmodified peptides due to the loss of the phosphate group (i.e., H_3PO_4 or HPO_3) under the CID conditions typically employed for peptide fragmentation (DeGnore and Qin 1998). Although partial loss of the phosphate moiety can be tolerated, or even exploited, for multidissociation acquisitions to better identify phosphopeptides (Schroeder et al. 2004), it typically involves manual inspection of the data to verify product ion assignments and neutral losses.

Alternative gas-phase fragmentation techniques such as electron capture dissociation (ECD) (Stensballe et al. 2000; McLafferty et al. 2001; Shi et al. 2001; Zubarev 2004; Meng et al. 2005) and electron-transfer dissociation (ETD) (McLuckey and Stephenson 1998; Syka et al. 2004; Mikesh et al. 2006; Molina et al. 2007; Chi et al. 2007) are viable alternatives for phosphoproteomics because the phosphate moiety remains intact during peptide fragmentation while sequence-specific c- and z-ions are generated which can be used for sequence determination. In addition, ion suppression effects of abundant non-phosphorylated peptides can severely affect the detection of the substoichiometric amounts of phosphopeptides, which is further compounded by the partial negative character of the attached phosphate group that affects ionization in the positive mode during liquid chromatography-tandem mass spectrometry (LC/MS/MS) analysis. Therefore, advanced fractionation and enrichment techniques, in combination with high-end mass analyzers capable of accurately measuring both precursor and fragment ions, enable a more rigorous phosphoproteomic approach and provide more reliable information regarding phosphorylation site mapping as well as the relative or absolute amount of phosphopeptides.

2.1 Gel-Based Approaches

The evolution of quantitative proteomics strategies over the last decade has involved two major analytical platforms: gel-based or liquid chromatography (LC)-based mass spectrometry analysis. In two-dimensional polyacrylamide gel electrophoresis (2D-PAGE), proteins are separated according to their isoelectric points as well as their molecular weight. Prefractionation steps such as additional isoelectric focusing in granulated gels or ion exchange can also be applied prior to the 2D-PAGE separation in order to reduce sample complexity (Gorg et al. 2002; Doud et al. 2004). After the 2D-PAGE separation, relative protein abundances between two distinct samples can be estimated by calculating the intensity of the corresponding spots using a densitometer. In the Difference Gel Electrophoresis (DiGE) approach, a modification of 2D-PAGE, two different samples are individually labeled with dyes (Cy2 is used to label an internal standard, whereas Cy3 and Cy5 are used to label the control and experimental samples, respectively) and the fluorescence intensity of the labeled proteins is used for quantification. Regardless of the method of quantification, the spots of interest can be excised and in-gel proteolytically digested to produce peptides that are subjected to liquid chromatography-tandem mass spectrometry (LC/MS/MS) or matrix-assisted laser desorption/ionization (MALDI)-time-of-flight (TOF) analysis for protein identification (Oda et al. 2003; Viswanathan et al. 2006; Tannu and Hemby 2006). The main drawback of this approach in terms of quantification is that a given spot intensity is often composed of several proteins, and thus any change in spot intensity cannot be unequivocally attributed to just one protein. In addition, posttranslational modifications (PTMs) such as phosphorylation will change the isoelectric point of the protein, potentially causing it to migrate into a different region of the gel than its unmodified form, thus impairing accurate quantification.

Gel-based approaches have been widely used in plant studies and have resulted in discovery of several RLK signaling components. For example, in a study of rice (*Oryza sativa*) using a 2D-PAGE approach, Cheng et al. identified a new leucine-rich repeat (LRR) RLK, OsRPK1, responding to salt stress (Cheng et al. 2009), while in a proteomic study of Arabidopsis, the DiGE approach was used to identify novel proteins involved in brassinosteroid (BR) signaling (Deng et al. 2007; Tang et al. 2008).

2.2 LC-Based Approaches

For the LC-based method, a "shotgun" or "bottom-up" strategy has been widely used for proteome-wide identification and quantification. In this approach, protein mixtures are digested by one or more proteases and the resulting peptides are analyzed by reversed-phase LC/MS/MS typically using CID of the charged peptides to generate product ion spectra of b- and y-ions that are identified by database searching using a translated genomic database (Yates et al. 1997). Increased peptide identification can be achieved with prior fractionation such as strong-cation exchange (Washburn et al. 2001) or isoelectric focusing. Compared to the in-gel approach, LC-based methods are better suited for analyzing hydrophobic or extremely acid or basic proteins. However, the accuracy and efficiency of identification and quantification depend significantly upon the separation capacity of peptide fractionation and the data-processing algorithms that are capable of analyzing complex data sets. Several studies using LC/MS/MS analysis have been conducted to identify RLKs and their phosphorylation sites, as discussed in detail below.

2.3 Phosphopeptide Enrichment

Complete characterization of protein phosphorylation essentially requires complete sequence coverage of target proteins, which is difficult to achieve using bottom-up LC/MS/MS analysis of either a single phosphoprotein or complex phosphoproteome proteolytic digests. This may be due to the use of a protease that may generate peptides too large to be adequately fragmented by CID, which may be alleviated by using additional proteases, or may result from the inherent sample complexity of protein mixtures that can preclude the detection of phosphopeptides. To enable more reliable phosphopeptide detection from simple and complex peptide mixtures, a variety of enrichment techniques have been reported for the specific capture of phosphopeptides and phosphoproteins (Goshe 2006) including immobilized metal affinity chromatography (IMAC) (Andersson and Porath 1986; Nuwaysir and Stults 1993; Ficarro et al. 2002, 2003; Nuhse et al. 2003; Tsai et al. 2008), titianium dioxide supports (Sano and Nakamura 2004b; Pinkse et al. 2004; Yu et al. 2007; Li et al. 2008), phospho-amino acid motif antibody affinity enrichment (Gronborg et al. 2002; Bergstrom Lind et al. 2008), and ion exchange chromatography (Dai et al. 2007; Motoyama et al. 2007; Han et al. 2008) among others (Hung et al. 2007; Alpert 2008) and their combinations (Nuhse et al. 2003; Sano and Nakamura 2004a; Cantin et al. 2007; Zhang et al. 2007; Mann et al. 2007; Schilling and Knapp 2008).

Although affinity methods have been demonstrated in the literature to successfully enrich for phosphopeptides, the extent of phosphopeptide identification with any given approach is highly dependent upon the sample complexity, quantity, range of phosphorylation stoichiometry, as well as the LC and MS instrumentation capabilites. Consequently, there is no clear consensus in the literature as to the "best" affinity isolation approach or method of detection and often multiple approaches are required for complete phosphorylation site coverage. Many plant RLK studies have utilized phosphopeptide enrichment and MS analysis to characterize phosphorylation sites including analysis of individual RLKs (Oh et al. 2009b; Wang et al. 2005a, 2008) and global phosphoproteomic approaches (Schulze 2010), both of which are discussed below.

3 The BRI1/BAK1 Model of RLK Phosphorylation

BRs are endogenous plant growth-promoting hormones that act at nanomolar concentrations to influence cellular expansion and proliferation (Clouse and Sasse 1998; Sakurai et al. 1999), and are essential for normal organ elongation, vascular differentiation, reproductive development, timing of senescence, and leaf development (Altmann 1999; Bishop 2003). BRs are perceived at the cell surface by BRASSINOSTEROID INSENSITIVE 1 (BRI1), an LRR RLK that binds BR directly in the extracellular domain (Kinoshita et al. 2005; Li and Chory 1997; Wang et al. 2001). Mutational analysis in both Arabidopsis and crop species has shown conclusively that the BRI1 receptor is required for normal BR perception and plant growth and that mutations in either the extracellular domain or intracellular kinase can lead to multiple developmental defects including extreme dwarfism, male sterility, and altered vascular development (Bishop 2003; Clouse et al. 1996; Li and Chory 1997).

We previously identified at least 11 sites of in vivo phosphorylation in the JM, CT, and kinase domains of BRI1 by immunoprecipitation from BL-treated Arabidopsis seedlings followed by LC/MS/MS analysis and showed that BRI1 phosphorylation *in planta* on specific residues was BL dependent (Wang et al. 2005a). For detection of all 11 sites, treatment of plant tissue with BL and IMAC enrichment of phosphopeptides was crucial. Functional characterization of each identified site by biochemical and genetic analyses demonstrated that the highly conserved activation loop residues, S1044 and T1049, were critical for kinase function in vitro and proper BR signaling *in planta* (Figs. 1 and 2), while multiple JM and CT residues were required for optimal substrate phosphorylation by the BRI1 cytoplasmic domain (Wang et al. 2005a). This procedure of in vivo and in vitro phosphorylation site identification by LC/MS/MS, coupled with functional analysis using site-directed mutagenesis of specific phosphorylation sites followed by expression in an LRR RLK mutant with a visible phenotype, is a standard approach we are applying to all of our subsequent LRR RLK functional phosphorylation site analyses.

BRI1 can exist in plant membranes as a monomer or ligand-independent homodimer (Hink et al. 2008; Russinova et al. 2004; Wang et al. 2005b), but full expression of BR signaling requires heterooligomerization of BRI1 with members

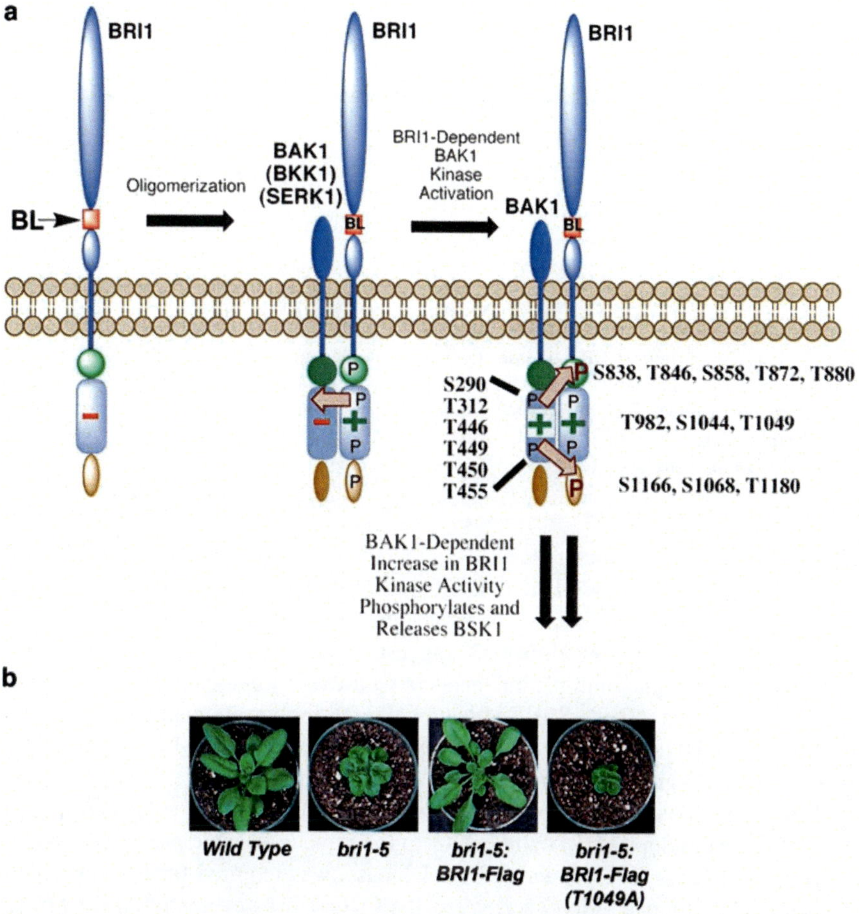

Fig. 1 BR-dependent phosphorylation of specific BRI1 residues is essential for BR signal transduction. (**a**) BR binding to the extracellular domain of the BRI1 receptor kinase results in phosphorylation of residues in the kinase activation loop including the essential T1049, promoting kinase activity and association with the BAK1 (or SERK1 or BKK1) co-receptor(s). BRI1 phosphorylates and activates BAK1 on kinase domain residues and BAK1 in turn transphosphorylates BRI1 on JM and CT residues, which increases BRI1 activity toward downstream substrates such as BSK1, activating BR signaling. A simple dimer model is shown but experimental data may also support a homodimer model for BRI1, which might then form a heterotetramer with BAK1. (**b**) Mutagenesis of the BRI1 T1049 activation loop residue demonstrates its critical role in BR signaling *in planta*. Images are adapted from Wang et al. (2005a) with permission

of the SOMATIC EMBROYGENESIS RECEPTOR KINASE (SERK) subfamily of LRR RLKs (Hecht et al. 2001), including BRI1-ASSOCIATED RECEPTOR KINASE (BAK1), also known as SERK3 (Li et al. 2002; Nam and Li 2002; Russinova et al. 2004). Moreover, BAK1 interacts with another LRR RLK, FLS2, and promotes its function in plant defense responses (Chinchilla et al. 2007; Heese et al. 2007). Thus, BAK1 functions in independent pathways by enhancing the

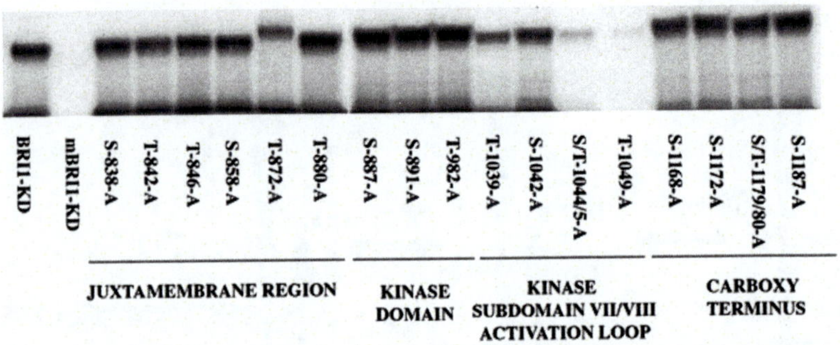

Fig. 2 Phosphorylation of BRI1 kinase domain activation loop residues is critical for in vitro kinase function. Mutagenesis of BRI1 phosphorylation sites demonstrates that specific kinase activation loop residues, including T1049, are essential for recombinant BRI1 cytoplasmic domain autophosphorylation in vitro. mBRI1KD is a K911E kinase inactive mutant. Adapted with permission from Wang et al. (2005a)

signaling output of distinct LRR RLK partners that bind different ligands. SERK4, alternatively named BAK1-LIKE (BKK1), also interacts with BRI1 in vivo (He et al. 2007), as does SERK1, which also heterodimerizes with BRI1 and enhances BR signaling (Karlova et al. 2006), suggesting that SERKs in general are co-receptors that regulate multiple independent pathways by association with different LRR RLKs. In this respect, SERKs functionally resemble mammalian ErbB2 RTKs that do not bind ligand but associate with and activate other ligand-binding members of the epidermal growth factor RTK family (Lemmon and Schlessinger 2010).

In our previous studies, expression of kinase- inactive and wild-type tagged versions of Arabidopsis BRI1 and BAK1 in the same transgenic plant showed that active BRI1 kinase, but not active BAK1 kinase, was required for BR-dependent association of the pair. Moreover, when BAK1-GFP was expressed in the *bri1-1* null mutant background, phosphorylation levels were dramatically reduced in BAK1-GFP suggesting BRI1 activates BAK1 kinase activity. A range of in vitro kinase assays also showed that BAK1 stimulates BRI1 kinase activity and that both BRI1 and BAK1 can transphosphorylate each other on specific residues, with BRI1 phosphorylating BAK1 kinase domain residues and BAK1 phosphorylating BRI1 on JM and CT sites (Wang et al. 2008). As with BRI1, we also identified multiple in vivo and in vitro phopshorylation sites for BAK1 and functionally characterized each site. Like BRI1 residue T1049, phosphorylation of the corresponding BAK1 activation loop residue T455 appears essential for BAK1 kinase function and BR signaling *in planta* (Wang et al. 2005a). Thus, this highly conserved residue may represent a fundamental site for kinase activation in LRR RLKs. Our studies combining LC/MS/MS analysis, functional characterization in mutant backgrounds and in vitro biochemical studies allowed us to develop a novel sequential transphosphorylation model of BRI1/BAK1 interaction (Fig. 1a) that demonstrates plant RLKs share some of the properties of both TGF-β and RTKs in mammals while retaining unique plant-specific features (Wang et al. 2008).

4 Other RLK Phosphorylation Sites Identified in Targeted Studies

In addition to BRI1 and BAK1, several other studies using LC/MS/MS analysis, in vitro mutagenesis or peptide mapping, have identified both in vivo and in vitro RLK phosphorylation sites. Karlova et al. used LC/MALDI-TOF/MS analysis to identify the components of the SERK1 complex, which included the detection of SERK1 phosphorylation sites (Karlova et al. 2006) and in a later investigation identified further in vivo SERK1 sites and in vitro phosphorylation sites for additional Arabidopsis SERKs (Karlova et al. 2009). In a study using both organic solvent- and detergent-based solubilization and extraction methods for Arabidopsis membrane proteome analysis, Mitra et al. performed a multidimensional LC separation using strong-cation exchange chromatography and reversed-phase LC/MS/MS analysis to identify at least 17 members of the LRR RLK family, including the identification of several phosphorylation sites in these 17 LRR RLKs (Mitra et al. 2007). In a follow-up study incorporating two-phase partitioning of microsomal fractions and subsequent chloroform extraction, at least 30 LRR RLKs with multiple peptide identifications and phosphorylation sites were detected (Mitra et al. 2009).

Multiple in vitro phosphorylated Ser and Thr residues of the Arabidopsis CRINKLY4 RLK were identified by a hybrid mass spectrometer consisting of a linear quadrupole ion trap and a Thermo Scientific LTQ-Orbitrap XL (Meyer et al. 2011). Other RLK phosphorylation sites studied include the LsyM RLK, which is phosphorylated in response to chitin binding (Petutschnig et al. 2010), the nodule autoregulation receptor kinase and symbiosis receptor kinase involved in legume nodule formation (Miyahara et al. 2008; Yoshida and Parniske 2005), the tomato pollen receptor kinase LePRK2 (Salem et al. 2011), the rice LRR RLK, Xa21 (Xu et al. 2006), the soybean calmodulin-binding RLK, GmCaMK1 (DeFalco et al. 2010), and the Arabidopsis RLCK BIK1, which associates with BAK1 and FLS2 and is phosphorylated on Thr237 in response to flagellin (Lu et al. 2010).

5 RLK Phosphorylation Sites Identified in Phosphoproteomic Studies

Technical advances in phosphopeptide enrichment and other methods to reduce sample complexity, coupled with increasing sensitivity and mass accuracy of MS instrumentation, have allowed a rapid increase in the number of plant phosphoproteomic studies, which have generated thousands of in vivo plant phosphorylation sites (Mithoe and Menke 2011; Peck 2006; Schulze 2010). Web-based plant phosphorylation databases are now available for a variety of species including *Brassica napus*, soybean and Arabidopsis (http://digbio.missouri.edu/p3db/statistics.php) (Gao et al. 2009); rice and Arabidopsis (https://database.riken.jp/sw/links/en/ria102i/) (Nakagami et al. 2010; Sugiyama et al. 2008); Medicago (http://www.

phospho.medicago.wisc.edu/db/index.php) (Grimsrud et al. 2010); and the PhosPhAt 3.0 Arabidopsis database that incorporates phosphorylation sites from 15 phosphoproteomic publications (http://phosphat.mpimp-golm.mpg.de/) (Durek et al. 2010; Heazlewood et al. 2008). Entering key words related to RLKs or specific RLK accession numbers into these databases returns hundreds of RLK phosphorylation sites. For example, entering the Arabidopsis Genome Initiative (AGI) locus numbers for 224 LRR RLKs into PhosPhAt 3.0 returned 91 LRR RLKs with at least one and more often with multiple in vivo phosphorylation sites.

Although the above resources are a great advance in RLK phosphorylation site analysis and offer opportunities for comparative studies of phosphorylation motifs, a word of caution about global phosphoproteomic approaches is also required. As mentioned earlier, we identified 11 in vivo phosphorylation sites of BRI1 from BR-treated plant tissue using immunoprecipitation followed by LC/MS/MS analysis. Multiple experiments and approaches were required to identify all 11 sites. Entering the BRI1 AGI number (At4g39400) into PhosPhAt returned only one phosphorylation site, Ser838, which we found to be the BRI1 site routinely identified in all of our experiments, suggesting a high stoichiometry of phosphorylation for that site. Sites of BRI1 phosphorylation with lower stoichiometry, which are critical for BR signaling, were not identified in these phosphoproteomic studies. Sample complexity and the requirement for treatment of tissue with a specific RLK ligand to maximize phosphorylation levels suggests that global approaches may not be able to identify many key sites in RLK signaling pathways. To address this issue, we have cloned more than 200 full-length Arabidopsis LRR RLK family members into plant transformation vectors with epitope tags (Gou et al. 2010), which will allow immunoprecipitation of individual LRR RLKs followed by LC/MS/MS analysis as described above for BRI1 and BAK1 (Clouse et al. 2008). A phosphorylation database of at least 50 LRR RLKs treated in this manner is being generated and will be compared to sites determined in the broad phosphoproteomic studies. When curated, these sites and their associated spectra will be available on our NSF Arabidopsis 2010 project website, http://www4.ncsu.edu/~sclouse/Clouse2010.htm.

6 In Vitro RLK Phosphorylation Site Database

We found that in vitro BRI1 and BAK1 kinase domain autophosphorylation sites were highly predictive of in vivo phosphorylation and thus identification of in vitro LRR RLK sites can be a valuable preliminary exercise in examining biochemical function of kinase domains (Oh et al. 2000; Wang et al. 2005a, 2008). As part of our NSF Arabidopsis 2010 project on LRR RLK phosphorylation (Clouse et al. 2008), we cloned and transformed 334 His- and/or Flag-tagged bacterial expression constructs for LRR-RLK kinase domains in BL21 *E. coli* cells and developed a high-throughput protocol for analyzing LRR RLK in vitro autophosphorylation sites using a Waters Q-ToF Premier mass spectrometer functioning in both data-dependent LC/MS/MS and data-independent LC/MS^E modes (see below). Analyses with both un-enriched and IMAC-enriched peptide samples for over 100 LRR RLKs, with four injections per LRR RLK, were performed (Mitra, Blackburn, Goshe, Clouse, unpublished). Hundreds of in vitro phosphorylation sites are now available in a preliminary database (http://www4.ncsu.edu/~sclouse/Clouse2010invitro.htm) and can be compared with in vivo sites available on PhosPhAt. In an initial screen of 47 LRR RLKs for which we have in vitro phosphorylation sites, 30 were not present in PhosPhAt and of the 17 that were, nine had in vivo phosphorylation sites that matched in vitro sites.

7 Tyrosine Phosphorylation in RLKs

Recent phosphoproteomic surveys indicate that Tyr phosphorylation is more abundant in plants than has been generally recognized and the suggestion is emerging that dual-specificity kinases probably play a major role in this modification (Mithoe and Menke 2011; Nakagami et al. 2010; de la Fuente van Bentem and Hirt 2009; Sugiyama et al. 2008). Typically, the soluble protein kinases with dual specificity are considered in this regard, but it is becoming clear that the membrane-associated receptor kinases may also play a role. Although the plant RLKs are categorized as Ser/Thr protein kinases (Afzal et al. 2008), the ability of recombinant RLKs to autophosphorylate on Tyr was demonstrated for BRI1, BAK1, and several other RLKs using anti-phosphoTyr antibodies (Oh et al. 2009b), and thus it appears that many RLKs may be dual-specificity kinases. Consequently, it is appropriate to compare and contrast Tyr phosphorylation in animal RTKs with our current understanding of dual-specific plant receptor kinases.

Ligand-dependent RTK autophosphorylation typically occurs on one or more residues in the activation loop, located between subdomains VII and VIII, and is usually essential for kinase activity (Adams 2003). In addition, autophosphorylation can also occur on multiple residues throughout the cytoplasmic domain, which can also regulate kinase activity but more typically are thought to generate the phosphorylation-dependent interaction motifs required for binding of downstream signaling components (some of which may be substrates of the kinase). An exception to the general model for mammalian RTK activation described above is the subfamily of receptors that include the epidermal growth factor receptors (EGFRs), also known as the ErbB1 and Her1 receptors (Schlessinger 2002). Ligand-induced dimerization of these receptors does not initially involve phosphorylation of residues within the activation loop; rather, formation of an asymmetric dimer of the cytoplasmic domains results in an allosteric activation whereby one kinase domain (referred to as the activator) stimulates the activity of the other kinase domain (the receiver). As a result, multiple Tyr residues in the CT domain are phosphorylated which initiates signal transduction.

The JM domain of the EGFR subfamily is essential for formation of the asymmetric dimer and thus functions as an activator of the kinase domain (Jura et al. 2009). Interestingly, the JM domain of Arabidopsis BRI1 is also an activator

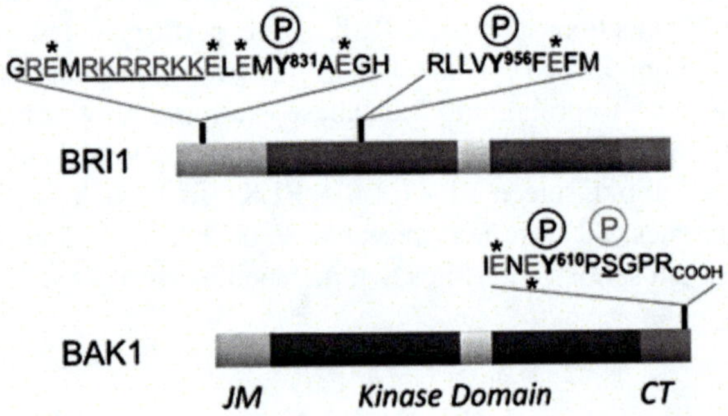

Fig. 3 Schematic representation of the cytoplasmic domains of Arabidopsis BRI1 and BAK1 showing the locations and sequences surrounding sites of Tyr phosphorylation confirmed to occur in vivo. In the sequences surrounding Tyr phosphorylation sites, acidic residues are shown with an *asterisk* and the stretch of basic residues in the N-terminal portion of the BRI1 JM domain is *underlined*. The basic enriched sequences in the JM domain are in common with the mammalian EGFRs. For simplicity, the many other sites of Ser/Thr phosphorylation (Wang et al. 2008) are not shown with the exception of BAK1-Ser612 (Karlova et al. 2009), which is located close to the Tyr-610 site and may contribute to phosphorylation of this site

of kinase activity, as recombinant cytoplasmic domains lacking the JM domain have dramatically reduced autophosphorylation in particular on Thr and Tyr residues (Oh et al. 2009b), and impaired function in vivo (Wang et al. 2005b). The JM domain is required for dimerization of the recombinant cytoplasmic domain of BRI1 (Jaillais et al. 2011) and likely plays a role in the ligand-independent homodimerization of BRI1 that occurs in vivo. Another feature in common between the EGFR RTKs and BRI1 (and also many other plant receptor kinases) is that the portion of the JM domain immediately following the transmembrane domain has an abundance of basic residues (Jura et al. 2009) (see Fig. 3 for the BRI1 sequence), suggesting some common mechanisms for dimerization and perhaps activation. Finally, another common feature between plant receptor kinases and mammalian RTKs is that both can potentially use endosomes for signaling. With BRI1, the protein can be found in both the plasma membrane and early endosomes, and results suggest that the endosomal location may enhance signaling (Geldner et al. 2007). Thus, signaling is not restricted to the receptors that are located in the plasma membrane.

Autophosphorylation is often studied in vitro using recombinant receptor kinase cytoplasmic domains, which when incubated with [γ-^{32}P]ATP in vitro become radiolabeled, as first reported with RLK5 (Horn and Walker 1994). With BRI1, incubation with [γ-^{32}P]ATP resulted in phosphorylation of Ser and Thr residues (Oh et al. 2000; Wang et al. 2005a, b) consistent with its classification as a Ser/Thr protein kinase (Vert et al. 2005). Similar results were obtained with other recombinant RLKs (Liu et al. 2002; Yoshida and Parniske 2005), but there were two exceptions as AtSERK1 (Shah et al. 2001) and petunia PRK1 (Mu et al. 1994) were reported to autophosphorylate on Ser, Thr, *and Tyr* residues. However, the occurrence of Tyr phosphorylation of these receptor kinases in vivo was not established nor was the significance of Tyr phosphorylation established. Consequently, the paradigm remained that plant RLKs are Ser/Thr protein kinases (Afzal et al. 2008).

Recently, the ability of recombinant RLKs to autophosphorylate on Tyr was demonstrated for BRI1, BAK1, and several other RLKs using anti-phosphoTyr antibodies (Oh et al. 2009b), and it now appears that many RLKs may be dual-specificity kinases. Moreover, the occurrence and significance in vivo has been established in at least two cases (see below). Nonetheless, many important questions remain unanswered; for example, if BRI1 can autophosphorylate on Tyr why was this not observed following preincubation with [γ-^{32}P]ATP in vitro? This paradox – that BRI1 contains phosphotyrosine but has not been observed to autophosphorylate on Tyr residues in vitro – has several potential explanations, including the possibility that Tyr autophosphorylation may be a cotranslational rather than posttranslational event. Cotranslational autophosphorylation on Tyr residues has been demonstrated for glycogen synthase kinase 3 (GSK3) (Cole et al. 2004; Lochhead et al. 2006) and dual-specificity tyrosine-phosphorylation-regulated protein kinase (DRKY) (Lochhead et al. 2005). Interestingly, the GSK3 and DRKY family protein kinases autophosphorylate on tyrosine residues as transitional intermediates, while the mature proteins can only catalyze phosphorylation on serine/threonine residues. This is a formal possibility that may be relevant to plant receptor kinases such as BRI1 and remains an area that needs additional experimental investigation. However, regardless of the mechanism involved, how does Tyr phosphorylation impact plant receptor kinase signaling? Possibilities include (1) direct electrostatic effects of the phosphate group and (2) recruitment of phosphotyrosine-binding proteins (Yaffe 2002). Very little is known about the latter category of proteins in plants, and this also remains as an important area for future study.

7.1 *Physiological Significance of Tyrosine Phosphorylation in BR Signaling*

Sites of Tyr phosphorylation have been identified in plant receptor kinases by a combination of mass spectrometry, site-directed mutagenesis, and modification- and sequence-specific antibodies (Karlova et al. 2009; Oh et al. 2009b, 2010). Once sites are identified, their functional significance can be studied by expression of site-directed mutants in transgenic plants; however, this approach will only work when the Tyr residue modified is not essential for kinase activity. For example, with BRI1, site-directed mutagenesis involving individual substitution of the ten Tyr residues in the cytoplasmic domain of the protein with Phe (to prevent phosphorylation at the site) identified four residues that are essential for kinase activity:

Tyr-956, -1052, -1057, and -1072 (Oh et al. 2009b). This result suggested that either these residues must be phosphorylated in order for the kinase to be active, or that the hydroxyl groups at these positions are required for activity. Whether phosphorylation occurs or not must be determined by mass spectrometry or cross reactivity with specific antibodies. Interestingly, with BRI1, two of the essential Tyr residues (Tyr-956 and -1072) were shown to be sites of phosphorylation (Oh et al. 2009b) but current thinking is that these modifications are not essential for activity; in fact, phosphorylation, to the extent that it occurs, probably inhibits kinase activity (Oh and Huber, 2010, unpublished). In contrast, substitution of Tyr-831 with Phe (to produce the Y831F mutant) had little effect on Ser/Thr autophosphorylation but dramatically reduced Tyr autophosphorylation, suggesting that Tyr-831 may be a major site of autophosphorylation and of potential regulatory significance. A similar mutagenesis analysis of the ten Tyr residues in the cytoplasmic domain of BAK1 identified Tyr-463 in the kinase domain as essential for catalytic activity, and Tyr-610 in the CT domain as a major potential site of Tyr phosphorylation that could play a regulatory role. The sites of Tyr phosphorylation in BRI1 and BAK1 that have been confirmed in vivo are presented schematically in Fig. 3. It is interesting to note that the sites are flanked by acidic residues (shown with an asterisk in Fig. 3), which may play a role in targeting (Miller et al. 2008). The feature in common among all three sites shown in Fig. 3 is a negative group at the +2 position (i.e., 2 residues C-terminal to the phosphorylation site). In both sites in BRI1, the acidic residue is a Glu, while in BAK1, negative charge may be contributed by phosphorylation of the Ser at this position. In the future, it will be interesting to determine whether phosphorylation of Ser-612 promotes phosphorylation of Tyr-610 as predicted.

Another reason to think that acidic residues may play a role in Tyr phosphorylation is that alignment of the JM domains of BRI1 orthologs reveals that the JM domains are, in general, much more hyper-variable than the kinase domains. However, alignment of the conserved Tyr residue at about the 831 position identifies conserved acidic residues at the −2 and +2 positions. Future studies will be required to determine whether the acidic residues are indeed essential positive elements of the Tyr phosphorylation motif. If so, the Tyr phosphorylation motif will be quite distinct from the motif that is thought to be required for phosphorylation of Ser-containing peptides. BRI1 will readily phosphorylate the SP11 synthetic peptide (sequence: GRJRRIASVEJJKK, where J is norleucine, a nonoxidizing analog of Met), and peptide variants have shown that both basic and hydrophobic residues are important (Oh et al. 2000). This may suggest that Ser/Thr- and Tyr-containing sequences that are phosphorylated by BRI1 may bind to distinct residues lining the active site. A final point to note with respect to the Tyr-831 site of BRI1 is that there is a Met residue at the −1 position. Methionine residues can be readily oxidized to Met sulfoxide, and can impact the phosphorylation of nearby Ser/Thr residues (Hardin et al. 2009). Whether Tyr phosphorylation can also be affected by Met oxidation remains to be determined but is certainly a possibility, and if so, could provide a link between receptor kinase function and ROS signaling.

The function of phosphorylation of a given residue can be most convincingly studied when that residue is not essential for kinase activity. Such is the case with BRI1-Tyr831 and BAK1-Tyr610. Prevention of phosphorylation at the Tyr-831 site in BRI1 substantially enhances rescue of the dwarf phenotype of the *bri1-5* Arabidopsis mutant. Plants expressing BRI1-Y831F are larger, have rounder leaves and flower earlier than plants expressing wild-type BRI1 (Oh et al. 2009b), suggesting that phosphorylation at this site attenuates BR signaling (Oh et al. 2009a). The possible physiological and molecular genetic bases underlying the Y831F phenotype are currently being studied but appear to involve changes in expression of a number of genes that are not generally considered to be BL regulated in wild-type plants (unpublished data). In contrast, expression of site-directed mutants of BAK1 at the Tyr-610 site revealed that phosphorylation of this residue is essential for enhanced BR signaling (Oh et al. 2010). Thus, plants expressing BAK1(Y610F)-Flag are severely dwarfed in comparison with plants expressing wild-type BAK1-Flag or the phosphomimetic BAK1(Y610E)-Flag. This phenotype is much more severe than the BAK1 loss-of-function phenotype, indicating potentially strong inhibition of BRI1 activity in this overexpressed mutant. Interestingly, while the expression of many BL-regulated genes are altered in the BAK1(Y610F)-Flag plants, the changes in gene expression extend far beyond the normal BL-regulatory network, indicating that Tyr phosphorylation events can have far-reaching, and difficult to predict, effects on cellular control mechanisms. How these effects are exerted at the molecular level is completely unknown at present, but a role for phosophotyrosine-binding proteins seems likely. In mammalian systems, one of the conserved mechanisms mediating phosphotyrosine signaling is binding of Src Homology 2 (SH2) domain-containing proteins (Pincus et al. 2008). There are also unconventional phosphotyrosine-binding proteins in animals such as the M2 isoform of pyruvate kinase (Christofk et al. 2008), and consequently there are likely to be many potential binding proteins in plants as well.

It is now clear that Tyr phosphorylation plays an important role with at least four proteins involved in BR signaling. As illustrated schematically in Fig. 4, in the absence of hormone ligand, BRI1 and BAK1 are separated and inactive in the plasma membrane, in part because of the association of BKI1 with BRI1 (Wang and Chory 2006). Meanwhile, the soluble BIN2 kinase (a GSK3 enzyme) is phosphorylated on Tyr-200 and therefore active and able to repress BL-regulated gene expression by phosphorylation of the transcription factors BRASSINAZOLE-RESISTANT1/2 (BZR1/2) (Wang et al. 2002). In the presence of the BL ligand, BRI1 and BAK1 interact and activate by auto- and/or transphosphorylation (Wang et al. 2008). Phosphorylation of BAK1 at the Tyr610 site is essential for its function in BR signaling (Oh et al. 2010), whereas phosphorylation of BRI1 at the Tyr831 and Tyr956 sites are thought to attenuate signaling (Oh et al. 2009b). There are two immediate downstream components that are rapidly phosphorylated by the BRI1: BAK1 signaling complex. One is BRI1 KINASE INHIBITOR 1 (BKI1), which is associated with the plasma membrane via an N-terminal polybasic region and interacts with and inhibits BRI1 through a C-terminal domain. A key step in the

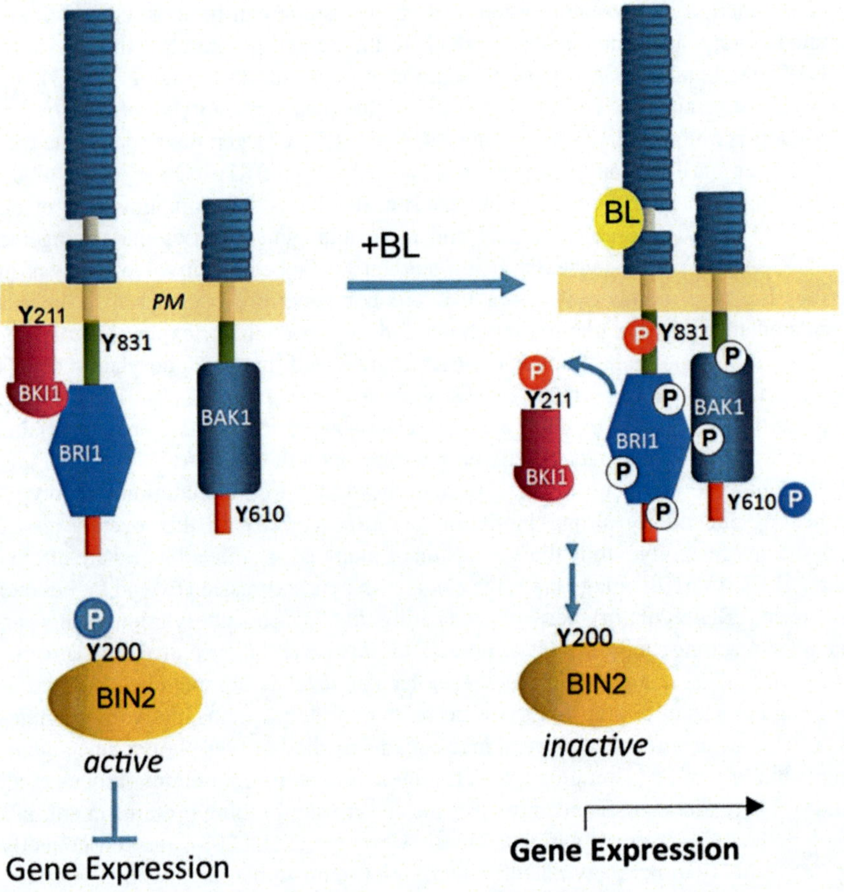

Fig. 4 Highly simplified schematic representation of the upstream components involved in BR signaling illustrating the regulatory role for Tyr phosphorylation of BRI1, BAK1, BKI1, and BIN2. See text for discussion

release of BKI1 inhibition is phosphorylation of Tyr-211, which is located within the membrane-targeting motif (Jaillais et al. 2011). The exact sequence of steps is not known, but it is likely that release of BKI1 preceeds formation of the BRI1: BAK1 heterodimer. Once the inhibitor is released and the receptor kinase heterodimer forms and fully activates, a signaling cascade is initiated that ultimately activates the PP1-related protein phosphatase BRI SUPPRESSOR 1 (BSU1), which dephosphorylates and inactivates BRASSINOSTEROID INSENSITIVE 2 (BIN2) at the phospho-Tyr200 site (Tang et al. 2010). Once the kinase is inactive, specific members of the PP2A family dephosphorylate BZR1 sites phosphorylated by BIN2, including Ser-173, the critical 14-3-3 protein binding residue, thereby activating BZR1(Tang et al. 2011), and allowing BL-regulated gene expression to commence.

8 Quantitative Analysis of RLK Phosphorylation Sites

Since phosphorylation is a dynamic event, it is crucial to monitor changes in phosphorylation in response to ligand over time in order to determine the contribution of individual sites to LRR RLK interactions and downstream signaling. Over the past decade, many stable isotope-labeling approaches, such as ICAT (Gygi et al. 1999), PhIAT (Goshe et al. 2001), SILAC (Ong et al. 2002), AQUA (Gerber et al. 2003), iTRAQ (Ross et al. 2004), and SILIP (Schaff et al. 2008) have been used to quantify both protein levels and phosphorylation sites (Jones et al. 2006; Huttlin et al. 2007; Nelson et al. 2007; Thelen and Peck 2007; Kaffarnik et al. 2009).

With the increase in analytical capabilities of LC/MS/MS instrumentation in terms of resolution, accuracy, and sensitivity coupled with the development of more rapid data collection and software tools, robust label-free quantitative protein abundance measurements can now be performed based on electrospray intensity that does not require labeling with an isotopic tag. The data-independent LC/MS^E approach developed by Waters Corporation is capable of measuring absolute protein abundance by simply spiking in appropriate internal standard proteins (Silva et al. 2006b) and delivers both quantitative and qualitative data in a single experiment (Silva et al. 2006a). To achieve this, alternating Q-ToF acquisition scans are acquired with normal (MS) and elevated collision energies (MS^E), allowing detection of both intact peptides and their product ions in the same experiment. ProteinLynx Global Server software coupled with accurate Q-ToF mass measurements, allow relative levels of phosphorylation at an individual residue to be monitored. We have used this approach to determine quantitative levels of phosphorylation of residue S-1166 in Arabidopsis BRI1 in the presence and absence of BAK1, which was a key feature in the development of the sequential transphosphorylation model of BRI1/BAK1 interaction (Wang et al. 2008). We have also used this approach in vivo by monitoring specific phosphorylation of Ser and Thr residues of BRI1 in response to BL (U. Kota, K. Blackburn, M. Goshe, S. Clouse, unpublished data). By quantitative analysis of individual RLK phosphorylation sites, one can assign relative intensities to specific phosphopeptides across different ligand treatments, time courses, and mutant vs. wild-type backgrounds, which will provide novel insight regarding RLK signaling function. A survey of quantitative MS approaches to plant phosphorylation analysis has been addressed in a recent comprehensive review (Schulze and Usadel 2010).

9 Conclusions

Significant advances in our understanding of RLK phosphorylation and its role in regulating the function and signaling output of these critical membrane proteins have been achieved over the past several years by both the targeted analysis of individual RLKs and by global proteomic approaches that vastly increase the number of available RLK phosphorylation sites for comparative analysis of sequences that may suggest specific phosphorylation motifs. Cooperation between laboratories skilled at advance MS technologies for phosphorylation site analysis with those studying the molecular genetics of RLK families, which provide RLK mutants with appropriate phenotypes for functional analysis of individual RLK phosphorylation sites, needs to continue and expand.

The impressive increase in plant phosphoproteomic studies would be enhanced by the development of methods to further reduced sample complexity and maximize the depth of RLK phosphorylation site coverage. Moreover, quantitative phosphoproteomic approaches conducted under a broad range of treatment conditions and developmental stages will help to classify RLK phosphorylation sites with respect to their signaling role in different physiological responses including hormonal regulation of growth, abiotic stress, and pathogen attack. The increasing availability of genetic resources for entire subfamilies of RLKs, such as constructs for bacterial and plant expression and multiple mutant lines, will allow high-throughput, family-wide biochemical analysis of kinase domain function including substrate identification using random peptide libraries and protein microarrays. The utility of broad discovery-based phosphoproteomic screens in generating targeted, smaller-scale, functional analyses of specific RLKs is clear, and these complementary approaches will increase our understanding of RLK function in regulating plant growth, development, and adaptation.

References

Adams JA (2003) Activation loop phosphorylation and catalysis in protein kinases: is there functional evidence for the autoinhibitor model? Biochemistry 42:601–607

Afzal AJ, Wood AJ, Lightfoot DA (2008) Plant receptor-like serine threonine kinases: roles in signaling and plant defense. Mol Plant-Microbe Interact 21:507–517

Alpert AJ (2008) Electrostatic repulsion hydrophilic interaction chromatography for isocratic separation of charged solutes and selective isolation of phosphopeptides. Anal Chem 80:62–76

Altmann T (1999) Molecular physiology of brassinosteroids revealed by the analysis of mutants. Planta 208:1–11

Andersson L, Porath J (1986) Isolation of phosphoproteins by immobilized metal (Fe^{3+}) affinity chromatography. Anal Biochem 154(1):250–254

Becraft PW (2002) Receptor kinase signaling in plant development. Annu Rev Cell Dev Biol 18:163–192

Bergstrom Lind S, Molin M, Savitski MM, Emilsson L, Astrom J, Hedberg L, Adams C, Nielsen ML, Engstrom A, Elfineh L, Andersson E, Zubarev RA, Pettersson U (2008) Immunoaffinity enrichments followed by mass spectrometric detection for studying global protein tyrosine phosphorylation. J Proteome Res 7:2897–2910

Bishop GJ (2003) Brassinosteroid mutants of crops. J Plant Growth Regul 22:325–335

Cantin GT, Shock TR, Park SK, Madhani HD, Yates JR 3rd (2007) Optimizing TiO2-based phosphopeptide enrichment for automated multidimensional liquid chromatography coupled to tandem mass spectrometry. Anal Chem 79:4666–4673

Cheng Y, Qi Y, Zhu Q, Chen X, Wang N, Zhao X, Chen H, Cui X, Xu L, Zhang W (2009) New changes in the plasma-membrane-associated proteome of rice roots under salt stress. Proteomics 9:3100–3114

Chi A, Huttenhower C, Geer LY, Coon JJ, Syka JEP, Bai DL, Shabanowitz J, Burke DJ, Troyanskaya OG, Hunt DF (2007) Analysis of phosphorylation sites on proteins from *Saccharomyces cerevisiae* by electron transfer dissociation (ETD) mass spectrometry. Proc Natl Acad Sci USA 104:2193–2198

Chinchilla D, Zipfel C, Robatzek S, Kemmerling B, Nurnberger T, Jones JD, Felix G, Boller T (2007) A flagellin-induced complex of the receptor FLS2 and BAK1 initiates plant defence. Nature 448:497–500

Christofk HR, Vander Heiden MG, Wu N, Asara JM, Cantley LC (2008) Pyruvate kinase M2 is a phosphotyrosine-binding protein. Nature 452:181–186

Clouse SD, Sasse JM (1998) BRASSINOSTEROIDS: essential regulators of plant growth and development. Annu Rev Plant Physiol Plant Mol Biol 49:427–451

Clouse SD, Langford M, McMorris TC (1996) A brassinosteroid-insensitive mutant in *Arabidopsis thaliana* exhibits multiple defects in growth and development. Plant Physiol 111:671–678

Clouse SD, Goshe MB, Huber SC, Li J (2008) Functional analysis and phosphorylation site mapping of leucine-rich repeat receptor-like kinases. In: Agrawal GK, Rakwal R (eds) Plant proteomics: technologies, strategies and applications. Wiley, Hoboken, NJ, pp 469–484

Cock JM, Vanoosthuyse V, Gaude T (2002) Receptor kinase signalling in plants and animals: distinct molecular systems with mechanistic similarities. Curr Opin Cell Biol 14:230–236

Cole A, Frame S, Cohen P (2004) Further evidence that the tyrosine phosphorylation of glycogen synthase-3 (GSK3) in mammalian cells is an autophosphorylation event. Biochem J 377:249–255

Dai J, Jin WH, Sheng QH, Shieh CH, Wu JR, Zeng R (2007) Protein phosphorylation and expression profiling by Yin-yang multidimensional liquid chromatography (Yin-yang MDLC) mass spectrometry. J Proteome Res 6:250–262

de la Fuente Van Bentem S, Hirt H (2009) Protein tyrosine phosphorylation in plants: more abundant than expected? Trends Plant Sci 14:71–76

de Lorenzo L, Merchan F, Laporte P, Thompson R, Clarke J, Sousa C, Crespi M (2009) A novel plant leucine-rich repeat receptor kinase regulates the response of *Medicago truncatula* roots to salt stress. Plant Cell 21:668–680

DeFalco TA, Chiasson D, Munro K, Kaiser BN, Snedden WA (2010) Characterization of GmCaMK1, a member of a soybean calmodulin-binding receptor-like kinase family. FEBS Lett 584:4717–4724

DeGnore JP, Qin J (1998) Fragmentation of phosphopeptides in an ion trap mass spectrometer. J Am Soc Mass Spectrom 9:1175

Deng Z, Zhang X, Tang W, Oses-Prieto JA, Suzuki N, Gendron JM, Chen H, Guan S, Chalkley RJ, Peterman TK, Burlingame AL, Wang ZY (2007) A proteomics study of brassinosteroid response in Arabidopsis. Mol Cell Proteomics 6:2058–2071

Doud MK, Schmidt MW, Hines D, Naumann C, Kocourek A, Kashani-Poor N, Zeidler R, Wolf DA (2004) Rapid prefractionation of complex protein lysates with centrifugal membrane adsorber units improves the resolving power of 2D-PAGE-based proteome analysis. BMC Genomics 5:25

Durek P, Schmidt R, Heazlewood JL, Jones A, MacLean D, Nagel A, Kersten B, Schulze WX (2010) PhosPhAt: the *Arabidopsis thaliana* phosphorylation site database. An update. Nucleic Acids Res 38:D828–D834

Ficarro SB, McCleland ML, Stukenberg PT, Burke DJ, Ross MM, Shabanowitz J, Hunt DF, White FM (2002) Phosphoproteome analysis by mass spectrometry and its application to *Saccharomyces cerevisiae*. Nat Biotechnol 20:301–305

Ficarro S, Chertihin O, Westbrook VA, White F, Jayes F, Kalab P, Marto JA, Shabanowitz J, Herr JC, Hunt DF, Visconti PE (2003) Phosphoproteome analysis of capacitated human sperm.

Evidence of tyrosine phosphorylation of a kinase-anchoring protein 3 and valosin-containing protein/p97 during capacitation. J Biol Chem 278:11579–11589

Gao J, Agrawal GK, Thelen JJ, Xu D (2009) P3DB: a plant protein phosphorylation database. Nucleic Acids Res 37:D960–D962

Geldner N, Hyman DL, Wang X, Schumacher K, Chory J (2007) Endosomal signaling of plant steroid receptor kinase BRI1. Genes Dev 21:1598–1602

Gerber SA, Rush J, Stemman O, Kirschner MW, Gygi SP (2003) Absolute quantification of proteins and phosphoproteins from cell lysates by tandem MS. Proc Natl Acad Sci USA 100:6940–6945

Gorg A, Boguth G, Kopf A, Reil G, Parlar H, Weiss W (2002) Sample prefractionation with Sephadex isoelectric focusing prior to narrow pH range two-dimensional gels. Proteomics 2:1652

Goshe MB (2006) Characterizing phosphoproteins and phosphoproteomes using mass spectrometry. Brief Funct Genomic Proteomic 4:363–376

Goshe MB, Conrads TP, Panisko EA, Angell NH, Veenstra TD, Smith RD (2001) Phosphoprotein isotope-coded affinity tag approach for isolating and quantitating phosphopeptides in proteome-wide analyses. Anal Chem 73:2578–2586

Gou X, He K, Yang H, Yuan T, Lin H, Clouse SD, Li J (2010) Genome-wide cloning and sequence analysis of leucine-rich repeat receptor-like protein kinase genes in *Arabidopsis thaliana*. BMC Genomics 11:19

Grimsrud PA, den Os D, Wenger CD, Swaney DL, Schwartz D, Sussman MR, Ane JM, Coon JJ (2010) Large-scale phosphoprotein analysis in *Medicago truncatula* roots provides insight into in vivo kinase activity in legumes. Plant Physiol 152:19–28

Gronborg M, Kristiansen TZ, Stensballe A, Andersen JS, Ohara O, Mann M, Jensen ON, Pandey A (2002) A mass spectrometry-based proteomic approach for identification of serine/threonine-phosphorylated proteins by enrichment with phospho-specific antibodies: identification of a novel protein, Frigg, as a protein kinase A substrate. Mol Cell Proteomics 1:517–527

Gygi SP, Rist B, Gerber SA, Turecek F, Gelb MH, Aebersold R (1999) Quantitative analysis of complex protein mixtures using isotope-coded affinity tags. Nat Biotechnol 17:994

Han G, Ye M, Zhou H, Jiang X, Feng S, Tian R, Wan D, Zou H, Gu J (2008) Large-scale phosphoproteome analysis of human liver tissue by enrichment and fractionation of phosphopeptides with strong anion exchange chromatography. Proteomics 8:1346–1361

Hardin SC, Larue CT, Oh MH, Jain V, Huber SC (2009) Coupling oxidative signals to protein phosphorylation via methionine oxidation in Arabidopsis. Biochem J 422:305–312

He K, Gou X, Yuan T, Lin H, Asami T, Yoshida S, Russell SD, Li J (2007) BAK1 and BKK1 regulate brassinosteroid-dependent growth and brassinosteroid-independent cell-death pathways. Curr Biol 17:1109–1115

Heazlewood JL, Durek P, Hummel J, Selbig J, Weckwerth W, Walther D, Schulze WX (2008) PhosPhAt: a database of phosphorylation sites in *Arabidopsis thaliana* and a plant-specific phosphorylation site predictor. Nucleic Acids Res 36:D1015–D1021

Hecht V, Vielle-Calzada JP, Hartog MV, Schmidt ED, Boutilier K, Grossniklaus U, de Vries SC (2001) The Arabidopsis SOMATIC EMBRYOGENESIS RECEPTOR KINASE 1 gene is expressed in developing ovules and embryos and enhances embryogenic competence in culture. Plant Physiol 127:803–816

Heese A, Hann DR, Gimenez-Ibanez S, Jones AM, He K, Li J, Schroeder JI, Peck SC, Rathjen JP (2007) The receptor-like kinase SERK3/BAK1 is a central regulator of innate immunity in plants. Proc Natl Acad Sci USA 104:12217–12222

Hink MA, Shah K, Russinova E, de Vries SC, Visser AJ (2008) Fluorescence fluctuation analysis of *Arabidopsis thaliana* somatic embryogenesis receptor-like kinase and brassinosteroid insensitive 1 receptor oligomerization. Biophys J 94:1052–1062

Horn MA, Walker JC (1994) Biochemical properties of the autophosphorylation of RLK5, a receptor-like protein kinase from *Arabidopsis thaliana*. Biochim Biophys Acta 1208:65–74

Hubbard SR (2004) Juxtamembrane autoinhibition in receptor tyrosine kinases. Nat Rev Mol Cell Biol 5:464–471

Hung CW, Kubler D, Lehmann WD (2007) pI-based phosphopeptide enrichment combined with nanoESI-MS. Electrophoresis 28:2044–2052

Huttlin EL, Hegeman AD, Harms AC, Sussman MR (2007) Comparison of full versus partial metabolic labeling for quantitative proteomics analysis in Arabidopsis thaliana. Mol Cell Proteomics 6:860–881

Jaillais Y, Hothorn M, Belkhadir Y, Dabi T, Nimchuk ZL, Meyerowitz EM, Chory J (2011) Tyrosine phosphorylation controls brassinosteroid receptor activation by triggering membrane release of its kinase inhibitor. Genes Dev 25:232–237

Jones AM, Bennett MH, Mansfield JW, Grant M (2006) Analysis of the defence phosphoproteome of *Arabidopsis thaliana* using differential mass tagging. Proteomics 6:4155–4165

Jura N, Endres NF, Engel K, Deindl S, Das R, Lamers MH, Wemmer DE, Zhang X, Kuriyan J (2009) Mechanism for activation of the EGF receptor catalytic domain by the juxtamembrane segment. Cell 137:1293–1307

Kaffarnik FA, Jones AM, Rathjen JP, Peck SC (2009) Effector proteins of the bacterial pathogen *Pseudomonas syringae* alter the extracellular proteome of the host plant, *Arabidopsis thaliana*. Mol Cell Proteomics 8:145–156

Karlova R, Boeren S, Russinova E, Aker J, Vervoort J, de Vries S (2006) The Arabidopsis SOMATIC EMBRYOGENESIS RECEPTOR-LIKE KINASE1 protein complex includes BRASSINOSTEROID-INSENSITIVE1. Plant Cell 18:626–638

Karlova R, Boeren S, van Dongen W, Kwaaitaal M, Aker J, Vervoort J, de Vries S (2009) Identification of *in vitro* phosphorylation sites in the *Arabidopsis thaliana* somatic embryogenesis receptor-like kinases. Proteomics 9:368–379

Kinoshita T, Cano-Delgado A, Seto H, Hiranuma S, Fujioka S, Yoshida S, Chory J (2005) Binding of brassinosteroids to the extracellular domain of plant receptor kinase BRI1. Nature 433:167–171

Lemmon MA, Schlessinger J (2010) Cell signaling by receptor tyrosine kinases. Cell 141:1117–1134

Li J, Chory J (1997) A putative leucine-rich repeat receptor kinase involved in brassinosteroid signal transduction. Cell 90:929–938

Li J, Wen J, Lease KA, Doke JT, Tax FE, Walker JC (2002) BAK1, an Arabidopsis LRR receptor-like protein kinase, interacts with BRI1 and modulates brassinosteroid signaling. Cell 110:213–222

Li Y, Xu X, Qi D, Deng C, Yang P, Zhang X (2008) Novel Fe3O4@TiO2 core-shell microspheres for selective enrichment of phosphopeptides in phosphoproteome analysis. J Proteome Res 7:2526–2538

Liu GZ, Pi LY, Walker JC, Ronald PC, Song WY (2002) Biochemical characterization of the kinase domain of the rice disease resistance receptor-like kinase XA21. J Biol Chem 277:20264–20269

Lochhead PA, Sibbet G, Morrice N, Cleghon V (2005) Activation-loop autophosphorylation is mediated by a novel transitional intermediate form of DYRKs. Cell 121:925–936

Lochhead PA, Kinstrie R, Sibbet G, Rawjee T, Morrice N, Cleghon V (2006) A chaperone-dependent GSK3β transitional intermediate mediates activation-loop autophosphorylation. Mol Cell 24:627–633

Lu D, Wu S, Gao X, Zhang Y, Shan L, He P (2010) A receptor-like cytoplasmic kinase, BIK1, associates with a flagellin receptor complex to initiate plant innate immunity. Proc Natl Acad Sci USA 107:496–501

Mann K, Olsen JV, Macek B, Gnad F, Mann M (2007) Phosphoproteins of the chicken eggshell calcified layer. Proteomics 7:106–115

McLafferty FW, Horn DM, Breuker K, Ge Y, Lewis MA, Cerda B, Zubarev RA, Carpenter BK (2001) Electron capture dissociation of gaseous multiply charged ions by Fourier-transform ion cyclotron resonance. J Am Soc Mass Spectrom 12:245–249

McLuckey SA, Stephenson JL Jr (1998) Ion/ion chemistry of high-mass multiply charged ions. Mass Spectrom Rev 17:369–407

Meng F, Forbes AJ, Miller LM, Kelleher NL (2005) Detection and localization of protein modifications by high resolution tandem mass spectrometry. Mass Spectrom Rev 24:126–134

Meyer MR, Lichti CF, Townsend RR, Rao AG (2011) Identification of *in vitro* autophosphorylation sites and effects of phosphorylation on the Arabidopsis CRINKLY4 (ACR4) receptor-like kinase intracellular domain: insights into conformation, oligomerization, and activity. Biochemistry 50(12):2170–2186

Mikesh LM, Ueberheide B, Chi A, Coon JJ, Syka JEP, Shabanowitz J, Hunt DF (2006) The utility of ETD mass spectrometry in proteomic analysis. Biochim Biophys Acta 1764:1811

Miller ML, Jensen LJ, Diella F, Jorgensen C, Tinti M, Li L, Hsiung M, Parker SA, Bordeaux J, Sicheritz-Ponten T, Olhovsky M, Pasculescu A, Alexander J, Knapp S, Blom N, Bork P, Li S, Cesareni G, Pawson T, Turk BE, Yaffe MB, Brunak S, Linding R (2008) Linear motif atlas for phosphorylation-dependent signaling. Sci Signal 1:ra2

Mithoe SC, Menke FL (2011) Phosphoproteomics perspective on plant signal transduction and tyrosine phosphorylation. Phytochemistry 72(10):997–1006

Mitra SK, Gantt JA, Ruby JF, Clouse SD, Goshe MB (2007) Membrane proteomic analysis of *Arabidopsis thaliana* using alternative solubilization techniques. J Proteome Res 6:1933–1950

Mitra SK, Walters BT, Clouse SD, Goshe MB (2009) An efficient organic solvent based extraction method for the proteomic analysis of Arabidopsis plasma membranes. J Proteome Res 8:2752–2767

Miyahara A, Hirani TA, Oakes M, Kereszt A, Kobe B, Djordjevic MA, Gresshoff PM (2008) Soybean nodule autoregulation receptor kinase phosphorylates two kinase-associated protein phosphatases *in vitro*. J Biol Chem 283:25381–25391

Molina H, Horn DM, Tang N, Mathivanan S, Pandey A (2007) Global proteomic profiling of phosphopeptides using electron transfer dissociation tandem mass spectrometry. Proc Natl Acad Sci USA 104:2199–2204

Motoyama A, Xu T, Ruse CI, Wohlschlegel JA, Yates JR 3rd (2007) Anion and cation mixed-bed ion exchange for enhanced multidimensional separations of peptides and phosphopeptides. Anal Chem 79:3623–3634

Mu J-H, Lee H-S, Kao T-H (1994) Characterization of a pollen-expressed receptor-like kinase gene of *Petunia inflata* and the activity of its encoded kinase. Plant Cell 6:709–721

Nakagami H, Sugiyama N, Mochida K, Daudi A, Yoshida Y, Toyoda T, Tomita M, Ishihama Y, Shirasu K (2010) Large-scale comparative phosphoproteomics identifies conserved phosphorylation sites in plants. Plant Physiol 153:1161–1174

Nam KH, Li J (2002) BRI1/BAK1, a receptor kinase pair mediating brassinosteroid signaling. Cell 110:203–212

Nelson CJ, Huttlin EL, Hegeman AD, Harms AC, Sussman MR (2007) Implications of ^{15}N-metabolic labeling for automated peptide identification in *Arabidopsis thaliana*. Proteomics 7:1279–1292

Nuhse TS, Stensballe A, Jensen ON, Peck SC (2003) Large-scale analysis of *in vivo* phosphorylated membrane proteins by immobilized metal ion affinity chromatography and mass spectrometry. Mol Cell Proteomics 2:1234–1243

Nuwaysir LM, Stults JT (1993) Electrospray ionization mass spectrometry of phosphopeptides isolated by on-line immobilized metal-ion affinity chromatography. J Am Soc Mass Spectrom 4:662–669

Oda Y, Owa T, Sato T, Boucher B, Daniels S, Yamanaka H, Shinohara Y, Yokoi A, Kuromitsu J, Nagasu T (2003) Quantitative chemical proteomics for identifying candidate drug targets. Anal Chem 75:2159

Oh MH, Ray WK, Huber SC, Asara JM, Gage DA, Clouse SD (2000) Recombinant brassinosteroid insensitive 1 receptor-like kinase autophosphorylates on serine and threonine residues and phosphorylates a conserved peptide motif *in vitro*. Plant Physiol 124:751–766

Oh MH, Clouse SD, Huber SC (2009a) Tyrosine phosphorylation in brassinosteroid signaling. Plant Signal Behav 4:1–4

Oh MH, Wang X, Kota U, Goshe MB, Clouse SD, Huber SC (2009b) Tyrosine phosphorylation of the BRI1 receptor kinase emerges as a component of brassinosteroid signaling in Arabidopsis. Proc Natl Acad Sci USA 106:658–663

Oh M-H, Wang X, Wu X, Zhao Y, Clouse SD, Huber SC (2010) Autophosphorylation of Tyr-610 in the receptor kinase BAK1 plays a role in brassinosteroid signaling and basal defense gene expression. Proc Natl Acad Sci USA 107:17827–17832

Ong S-E, Blagoev B, Kratchmarova I, Kristensen DB, Steen H, Pandey A, Mann M (2002) Stable isotope labeling by amino acids in cell culture, SILAC, as a simple and accurate approach to expression proteomics. Mol Cell Proteomics 1:376

Pawson T (2004) Specificity in signal transduction: from phosphotyrosine-SH2 domain interactions to complex cellular systems. Cell 116:191–203

Peck SC (2006) Phosphoproteomics in Arabidopsis: moving from empirical to predictive science. J Exp Bot 57:1523–1527

Petutschnig EK, Jones AM, Serazetdinova L, Lipka U, Lipka V (2010) The lysin motif receptor-like kinase (LysM-RLK) CERK1 is a major chitin-binding protein in *Arabidopsis thaliana* and subject to chitin-induced phosphorylation. J Biol Chem 285:28902–28911

Pincus D, Letunic I, Bork P, Lim WA (2008) Evolution of the phospho-tyrosine signaling machinery in premetazoan lineages. Proc Natl Acad Sci USA 105:9680–9684

Pinkse MW, Uitto PM, Hilhorst MJ, Ooms B, Heck AJ (2004) Selective isolation at the femtomole level of phosphopeptides from proteolytic digests using 2D-NanoLC-ESI-MS/MS and titanium oxide precolumns. Anal Chem 76:3935–3943

Rahimi RA, Leof EB (2007) TGF-beta signaling: a tale of two responses. J Cell Biochem 102:593–608

Ross PL, Huang YN, Marchese JN, Williamson B, Parker K, Hattan S, Khainovski N, Pillai S, Dey S, Daniels S, Purkayastha S, Juhasz P, Martin S, Bartlet-Jones M, He F, Jacobson A, Pappin DJ (2004) Multiplexed protein quantitation in *Saccharomyces cerevisiae* using amine-reactive isobaric tagging reagents. Mol Cell Proteomics 3:1154

Russinova E, Borst JW, Kwaaitaal M, Cano-Delgado A, Yin Y, Chory J, de Vries SC (2004) Heterodimerization and endocytosis of Arabidopsis brassinosteroid receptors BRI1 and AtSERK3 (BAK1). Plant Cell 16:3216–3229

Sakurai A, Yokota T, Clouse SD (eds) (1999) Brassinosteroids: steroidal plant hormones. Springer, Tokyo

Salem T, Mazzella A, Barberini ML, Wengier D, Motillo V, Parisi G, Muschietti J (2011) Mutations in two putative phosphorylation motifs in the tomato pollen receptor kinase LePRK2 show antagonistic effects on pollen tube length. J Biol Chem 286:4882–4889

Sano A, Nakamura H (2004a) Chemo-affinity of titania for the column-switching HPLC analysis of phosphopeptides. Anal Sci 20:565–566

Sano A, Nakamura H (2004b) Titania as a chemo-affinity support for the column-switching HPLC analysis of phosphopeptides: application to the characterization of phosphorylation sites in proteins by combination with protease digestion and electrospray ionization mass spectrometry. Anal Sci 20:861–864

Schaff JE, Mbeunkui F, Blackburn K, Bird DM, Goshe MB (2008) SILIP: a novel stable isotope labeling method for in planta quantitative proteomic analysis. Plant J 56:840–854

Schilling M, Knapp DR (2008) Enrichment of phosphopeptides using biphasic immobilized metal affinity-reversed phase microcolumns. J Proteome Res 7:4164–4172

Schlessinger J (2002) Ligand-induced, receptor-mediated dimerization and activation of EGF receptor. Cell 110:669–672

Schroeder MJ, Shabanowitz J, Schwartz JC, Hunt DF, Coon JJ (2004) A neutral loss activation method for improved phosphopeptide sequence analysis by quadrupole ion trap mass spectrometry. Anal Chem 76:3590

Schulze WX (2010) Proteomics approaches to understand protein phosphorylation in pathway modulation. Curr Opin Plant Biol 13(3):280–287

Schulze WX, Usadel B (2010) Quantitation in mass-spectrometry-based proteomics. Annu Rev Plant Biol 61:491–516

Shah K, Vervoort J, de Vries SC (2001) Role of threonines in the *Arabidopsis thaliana* somatic embryogenesis receptor kinase 1 activation loop in phosphorylation. J Biol Chem 276:41263–41269

Shi SD, Hemling ME, Carr SA, Horn DM, Lindh I, McLafferty FW (2001) Phosphopeptide/phosphoprotein mapping by electron capture dissociation mass spectrometry. Anal Chem 73 (1):19

Shiu SH, Bleecker AB (2001) Plant receptor-like kinase gene family: diversity, function, and signaling. Sci STKE 2001:re22

Silva JC, Denny R, Dorschel C, Gorenstein MV, Li GZ, Richardson K, Wall D, Geromanos SJ (2006a) Simultaneous qualitative and quantitative analysis of the *Escherichia coli* proteome: a sweet tale. Mol Cell Proteomics 5:589–607

Silva JC, Gorenstein MV, Li GZ, Vissers JP, Geromanos SJ (2006b) Absolute quantification of proteins by LCMSE: a virtue of parallel MS acquisition. Mol Cell Proteomics 5:144–156

Stensballe A, Jensen ON, Olsen JV, Haselmann KF, Zubarev RA (2000) Electron capture dissociation of singly and multiply phosphorylated peptides. Rapid Commun Mass Spectrom 14:1793

Sugiyama N, Nakagami H, Mochida K, Daudi A, Tomita M, Shirasu K, Ishihama Y (2008) Large-scale phosphorylation mapping reveals the extent of tyrosine phosphorylation in Arabidopsis. Mol Syst Biol 4:193

Syka JE, Coon JJ, Schroeder MJ, Shabanowitz J, Hunt DF (2004) Peptide and protein sequence analysis by electron transfer dissociation mass spectrometry. Proc Natl Acad Sci USA 101:9528–9533

Tang W, Deng Z, Oses-Prieto JA, Suzuki N, Zhu S, Zhang X, Burlingame AL, Wang ZY (2008) Proteomics studies of brassinosteroid signal transduction using prefractionation and two-dimensional DIGE. Mol Cell Proteomics 7:728–738

Tang W, Deng Z, Wang ZY (2010) Proteomics shed light on the brassinosteroid signaling mechanisms. Curr Opin Plant Biol 13:27–33

Tang W, Yuan M, Wang R, Yang Y, Wang C, Oses-Prieto JA, Kim TW, Zhou HW, Deng Z, Gampala SS, Gendron JM, Jonassen EM, Lillo C, Delong A, Burlingame AL, Sun Y, Wang ZY (2011) PP2A activates brassinosteroid-responsive gene expression and plant growth by dephosphorylating BZR1. Nat Cell Biol 13:124–131

Tannu NS, Hemby SE (2006) Two-dimensional fluorescence difference gel electrophoresis for comparative proteomics profiling. Nat Protoc 1:1732

Thelen JJ, Peck SC (2007) Quantitative proteomics in plants: choices in abundance. Plant Cell 19:3339–3346

Tor M, Lotze MT, Holton N (2009) Receptor-mediated signalling in plants: molecular patterns and programmes. J Exp Bot 60:3645–3654

Torii KU (2004) Leucine-rich repeat receptor kinases in plants: structure, function, and signal transduction pathways. Int Rev Cytol 234:1–46

Tsai CF, Wang YT, Chen YR, Lai CY, Lin PY, Pan KT, Chen JY, Khoo KH, Chen YJ (2008) Immobilized metal affinity chromatography revisited: pH/acid control toward high selectivity in phosphoproteomics. J Proteome Res 7:4058–4069

Vert G, Nemhauser JL, Geldner N, Hong F, Chory J (2005) Molecular mechanisms of steroid hormone signaling in plants. Annu Rev Cell Dev Biol 21:177–201

Viswanathan S, Unlu M, Minden JS (2006) Two-dimensional difference gel electrophoresis. Nat Protoc 1:1351

Wang X, Chory J (2006) Brassinosteroids regulate dissociation of BKI1, a negative regulator of BRI1 signaling, from the plasma membrane. Science 313:1118–1122

Wang ZY, Seto H, Fujioka S, Yoshida S, Chory J (2001) BRI1 is a critical component of a plasma-membrane receptor for plant steroids. Nature 410:380–383

Wang ZY, Nakano T, Gendron J, He J, Chen M, Vafeados D, Yang Y, Fujioka S, Yoshida S, Asami T, Chory J (2002) Nuclear-localized BZR1 mediates brassinosteroid-induced growth and feedback suppression of brassinosteroid biosynthesis. Dev Cell 2:505–513

Wang X, Goshe MB, Soderblom EJ, Phinney BS, Kuchar JA, Li J, Asami T, Yoshida S, Huber SC, Clouse SD (2005a) Identification and functional analysis of *in vivo* phosphorylation sites of the Arabidopsis BRASSINOSTEROID-INSENSITIVE1 receptor kinase. Plant Cell 17:1685–1703

Wang X, Li X, Meisenhelder J, Hunter T, Yoshida S, Asami T, Chory J (2005b) Autoregulation and homodimerization are involved in the activation of the plant steroid receptor BRI1. Dev Cell 8:855–865

Wang X, Kota U, He K, Blackburn K, Li J, Goshe MB, Huber SC, Clouse SD (2008) Sequential transphosphorylation of the BRI1/BAK1 receptor kinase complex impacts early events in brassinosteroid signaling. Dev Cell 15:220–235

Washburn MP, Wolters D, Yates JR 3rd (2001) Large-scale analysis of the yeast proteome by multidimensional protein identification technology. Nat Biotechnol 19:242

Xu WH, Wang YS, Liu GZ, Chen X, Tinjuangjun P, Pi LY, Song WY (2006) The autophosphorylated Ser686, Thr688, and Ser689 residues in the intracellular juxtamembrane domain of XA21 are implicated in stability control of rice receptor-like kinase. Plant J 45:740–751

Yaffe MB (2002) Phosphotyrosine-binding domains in signal transduction. Nat Rev Mol Cell Biol 3:177–186

Yates JR 3rd, McCormack AL, Schieltz D, Carmack E, Link A (1997) Direct analysis of protein mixtures by tandem mass spectrometry. J Protein Chem 16:495

Yoshida S, Parniske M (2005) Regulation of plant symbiosis receptor kinase through serine and threonine phosphorylation. J Biol Chem 280:9203–9209

Yu LR, Zhu Z, Chan KC, Issaq HJ, Dimitrov DS, Veenstra TD (2007) Improved titanium dioxide enrichment of phosphopeptides from HeLa cells and high confident phosphopeptide identification by cross-validation of MS/MS and MS/MS/MS spectra. J Proteome Res 6:4150–4162

Zhang X, Ye J, Jensen ON, Roepstorff P (2007) Highly efficient phosphopeptide enrichment by calcium phosphate precipitation combined with subsequent IMAC enrichment. Mol Cell Proteomics 6(11):2032–2042

Zubarev RA (2004) Electron-capture dissociation tandem mass spectrometry. Curr Opin Biotechnol 15:12–16

Receptor Trafficking in Plants

Martina Beck and Silke Robatzek

Abstract Signalling through receptor-like kinases (RLKs) from the plasma membrane (PM) is important for cell-to-cell communication and for responding to environmental cues. RLKs not only reside at the PM but also localize to endosomes, and existing data provide good evidence that RLK signalling can occur from both locations. Constitutive as well as ligand-induced endocytosis of RLKs has been demonstrated, underlining the complexity of membrane trafficking pathways in plants for recycling RLKs back to the PM or targeting them for degradation.

1 Trafficking Pathways

Membrane compartmentalization and membrane trafficking is pivotal for plant cells. Secretory vesicles are required to deliver newly synthesized proteins to the plasma membrane (PM). Despite being questioned for decades, it now has been accepted that plants employ endocytic vesicles to uptake PM, PM-resident proteins and extracellular molecules, and that the types of endocytic vesicles exceed the complexity of those known from animal models (Jürgens 2004). Endosomal trafficking allows sorting of proteins to various destinations including the PM and the vacuole. If we are to understand protein functions, in particular the functions of signalling proteins such as RLKs, it is important to know their subcellular localization and dynamics.

1.1 Route to the Plasma Membrane

Newly synthesized proteins enter the endomembrane system from the endoplasmic reticulum (ER), where they can undergo maturation processes such as folding and

M. Beck • S. Robatzek (✉)
The Sainsbury Laboratory, Norwich Research Park, Norwich NR4 7UH, UK
e-mail: robatzek@TSL.ac.uk

F. Tax and B. Kemmerling (eds.), *Receptor-like Kinases in Plants*,
Signaling and Communication in Plants 13,
DOI 10.1007/978-3-642-23044-8_13,

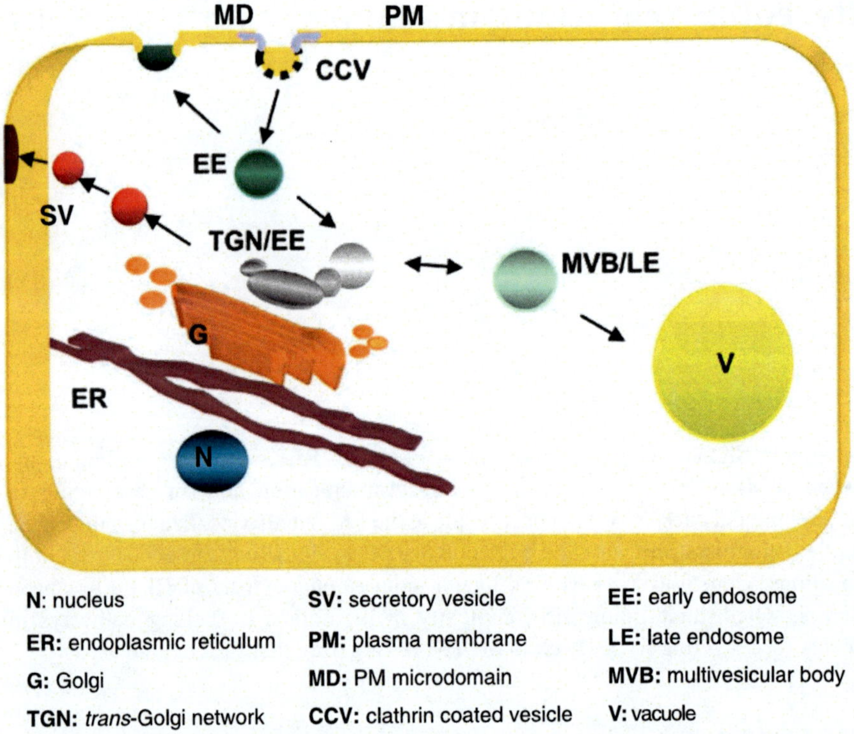

Fig. 1 Graphic depiction of membrane trafficking routes and membrane components in plant cells. Endocytosis starts at the PM with invagination of CC pits and budding of CCV inside the cell. CCVs lose their clathrin coats after internalization and enter either into a recycling pathway back to the PM or a degradation pathway to the vacuole via EE and MVB/LE. From the MVB, cargo can also be sorted back to the TGN/EE or back to the PM. New proteins for the PM are synthesized in the ER, sorted and modified in the Golgi and then transported via SV to the PM

glycosylation, before being transported to the Golgi apparatus (Fig. 1). Plants possess numerous Golgi compartments per cell, which is in contrast to animal cells typically with a single Golgi apparatus per cell. The plant Golgi apparatus consists of numerous Golgi stacks, and cis and trans sides can be distinguished by functional and morphological differences (Jürgens 2004). The proteins can be further modified and move from the cis- to the trans-cisternae, finally entering the trans-Golgi network (TGN). The TGN in plants represents a distinct organelle on its own and sorting events take place to target proteins either to the cell surface or to the vacuole (Viotti et al. 2010). The secretory and endocytic pathways converge at the TGN, a unique feature of the plant's membrane trafficking system (Dettmer et al. 2006; Dhonukshe et al. 2007; Lam et al. 2007a; Viotti et al. 2010).

1.2 A Way into the Cell: Endocytosis

The first indications for the existence of endocytosis in plants were obtained two decades ago. It had long been postulated that endocytosis in plants cannot occur because of the high turgor pressure, which forces the PM against the plant cell wall. Electron-dense tracers have been applied to prove uptake via the PM into membrane-bound intracellular structures (Tanchak and Fowke 1987; Hillmer et al. 1988; Galway et al. 1993), and clathrin-coated pits associated with the PM have been identified in plant cells (Robinson and Hillmer 1990; Dhonukshe et al. 2007). Detailed observations of clathrin-coated vesicles (CCV) at the ultra-structural level revealed that plant CCVs are much smaller (50–80 nm in diameter) compared to animal CCVs (120 nm in diameter), possibly an adaptation in response to turgor pressure (Meckel et al. 2004). In the past years, a broad range of molecular markers for endosomes has been developed (Ueda et al. 2004; Uemura et al. 2004; Voigt et al. 2005; Jaillais et al. 2006; Ortiz-Zapater et al. 2006; Lam et al. 2007a, b; Geldner et al. 2009). Together with vital dyes such as FM1-43 and FM4-64 that can be used to trace the endocytic pathways (Vida and Emr 1995; Emans et al. 2002; Bolte et al. 2004), these markers are widely used to characterize endomembrane compartments and to analyze endocytic trafficking.

Endocytosis begins with the invagination of the PM, the retrieval of PM material including lipids, membrane-localized or -associated proteins and extracellular cargoes and internalization into endocytic vesicles (Fig. 1). Upon pinching-off, the vesicles mature into early endosomes, which can be associated with the TGN. Early endosomes can enter the recycling pathway and be targeted back to the PM or they can also fuse with other endosomal compartments and mature into multivesicular bodies (MVBs) or prevacuolar compartments (PVCs). These are functional homologues of late endosomes in animal cells, where cargoes are sorted within internal vesicles of the MVBs. In plants, the MVB appears to be an intermediate sorting compartment, from which vesicles can either be recycled back to the PM, enter into retrograde trafficking to the TGN, or be delivered to the lumen of the lytic vacuole for degradation (Robinson et al. 2008). In addition, MVBs can also be part of the secretory pathway by directing exosomes to the PM and secreting cargoes of intraluminal vesicles in the extracellular space (Mo et al. 2006).

In plants, clathrin seems to be the major coat protein involved in endocytic vesicle formation. Both TGN and PM are sites of the formation of clathrin-coated vesicles (Barth and Holstein 2004). The clathrin coat is composed of a three-legged structure referred to as a triskelion, each leg consisting of a clathrin heavy chain and a clathrin light chain. Triskelia are assembled at the vesicle formation sites at the PM and form a cage surrounding the invaginated membrane. Adaptor proteins (APs) together with accessory proteins, recruit PM-resident proteins into the forming CCVs. The adaptors recognize and specifically bind phospholipids and cytoplasmic domains of target proteins linking them to the clathrin molecules. Each AP complex is composed of four subunits called adaptins (Boehm and Bonifacino

2002). Four different types of AP complexes have been identified, and each of them is thought to be responsible for the recruitment of clathrin to specific donor compartments (Robinson and Bonifacino 2001; Kirchhausen 2009). Clathrin-mediated endocytosis can still occur in the absence of AP2 in animal cells revealing a cargo-specific function of AP complexes (Motley et al. 2003).

Recently, other molecules which can function as adaptor proteins, have been identified in a variety of eukaryotic cells. These include members of the Epsin family such as Epsin1, which plays a critical role in endocytosis (Chen et al. 1998). Plant Epsin and Epsin-related proteins have also been studied and shown to interact with clathrin, adaptins and phosphatidylinositol-3-phosphate [PtdIns(3)P] – the major lipid component of endosomes – and are also involved in membrane trafficking and cargo delivery (Song et al. 2006; Lee et al. 2007). Altogether, this demonstrates that the responsible genes required for clathrin-mediated endocytosis and adapter complexes are present in plants. Current work implicates clathrin-mediated endocytosis as the major mechanism for endocytic trafficking in plants (Dhonukshe et al. 2007).

2 Receptor Kinases

Membrane trafficking and endocytosis is important for PM-resident receptors in animals as well as in plants (Geldner and Robatzek 2008). Within plant genomes, a large number of genes coding for RLKs can be identified (Walker and Zhang 1990; Shiu and Bleecker 2003). They consist of a signal peptide, an extracellular region (ectodomain), a single transmembrane spanning domain and an intracellular kinase domain (Fig. 2). The presence of a signal peptide and a single transmembrane domain predicts PM localization, which has been revealed for a number of RLKs (Table 1). Plant RLKs have evolved a diverse complement of ectodomains including leucine-rich repeats (LRR) and self-incompatibility (S) domains (Cock et al. 2002; Shiu and Bleecker 2001). Often, conserved glycosylation motifs can be found within the ectodomains, indicative of maturation processes within the ER and the Golgi. In addition, RLKs carry typical endocytosis motifs in their cytosolic domains, and taken together, these features demonstrate that RLKs are subject to complex membrane trafficking pathways.

RLKs regulate a plethora of developmental processes and stress responses. Examples of well-characterized RLKs with known functions and subcellular localizations in plant development are the receptors for the plant hormone brassinosteroid BRI1 (Brassinosteroid Insensitive 1) and its co-receptor BAK1 (BRI1-Associated Kinase 1), as well as members of the SRK family (S-locus Receptor Kinase) mediating self-incompatibility (Russinova et al. 2004; Geldner et al. 2007; Ivanov and Gaude 2009). RLKs involved in plant pathogen defence are FLS2 (Flagellin Insensitive 2), which detects bacterial flagellin and forms a complex with BAK1; EFR (EF-Tu Receptor), which recognizes bacterial EF-Tu; and XA21, which confers resistance to Xanthomonas oryzae pv. oryzae race 6 upon

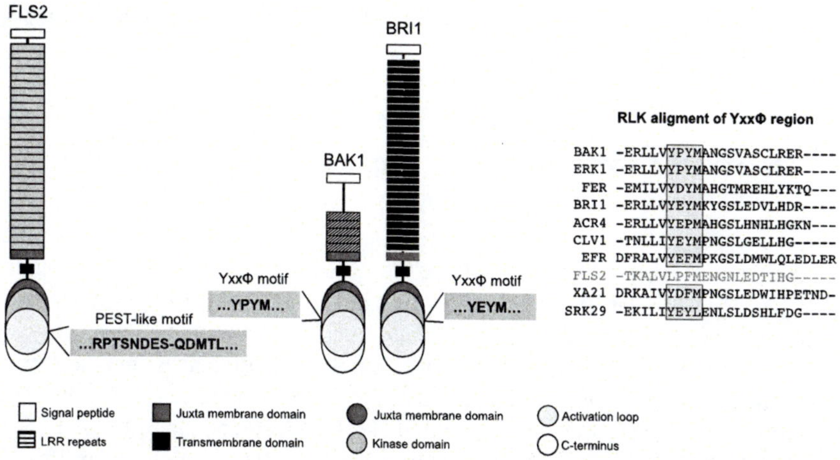

Fig. 2 Schematic representation of RLK domains and endocytosis motifs. The receptor for bacterial flagellin, FLS2, carries a PEST-like motif (sequence in *grey box*) in its kinase domains important for its endocytosis. By contrast, most RLKs such as BAK1 and BRI1 carry the tyrosine-based YxxΦ endocytic motif (sequences in *grey boxes*) at a conserved position in the kinase domain (alignment)

Table 1 Subcellular localization of RLKs. Listed are RLKs with demonstrated localizations and their function

RLK	Localization	Function	References
ACR4	Endosomes	Epidermal cell development	Gifford et al. (2005)
BAK1/SERK3	PM, endosomes	Brassinosteroid signalling, pathogen defence, co-receptor of BRI1 and FLS2	Russinova et al. (2004)
BRI1	PM, endosomes	Brassinosteroid signalling	Russinova et al. (2004), Geldner et al. (2005)
CLV1	PM	Stem-cell regulation	Bleckmann et al. (2010)
EFR	PM, ER	Pathogen defence	Häweker et al. (2010)
EVR	PM, endosomes	Organ abscission	Leslie et al. (2010)
FER	DRM, PM	Root hair growth, pathogen defence, female fertility	Keinath et al. (2010)
FLS2	DRM, PM, endosomes	Pathogen defence	Robatzek et al. (2006), Keinath et al. (2010)
SERK1	PM, endosomes	Anther development, organ abscission	Kwaaitaal et al. (2005)
SRK3	PM, MVB	Self-incompatibility	Ivanov and Gaude (2009)
XA21	PM, endosomes, ER	Pathogen defence	Chen et al. (2010)

perception of the secreted peptide Ax21 (Zipfel et al. 2004, 2006; Chinchilla et al. 2007a; Lee et al. 2009).

2.1 Receptor Localization at the Plasma Membrane

Delivery to the PM, targeted by the N-terminal signal peptide, is critical for the function of most RLKs. As RLKs undergo the steps of maturation, they are thought to be passaged through the ER and the Golgi. This includes N-glycosylation, a highly conserved co-translational modification of secreted proteins. N-glycosylation is essential for vesicle-mediated distribution of glycoproteins within the secretory pathway. Several conserved N-glycosylation sites, characterized by Nx(S/T) consensus motifs, can be identified within the LRR-ectodomains of e.g. FLS2, BAK1 and EFR (Häweker et al. 2010; Robatzek et al. 2007). Genetic and chemical interference of N-glycosylation causes mis-localization of fluorescent-tagged EFR and FLS2 in endomembrane structures, demonstrating that the transport of RLKs to the PM depends on proper protein decoration (Häweker et al. 2010; Nekrasov et al. 2009). The importance of correct protein maturation becomes evident from EFR and Xa21 interactions with ER chaperones, as well as genetic dependence of BRI1 and EFR on specific ER quality components (Park et al. 2010; Jin et al. 2007, 2009; Li et al. 2009; Nekrasov et al. 2009; Saijo et al. 2009; Lu et al. 2009). These data reveal that RLKs traffic through the ER and the Golgi for secretion to the PM.

2.2 Endocytosis Motifs in Receptors

The cytoplasmic domains of transmembrane proteins often carry endocytic or sorting motifs, short amino acid sequences recognized by adapter and sorting proteins, which in turn control trafficking and subcellular localization of these proteins. The tyrosine-based YXXΦ motif (whereby Φ represents an amino acid with a hydrophobic side chain) was first identified to be required for rapid internalization of low-density lipoprotein in animal cells (Davis et al. 1986). Protein sorting from the endosomal to the lysosomal pathway is conferred by a so-called PEST motif, a sequence region rich in proline (P), glutamic acid (E), serine (S) and threonine (T). The PEST motif is associated with proteins that have a short intracellular half-life, and acts as a signal peptide for protein degradation (Rogers et al. 1986). Modification of PEST motifs by ubiquitin is important for trafficking of proteins to MVBs and further routing for degradation (Babst et al. 1997; Seaman 2008). Current knowledge points to PM recycling of endosomal non-ubiquitinated proteins, while ubiquitination is required for both internalization and targeting to vacuoles (Seaman 2008).

Many RLKs including BRI1, BAK1 and EFR carry the YXXΦ endocytic motif, often at a conserved position in their kinase domains (Fig. 2), and therefore endocytosis of RLKs could be considered a more general phenomenon (Geldner and Robatzek 2008). Interestingly, the FLS2 cytosolic domain is devoid of any YXXΦ motif, an exception compared to all inspected RLKs (Geldner and Robatzek 2008; Fig. 2). FLS2 carries instead a PEST-like motif important for FLS2 endocytosis (Robatzek et al. 2006). Except for a few RLKs, which were shown to localize to endosomes (Table 1), a functional link between endocytic motifs and subcellular localization remains to be addressed for most of the RLKs.

3 Chemical Interference with RLK Trafficking

How can one dissect RLK trafficking and the endosomal compartments involved in the trafficking routes of membrane vesicles? One approach is to use specific inhibitors, which are known to block or interfere with distinct trafficking pathways (Drakakaki et al. 2009), and can provide insights into the dynamic subcellular localization and the role of RLK trafficking.

3.1 Filipin: Sterol Labelling and Interfering with Sterol-Dependent Endocytosis

Filipin is a polyene sterol-binding probe used for fluorescent detection of plant 3-β-hydroxy sterols that can be applied to study the distribution and trafficking of sterols. High concentrations of filipin and/or prolonged exposure to filipin effectively induce cross-linking of sterols, which in turn interferes with sterol distribution and function. Close interactions between sterol-enriched domains, vesicular recycling and endocytosis were reported for plants (Grebe et al. 2003; Boutté and Grebe 2009), fungi (Valdez-Taubas and Pelham 2003; Alvarez et al. 2007; Fischer et al. 2008; Takeshita et al. 2008) and algae (Klima and Foissner 2008). In fact, sterol-enriched domains recycle together with endocytic vesicles (Grebe et al. 2003).

3.2 Endosidin 1: Interfering with Early Endosomes

Endosidin 1 (ES1) was identified in a chemical screen of almost 2,000 compounds in a search for new inhibitors influencing pollen tube growth in Arabidopsis (Robert et al. 2008). As the growth of pollen tubes depends on membrane trafficking, detailed studies revealed that ES1 specifically affected early endosomes. ES1 treatment results in the formation of endomembrane, so-called "endosidin bodies", enriched with two early/TGN endosomal marker proteins (SYP61 and VHA-a1). Also, ES1 treatment affects the trafficking of endosomal AUX 1 and PIN2 (AUXIN

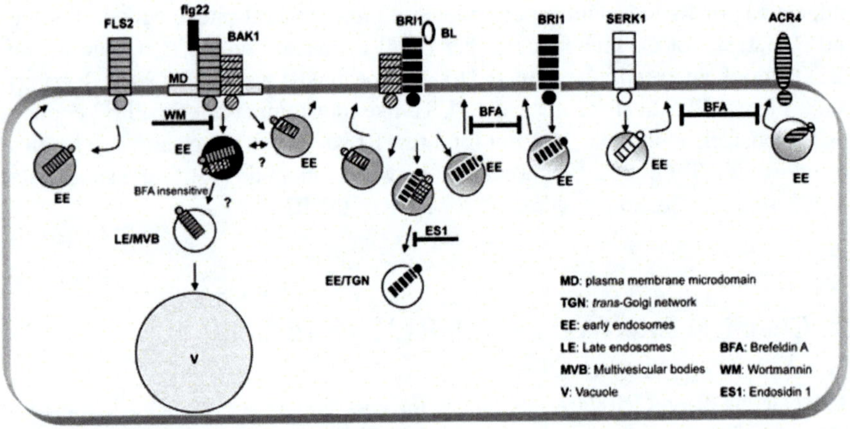

Fig. 3 Chemical dissection of RLK trafficking. FLS2 endocytosis after ligand perception can be blocked by WM, but then follows a BFA-insensitive route. BRI1 vesicles are sensitive to BFA and are constitutive recycled between the PM and RE, BRI1 can also be found on ES1-sensitive vesicles, which likely reflects EE/TGN. SERK1 and ACR4 are PM localized and constitutive recycled between BFA-sensitive vesicles and PM

TRANSPORTER 1 and the auxin efflux carrier PINFORMED 2, respectively), as well as BRI1 in Arabidopsis roots (Robert et al. 2008; Fig. 3). ES1 treatment also results in a seedling phenotype similar to BRI1 loss-of-function, which indicates the involvement of early endosomes in brassinosteroid signalling.

3.3 *Tyrphostin A23: Clathrin-Dependent Endocytosis*

A23 belongs to the Tyrphostins, which are tyrosine analogues developed as inhibitors of tyrosine kinases. Endocytosis of the human Transferrin Receptor (hTfR) expressed in animal cells, as well as Arabidopsis protoplasts, was specifically abolished by Tyrphostin A23 but not by other analogues such as Tyrphostin A51 (Perez-Gomez and Moore 2007). Tyrphostin A23 inhibits the interaction between YxxΦ motifs and the μ-subunit of the AP2 complex. Similarly, Tyrphostin A23 but not A51 affected endocytosis of several PM-resident proteins with polar and apolar distributions in Arabidopsis root tip cells (Dhonukshe et al. 2007). However, Tyrphostin A23 does not inhibit the uptake of the widely used endocytic tracer FM4-64. This is in contrast to the reported inhibition of FM4-64 internalization in protoplasts upon overexpression of the C-terminal part of clathrin heavy chain (the so-called clathrin hub fragment), which prevents proper triskelion assembly (Liu et al. 1995, Dhonukshe et al. 2007). This suggests that endocytosis in plants depends on clathrin formation and/or interaction with the μ-subunit of the AP2 complex.

3.4 Brefeldin A: Blocking the Recycling Pathway

One of the most popular drugs used to address vesicle-mediated protein trafficking is Brefeldin A (BFA), a macrocyclic lactone of fungal origin. Its specific target in plants is GNOM, an exchange factor for ARF GTPases (ARF-GEF), which localizes to endosomal compartments and blocks PM recycling in Arabidopsis roots (Geldner et al. 2003; Robinson et al. 2008). BFA treatment leads to the formation of large aggregations of TGN/early endosomal compartments, the so-called BFA-bodies, consisting of endocytosed proteins from the PM, which cannot be recycled back to the PM and are therefore "trapped" into BFA-bodies. However, even high concentrations of BFA can be washed out and the wild-type cellular phenotype can be recovered, an important tool for investigations of recycling processes at the PM.

BFA treatment triggers hTfR accumulation into larger patches inside of protoplasts, reminiscent of BFA-bodies, which reveals that the receptor cycles between the PM and endosomal compartments thereby showing similar trafficking properties in animal and plant cells (Ortiz-Zapater et al. 2006). Application of BFA also affects trafficking of RLKs in plants (Fig. 3). For example, BRI1 localizes to BFA-bodies in Arabidopsis root cells, demonstrating BRI1 trafficking through early/recycling endosomes (see below; Russinova et al. 2004). By contrast, BFA treatment showed no effect on ligand-induced endocytosis of FLS2 (see below; Robatzek et al. 2006). This suggests that inducibly endocytosed vesicles of FLS2 are not recycled back to the PM, but more likely follow a BFA-insensitive pathway to late endosomes for vacuolar degradation of the receptor.

3.5 Wortmannin: Addressing the Role of MVBs/PVCs

Wortmannin is a specific inhibitor of phosphatidylinositol 3-kinase responsible for the production of [PtdIns(3)P], a lipid characteristic of endosomal membranes. Treatment with Wortmannin causes an obvious swelling of MVBs in both mammalian and plant cells, and has therefore been used in several studies to follow membrane trafficking in the plant secretory and endocytic pathways (Reaves et al. 1996; Bright et al. 2001; Tse et al. 2004; Jaillais et al. 2006; Lam et al. 2007a). The enlargement of MVBs is a result of MVB/PVC fusion into small vacuoles (Wang et al. 2009), and Wortmannin-induced vacuolation is a useful tool for the identification of MVBs/PVCs (Tse et al. 2004; Miao et al. 2006; Lam et al. 2007a). Another effect of Wortmannin is the inhibition of endocytic uptake. FLS2 endocytosis is disturbed upon Wortmannin treatment (see below; Fig. 3; Robatzek et al. 2006), and reduces downstream MAPK signalling activated by flg22, the 22-amino acid peptide elicitor-active surrogate of bacterial flagellin (Chinchilla et al. 2007b). This provides evidence for FLS2 trafficking via MVBs/

PVCs and indicates a link between FLS2 endocytosis and activation of downstream signalling kinases.

4 Plasma Membrane Microdomains in Receptor-Mediated Endocytosis

Current work reports on RLKs localizing to PM microdomains, previously referred to as "lipid rafts". PM microdomains are enriched in cholesterol and sphingolipids, representing the liquid-ordered phase of membranes (Zappel and Panstruga 2008). In animals, they play a role in compartmentalization of cell signalling events and the subcellular trafficking of membrane-resident proteins (Menon 2002). PM microdomains can be biochemically purified as detergent-resistant membranes (DRMs; Borner et al. 2005; Minami et al. 2009). Plant DRMs are enriched with phytosphingolipids and phytosterols as well as flotillin-like proteins. The latter are known markers of PM microdomains in animals, also present in plants and localized to the PM (Borner et al. 2005). Structural sterols are implicated in the polar localization of PIN proteins and in the polar growth of tip-growing plant cells (Willemsen et al. 2003; Ovecka et al. 2010). Also, structural sterols and PM microdomains are involved in the recycling of plant proteins, and sterols are internalized into endocytic compartments (Grebe et al. 2003; Willemsen et al. 2003, Ovecka et al. 2010).

4.1 Receptor Localization to PM Microdomains

DRMs are not only enriched in structural sterols but also contain numerous protein kinases and other signalling proteins such as small GTPases (Mongrand et al. 2004; Borner et al. 2005; Morel et al. 2006; Lefebvre et al. 2007; Minami et al. 2010, Keinath et al. 2010; Sorek et al. 2010). A number of RLKs were recently reported to localize to PM microdomains. FLS2 was enriched in DRMs after activation of the receptor by its ligand, flg22 (Keinath et al. 2010). Other RLKs including FERONIA (FER) were also identified from flg22-induced PM microdomains. The localization of BAK1, clathrin heavy chain, actin and remorins to DRMs, further supports the role of PM microdomains for endocytic uptake in plants (Morel et al. 2006; Keinath et al. 2010). Heterodimerization of FLS2 and BAK1 may cause the reported decrease in lateral mobility of PM-resident FLS2 upon flg22 application (Ali et al. 2007). This high lateral mobility points to continuous receptor trafficking, and the reduction in mobility of activated FLS2 may be a prerequisite for establishing signalling cascades. PM microdomain localization of receptors appears to be important for downstream signalling events, consistent with previous evidence for signalling of Toll-like receptors in animals (Triantafilou et al. 2004). It is therefore quite possible that PM microdomains serve as subcellular signalling platforms.

4.2 The CrRLK FERONIA

FER is a member of the CrRLK family, receptor kinases related to Catharanthus RLK1, and was first identified as a regulator of female fertility in Arabidopsis important for terminating pollen tube growth (Escobar-Restrepo et al. 2007), but recently was demonstrated to function in a variety of processes. These include a role in cell elongation, in which FER participates together with two other RLKs of the same subfamily, HERKULES1 and THESEUS1 (Guo et al. 2009), in root hair elongation (Duan et al. 2010), in brassinosteroid and ethylene responsiveness (Deslauriers and Larsen 2010), and in plant defence (Keinath et al. 2010; Kessler et al. 2010). Interestingly, FER acts upstream of ROP signalling mediating reactive oxygen species (ROS) production by NADPH oxidases in root hair elongation as well as flg22-elicited plant defence, thus providing evidence for FER as a regulator of a common downstream messenger. FER is known to reside in the PM (Escobar-Restrepo et al. 2007), and was recently isolated from flg22-induced DRMs (Keinath et al. 2010). This points to endocytic trafficking of FER and may indicate a role as sensor of cell surface integrity similar to THESEUS1.

5 Receptor Kinase Localization to Endosomes

To date, reports of the subcellular localization and dynamics of RLKs are limited, mostly because this requires the generation of functional fluorescent-tagged fusion proteins, transgenic expression and advanced microscopic tools or methods. Nevertheless, data are available for members of different RLK subfamilies indicating that endocytic trafficking is often associated with the dynamic localization of RLKs (Geldner and Robatzek 2008). Moreover, internalization of the hTfR, a typical clathrin-specific cargo, has been demonstrated in plants by heterologous expression of this receptor in protoplasts (Ortiz-Zapater et al. 2006). This provides compelling evidence of clathrin-mediated endocytosis as well as receptor-mediated endocytosis in plants.

5.1 SERK1

The Arabidopsis thaliana Somatic Embryogenesis Receptor Kinases (SERKs) represent a small family of five closely related RLKs with an ectodomain consisting of an N-terminal leucine zipper and five LRRs and serve as co-regulators of multiple RLK pathways. SERK1 functions in anther development and brassinosteroid signalling (Albrecht et al. 2008). A SERK1 fluorescent protein fusion was expressed throughout diverse cell types, and detected at the PM and in small vesicle-like compartments (Kwaaitaal et al. 2005). Its rapid internalization into BFA-bodies shows that SERK1 is constitutively recycled between the PM and detected vesicles

(Fig. 3). Like many RLKs, SERK1 interacts with the Kinase-Associated Protein Phosphatase (KAPP; Stone et al. 1994; Torii, 2000; Gomez-Gomez et al. 2001; Rienties et al. 2005), which appears important for signal transduction (Li et al. 1999). Interestingly, KAPP is required for SERK1 endocytosis (Shah et al. 2002). Both proteins are present at the PM and in intracellular vesicles, but interaction occurs intracellularly in a phosphorylation-dependent manner. Thus, it seems possible that dephosphorylation of RLKs by KAPP and receptor internalization are linked mechanisms for down-regulation of signalling. Recently, SERK1 has been implicated in organ abscission (Lewis et al. 2010), which is also under the control of the RLK Evershed (EVR), shown to localize to the PM and to FM4-64 co-labelled vesicles indicative of endocytosis (Leslie et al. 2010).

5.2 *BRI1 and BAK1*

BRI1 is the receptor for brassinosteroids, steroidal plant hormones that promote cell expansion and division, and mediates brassinosteroid ligand perception through its LRR ectodomain (He et al. 2000). Activated BRI1 forms a complex with BAK1, a member of the SERK subfamily (also called SERK3), which is required for brassinosteroid signalling. Fluorescently labelled BRI1 localizes to the PM and endosomal compartments, confirmed by co-localization of BRI1 vesicles with the endocytic tracer FM4-64 (Russinova et al. 2004). BRI1 undergoes constitutive endocytic recycling in a BFA-sensitive manner and does not change upon exogenous addition of its ligand (Fig. 3). Similarly, BAK1 constitutively recycles between the PM and endocytic compartments (Aker and de Vries 2008). However, the interaction of BRI1 and its co-receptor BAK1 is restricted to only small domains of the PM and to vesicles near the PM or in the cytoplasm (Russinova et al. 2004). Co-expression of BRI1 and BAK1 resulted in enhanced internalization of endocytic vesicles containing either one or both of the receptors, which suggest receptor-specific sorting (Russinova et al. 2004). Interestingly, endosomal BRI1 is not associated with BKI1, an inhibitory protein released from the PM upon BRI1 activation (Wang et al. 2006), which suggests that BRI1 is active in endosomes. Endosomal signalling of BRI1 is further supported by enhanced brassinosteroid signalling in both cultured cells and Arabidopsis seedlings in the presence of BFA (Russinova et al. 2004; Geldner et al. 2007).

5.3 *CRINKLY4*

Intracellular localization at vesicles was also described for the non-LRR-RLK Arabidopsis CRINKLY4 (ACR4), which is important in epidermal cell development (Gifford et al. 2005). Fluorescent-tagged ACR4 was observed at the PM and endosomal vesicles, based on the partial co-labelling with FM4-64 stained endosomes (Gifford et al. 2005; Tian et al. 2007). ACR4 undergoes recycling, because ACR4 vesicles are internalized into BFA-bodies (Fig. 3).

5.4 *SRK*

Endosomal localization of receptor kinases also plays a role during the self-incompatibility response. SRK3 and its ligand, S-locus Cysteine Rich (SCR) protein are expressed in anthers and the pollen coat and prevent self-fertilization in Brassica (Gaude et al. 1993). SRK3 mainly localizes to endosomal compartments and is only weakly detected at the PM (Ivanov and Gaude 2009). Nevertheless, perception and recognition of the SCR ligand occurs at the PM, inducing SRK3 endocytosis. Internalized SRK3 is further routed to sorting endosomes (MVBs/PVCs), where it is deactivated by its negative regulator THL1 (Thioredoxin-H-Like) and degraded into the vacuole (Ivanov and Gaude 2009). Whether SRK3 endocytosis is needed for signalling or triggers termination of the signal remains to be addressed.

5.5 *FLS2*

FLS2 encodes an LRR-RLK responsible for the detection of bacterial flagellin, a conserved pathogen-associated molecular pattern (PAMP), through its elicitor-active peptide flg22 (Gomez-Gomez and Boller 2002; Chinchilla et al. 2006), and is a critical component of plant immunity to bacterial infection (Zipfel et al. 2004). A functional fluorescent FLS2 fusion protein resides at the PM and becomes internalized into highly mobile vesicles specifically upon addition of flg22, the first example of ligand-induced receptor-mediated endocytosis in plants (Robatzek et al. 2006). FLS2-containing vesicles can be detected at the cell periphery and intracellularly in different cell types (Fig. 4). Flg22-induced internalization of FLS2 depends on a PEST-like sequence as FLS2 lacks a typical YxxΦ endocytic motif (see above), which indicates endocytic trafficking through MVBs (Hurley 2008). This is further supported by the inhibition of flg22-induced FLS2 endocytosis in the presence of Wortmannin (Fig. 3). Wortmannin treatment reduces flg22-elicited activation of signalling MAP kinases, providing the first evidence of FLS2 endocytosis in downstream signalling (Chinchilla et al. 2007b). Flg22-induced FLS2 endocytosis and downstream signalling are impaired by a point mutation of threonine 867, a highly conserved residue in the juxta-membrane domain of RLKs (Robatzek et al. 2006), further pointing to a link between FLS2 endocytosis and signalling.

Like BRI1, FLS2 forms an inducible complex with BAK1 upon ligand binding (Chinchilla et al. 2007a; Heese et al. 2007). Although BAK1 is necessary for flg22-induced FLS2 endocytosis, subcellular localization of the FLS2-BAK1 interaction remains to be shown (Fig. 3). As FLS2 undergoes ligand-dependent internalization

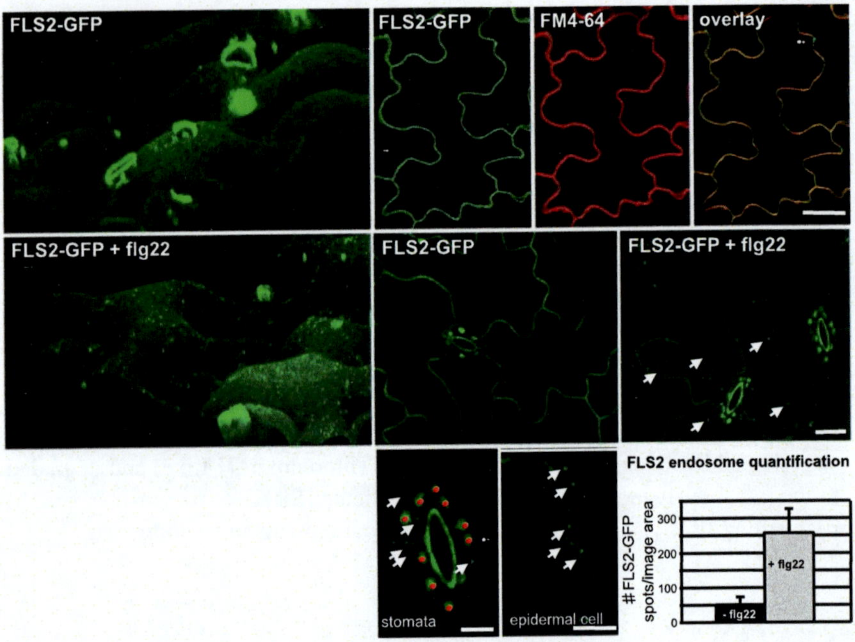

Fig. 4 FLS2 subcellular localization. Micrographs show confocal images. *Left panels*: FLS2-GFP localization in Arabidopsis seedling cotyledons before and after flg22 treatment (3D projection of epidermal tissue). *Right panels*: FLS2 is localized to PM (FM4-64 stained), and after flg22 perception FLS2 gets internalized into vesicles in epidermal leaves cells and stomata (indicated by *arrows*). *Lower panels*: detailed images of FLS2-GFP endosomes in guard cells of stomata and epidermal cells (*arrows*; *red spots* distinguish autofluorescence from chloroplasts). *Bars* represent 10 μm or 5 μm, respectively. *Bottom right panel*: Quantification of FLS2-GFP endosomes in the absence and presence of flg22 ligand

and BAK1 constitutive recycling, one can speculate that FLS2 and BAK1 are first endocytosed together and then sorted to their individual destinations.

5.6 XA21

Plant receptor trafficking is mostly studied in Arabidopsis but there is also evidence for RLK endocytosis in rice. XA21 is an LRR-RLK responsible for the detection of the PAMP Ax21 produced by Xanthomonas oryzae pv. oryzae, thereby conferring resistance to this bacterial pathogen (Lee et al. 2009). Overexpression of XA21 fused to a fluorescent protein showed localization to the PM of rice root cells and protoplasts (Chen et al. 2010). Although no constitutive partitioning between the PM and endosomal compartments was observed, XA21 accumulated into BFA-bodies indicative of endocytic trafficking. Autophosphorylation and kinase activity of Xa21 appeared not to affect Xa21 subcellular localization (Chen et al. 2010).

6 Receptor Signalling from Endosomes?

To date, in animals the phenomenon of "signalling endosomes" is well described (Howe et al. 2001). Although endocytosis was long regarded as a component of degradation or recycling of cell surface receptors, current knowledge clearly points to endosomal compartments contributing to signal transduction. Receptors can transmit signals from endosomes that are different from those arising at the PM, resulting in distinct physiological responses, as e.g. reported for the Nerve Growth Factor Receptor (NGFR) and the Epidermal Growth Factor Receptor (EGFR; Howe et al. 2001; Murphy et al. 2009). Nerve Growth Factor (NGF) and its receptor NGFR are sequestered into mobile CCVs, which are the origin of MAPK signal activation (Howe et al. 2001). EGFR mis-localization to late endosomes caused prolonged EGFR activity and sustained MAPK signalling (Taub et al. 2007).

There is also accumulating evidence for endosomes as a site for receptor signalling in plants. Elevated endosomal localization of BRI1 upon overexpression increased downstream signalling and transcriptional responses, indicative of endosomes playing a role in receptor signalling (Russinova et al. 2004). Furthermore, accumulation of BRI1 into BFA-bodies resulted in enhanced brassinosteroid signalling, strong evidence that plant RLKs can convey signals from endosomes (Geldner et al. 2007). Recent findings demonstrated that MPK6, a signalling MAP kinase, which is involved in transduction of abiotic and biotic stress responses, localizes to early endosomes in Arabidopsis (Müller et al. 2010). In animal cells, it is well known that tightly regulated spatial subcellular organization of MAPK is essential for proper signalling, and is one major factor to gain signal specificity. Association of MAPKs with PM, CCVs and endosomes have been shown to strongly influence cellular behaviour including the regulation of endocytic trafficking and immune responses (Fehrenbacher et al. 2009). Such regulation of MAPK may be also relevant in plants. Moreover, RLK internalization and endosomal signalling may be essential in plants due to the large number of RLKs that must be displayed at the PM, as well as the hindrance imposed by the large centrally located vacuole to access of signals to the nucleus (Robatzek and Geldner 2008). Despite the recent advances reviewed here, most changes in RLK trafficking and their roles in plant development and stress responses remain elusive, and further study is needed to fully understand signalling processes coupled with endocytosis in plants.

7 Conclusions

In animals, receptor kinase signalling is relevant from both, the cell surface and trafficking endosomes. In fact, endosomal signalling is an integral part of the signal transduction cascades for a number of well-studied receptor kinases such as the epidermal growth factor receptor (EGFR). Knowledge obtained in the recent past now also provides good evidence for active RLK signalling from endosomal compartments in plants. This includes constitutive and ligand-induced receptor-mediated endocytosis based on pioneering work of the receptors BRI1 and FLS2 functioning in plant growth and defence, respectively. Receptor kinase signalling from endosomes can serve to extend the signalling surface to amplify the signal for a robust and sustained response, to limit signalling spatially within the cell and over time. This plasticity in turn allows fine-tuning and efficient control over signalling pathways, and may have been a requirement for the development of complex multicellular organisms in order to integrate multiple signals from neighbouring cells as well as from the environment. However, several important questions need to be addressed whether we are to understand RLK endocytosis and its interception with signalling pathways: What are the subcellular localizations and dynamics of RLK family members? What is the subcellular localization of the immediate downstream components of RLKs? Which are the molecular components regulating RLK endocytosis? These challenging questions illustrate the need for advancing cell biological tools that in future would allow monitoring trafficking of active signalling receptors. A combination of cell biological, biochemical and genetic approaches over the next years will likely lead to the discovery of novel components involved in RLK trafficking, therefore allowing mechanistic insights into plant endocytosis and signalling.

References

Albrecht C, Russinova E, Kemmerling B, Kwaaitaal M, de Vries SC (2008) Arabidopsis SOMATIC EMBRYOGENESIS RECEPTOR KINASE proteins serve brassinosteroid-dependent and -independent signaling pathways. Plant Physiol 148:611–619

Ali GS, Prasad KV, Day I, Reddy AS (2007) Ligand-dependent reduction in the membrane mobility of FLAGELLIN SENSITIVE2, an Arabidopsis receptor-like kinase. Plant Cell Physiol 48:1601–1611

Alvarez FJ, Douglas LM, Konopka JB (2007) Sterol-rich plasma membrane domains in fungi. Eukaryot Cell 6:755–763

Babst M, Sato TK, Banta LM, Emr SD (1997) Endosomal transport function in yeast requires a novel AAA-type ATPase, Vps4p. EMBO J 16:1820–1831

Barth M, Holstein SE (2004) Identification and functional characterization of Arabidopsis AP180, a binding partner of plant alphaC-adaptin. J Cell Sci 15:2051–2062

Bleckmann A, Weidtkamp-Peters S, Seidel CA, Simon R (2010) Stem cell signaling in Arabidopsis requires CRN to localize CLV2 to the plasma membrane. Plant Physiol 152:166–176

Boehm M, Bonifacino JS (2002) Genetic analyses of adaptin function from yeast to mammals. Gene 286:175–186

Bolte S, Talbot C, Boutte Y, Catrice O, Read ND, Satiat-Jeunemaitre B (2004) FM-dyes as experimental probes for dissecting vesicle trafficking in living plant cells. J Microsc 214:159–173

Borner GH, Sherrier DJ, Weimar T, Michaelson LV, Hawkins ND, Macaskill A, Napier JA, Beale MH, Lilley KS, Dupree P (2005) Analysis of detergent-resistant membranes in Arabidopsis. Evidence for plasma membrane lipid rafts. Plant Physiol 137:104–116

Boutté Y, Grebe M (2009) Cellular processes relying on sterol function in plants. Curr Opin Plant Biol 12:705–713

Bright NA, Lindsay MR, Stewart A, Luzio JP (2001) The relationship between lumenal and limiting membranes in swollen late endocytic compartments formed after wortmannin treatment or sucrose accumulation. Traffic 2:631–642

Chen H, Fre S, Slepnev VI, Capua MR, Takei K, Butler MH, Di Fiore PP, De Camilli P (1998) Epsin is an EH-domain-binding protein implicated in clathrin-mediated endocytosis. Nature 394:793–797

Chen F, Gao MJ, Miao YS, Yuan YX, Wang MY, Li Q, Mao BZ, Jiang LW, He ZH (2010) Plasma membrane localization and potential endocytosis of constitutively expressed XA21 proteins in transgenic rice. Mol Plant 3:917–926

Chinchilla D, Bauer Z, Regenass M, Boller T, Felix G (2006) The Arabidopsis receptor kinase FLS2 binds flg22 and determines the specificity of flagellin perception. Plant Cell 18:465–476

Chinchilla D, Zipfel C, Robatzek S, Kemmerling B, Nürnberger T, Jones JD, Felix G, Boller T (2007a) A flagellin-induced complex of the receptor FLS2 and BAK1 initiates plant defence. Nature 448:497–500

Chinchilla D, Boller T, Robatzek S (2007b) Flagellin signalling in plant immunity. Adv Exp Med Biol 598:358–371

Cock JM, Vanoosthuyse V, Gaude T (2002) Receptor kinase signalling in plants and animals: distinct molecular systems with mechanistic similarities. Curr Opin Cell Biol 14:230–236

Davis CG, Lehrman MA, Russell DW, Anderson RG, Brown MS, Goldstein JL (1986) The J.D. mutation in familial hypercholesterolemia: amino acid substitution in cytoplasmic domain impedes internalization of LDL receptors. Cell 45:15–24

Deslauriers SD, Larsen PB (2010) FERONIA is a key modulator of brassinosteroid and ethylene responsiveness in Arabidopsis hypocotyls. Mol Plant 3:626–640

Dettmer J, Hong-Hermesdorf A, Stierhof YD, Schumacher K (2006) Vacuolar H1 ATPase activity is required for endocytic and secretory trafficking in Arabidopsis. Plant Cell 18:715–730

Dhonukshe P, Aniento F, Hwang I, Robinson DG, Mravec J, Stierhof YD, Friml J (2007) Clathrin-mediated constitutive endocytosis of PIN auxin efflux carriers in Arabidopsis. Curr Biol 17:520–527

Drakakaki G, Robert S, Raikhel NV, Hicks GR (2009) Chemical dissection of endosomal pathways. Plant Signal Behav 4:57–62

Duan Q, Kita D, Li C, Cheung AY, Wu HM (2010) FERONIA receptor-like kinase regulates RHO GTPase signaling of root hair development. Proc Natl Acad Sci USA 107:17821–17826

Emans N, Zimmermann S, Fischer R (2002) Uptake of a fluorescent marker in plant cells is sensitive to brefeldin A and wortmannin. Plant Cell 14:71–86

Fehrenbacher N, Bar-Sagi D, Philips M (2009) Ras/MAPK signaling from endomembranes. Mol Oncol 3:297–307

Fischer R, Zekert N, Takeshita N (2008) Polarized growth in fungi–interplay between the cytoskeleton, positional markers and membrane domains. Mol Microbiol 68:813–826

Galway ME, Rennie PJ, Fowke LC (1993) Ultrastructure of the endocytotic pathway in glutaraldehyde-fixed and high-pressure frozen/freeze-substituted protoplasts of white spruce (*Picea glauca*). J Cell Sci 106:847–858

Gaude T, Friry A, Heizmann P, Mariac C, Rougier M, Fobis I, Dumas C (1993) Expression of a self-incompatibility gene in a self-compatible line of *Brassica oleracea*. Plant Cell 5:75–86

Geldner N, Robatzek S (2008) Plant receptors go endosomal: a moving view on signal transduction. Plant Physiol 147:1565–1574

Geldner N, Anders N, Wolters H, Keicher J, Kornberger W, Muller P, Delbarre A, Ueda T, Nakano A, Jürgens G (2003) The Arabidopsis GNOM ARF-GEF mediates endosomal recycling, auxin transport, and auxin-dependent plant growth. Cell 112:219–230

Geldner N, Hyman DL, Wang X, Schumacher K, Chory J (2007) Endosomal signaling of plant steroid receptor kinase BRI1. Genes Dev 21:1598–1602

Geldner N, Dénervaud-Tendon V, Hyman DL, Mayer U, Stierhof YD, Chory J (2009) Rapid, combinatorial analysis of membrane compartments in intact plants with a multicolor marker set. Plant J 59:169–178

Gifford ML, Robertson FC, Soares DC, Ingram GC (2005) ARABIDOPSIS CRINKLY4 function, internalization, and turnover are dependent on the extracellular crinkly repeat domain. Plant Cell 17:1154–1166

Gómez-Gómez L, Boller T (2002) Flagellin perception: a paradigm for innate immunity. Trends Plant Sci 7:251–256

Gómez-Gómez L, Bauer Z, Boller T (2001) Both the extracellular leucine-rich repeat domain and the kinase activity of FSL2 are required for flagellin binding and signaling in Arabidopsis. Plant Cell 13:1155–1163

Grebe M, Xu J, Möbius W, Ueda T, Nakano A, Geuze HJ, Rook MB, Scheres B (2003) Arabidopsis sterol endocytosis involves actin-mediated trafficking via ARA6-positive early endosomes. Curr Biol 13:1378–1387

Guo H, Ye H, Li L, Yin Y (2009) A family of receptor-like kinases are regulated by BES1 and involved in plant growth in *Arabidopsis thaliana*. Plant Signal Behav 4:784–786

Häweker H, Rips S, Koiwa H, Salomon S, Saijo Y, Chinchilla D, Robatzek S, von Schaewen A (2010) A Pattern recognition receptors require N-glycosylation to mediate plant immunity. J Biol Chem 285:4629–4636

He Z, Wang ZY, Li J, Zhu Q, Lamb C, Ronald P, Chory J (2000) Perception of brassinosteroids by the extracellular domain of the receptor kinase BRI1. Science 288:2360–2363

Heese A, Hann DR, Gimenez-Ibanez S, Jones AM, He K, Li J, Schroeder JI, Peck SC, Rathjen JP (2007) The receptor-like kinase SERK3/BAK1 is a central regulator of innate immunity in plants. Proc Natl Acad Sci USA 104:12217–12222

Hillmer S, Freundt H, Robinson DG (1988) The partially coated reticulum and its relationship to the Golgi apparatus in higher plant cells. Eur J Cell Boil 47:206–212

Howe CL, Valletta JS, Rusnak AS, Mobley WC (2001) NGF signaling from clathrin-coated vesicles: evidence that signaling endosomes serve as a platform for the Ras-MAPK pathway. Neuron 32:801–814

Ivanov R, Gaude T (2009) Endocytosis and endosomal regulation of the S-receptor kinase during the self-incompatibility response in *Brassica oleracea*. Plant Cell 21:2107–2117

Jaillais Y, Fobis-Loisy I, Miege C, Rollin C, Gaude T (2006) AtSNX1 defines an endosome for auxin-carrier trafficking in Arabidopsis. Nature 443:106–109

Jin H, Yan Z, Nam KH, Li J (2007) Allele-specific suppression of a defective brassinosteroid receptor reveals a physiological role of UGGT in ER quality control. Mol Cell 26:821–830

Jin H, Hong Z, Su W, Li J (2009) A plant-specific calreticulin is a key retention factor for a defective brassinosteroid receptor in the endoplasmic reticulum. Proc Natl Acad Sci USA 106:13612–13617

Jurgens G (2004) Membrane trafficking in plants. Annu Rev Cell Dev Biol 20:481–504

Keinath NF, Kierszniowska S, Lorek J, Bourdais G, Kessler SA, Shimosato-Asano H, Grossniklaus U, Schulze WX, Robatzek S, Panstruga R (2010) PAMP (Pathogen-associated Molecular Pattern)-induced changes in plasma membrane compartmentalization reveal novel components of plant immunity. J Biol Chem 285:39140–39149

Kessler SA, Shimosato-Asano H, Keinath NF, Wuest SE, Ingram G, Panstruga R, Grossniklaus U (2010) Conserved molecular components for pollen tube reception and fungal invasion. Science 330:968–971

Kirchhausen T (2009) Imaging endocytic clathrin structures in living cells. Trends Cell Biol 19:596–605

Klima A, Foissner I (2008) FM dyes label sterol-rich plasma membrane domains and are internalized independently of the cytoskeleton in characean internodal cells. Plant Cell Physiol 49:1508–1521

Kwaaitaal MA, de Vries SC, Russinova E (2005) Arabidopsis thaliana Somatic Embryogenesis Receptor Kinase 1 protein is present in sporophytic and gametophytic cells and undergoes endocytosis. Protoplasma 226:55–65

Lam SK, Siu CL, Hillmer S, Jang S, An G, Robinson DG, Jiang L (2007a) Rice SCAMP1 defines clathrin-coated, trans-Golgi-located tubular-vesicular structures as an early endosome in tobacco BY-2 cells. Plant Cell 19:296–319

Lam SK, Tse YC, Robinson DG, Jiang L (2007b) Tracking down the elusive early endosome. Trends Plant Sci 12:497–505

Lee GJ, Kim H, Kang H, Jang M, Lee DW, Lee S, Hwang I (2007) EpsinR2 interacts with clathrin, adaptor protein-3, AtVTI12, and phosphatidylinositol-3-phosphate. Implications for EpsinR2 function in protein trafficking in plant cells. Plant Physiol 143:1561–1575

Lee SW, Han SW, Sririyanum M, Park CJ, Seo YS, Ronald PC (2009) A type I-secreted, sulfated peptide triggers XA21-mediated innate immunity. Science 326:850–853

Lefebvre B, Furt F, Hartmann MA, Michaelson LV, Carde JP, Sargueil-Boiron F, Rossignol M, Napier JA, Cullimore J, Bessoule JJ, Mongrand S (2007) Characterization of lipid rafts from *Medicago truncatula* root plasma membranes: a proteomic study reveals the presence of a raft-associated redox system. Plant Physiol 144:402–418

Leslie ME, Lewis MW, Youn JY, Daniels MJ, Liljegren SJ (2010) The EVERSHED receptor-like kinase modulates floral organ shedding in Arabidopsis. Development 137:467–476

Lewis MW, Leslie ME, Fulcher EH, Darnielle L, Healy PN, Youn JY, Liljegren SJ (2010) The SERK1 receptor-like kinase regulates organ separation in Arabidopsis flowers. Plant J 62:817–828

Li J, Smith GP, Walker JC (1999) Kinase interaction domain of kinase-associated protein phosphatase, a phosphoprotein-binding domain. Proc Natl Acad Sci USA 96:7821–7826

Li J, Zhao-Hui C, Batoux M, Nekrasov V, Roux M, Chinchilla D, Zipfel C, Jones JD (2009) Specific ER quality control components required for biogenesis of the plant innate immune receptor EFR. Proc Natl Acad Sci USA 106:15973–15978

Liu SH, Wong ML, Craik CS, Brodsky FM (1995) Regulation of clathrin assembly and trimerization defined using recombinant triskelion hubs. Cell 83:257–267

Lu X, Tintor N, Mentzel T, Kombrink E, Boller T, Robatzek S, Schulze-Lefert P, Saijo Y (2009) Uncoupling of sustained MAMP receptor signaling from early outputs in an Arabidopsis endoplasmic reticulum glucosidase II allele. Proc Natl Acad Sci USA 106:22522–22527

Meckel T, Hurst AC, Thiel G, Homann U (2004) Endocytosis against high turgor: intact guard cells of Vicia faba constitutively endocytose fluorescently labelled plasma membrane and GFP-tagged K^+-channel KAT1. Plant J 39:182–193

Menon AK (2002) Introduction: lipid transport–an overview. Semin Cell Dev Biol 13:159–162

Miao Y, Yan PK, Kim H, Hwang I, Jiang L (2006) Localization of green fluorescent protein fusions with the seven Arabidopsis vacuolar sorting receptors to prevacuolar compartments in tobacco BY-2 cells. Plant Physiol 142:945–962

Minami A, Fujiwara M, Furuto A, Fukao Y, Yamashita T, Kamo M, Kawamura Y, Uemura M (2009) Alterations in detergent-resistant plasma membrane microdomains in *Arabidopsis thaliana* during cold acclimation. Plant Cell Physiol 50:341–359

Minami A, Furuto A, Uemura M (2010) Dynamic compositional changes of detergent-resistant plasma membrane microdomains during plant cold acclimation. Plant Signal Behav 5:1115–1118

Mo B, Tse YC, Jiang L (2006) Plant prevacuolar/endosomal compartments. Int Rev Cytol 253:95–129

Mongrand S, Morel J, Laroche J, Claverol S, Carde JP, Hartmann MA, Bonneu M, Simon-Plas F, Lessire R, Bessoule JJ (2004) Lipid rafts in higher plant cells: purification and characterization of Triton X-100-insoluble microdomains from tobacco plasma membrane. J Biol Chem 279:36277–36286

Morel J, Claverol S, Mongrand S, Furt F, Fromentin J, Bessoule JJ, Blein JP, Simon-Plas F (2006) Proteomics of plant detergent-resistant membranes. Mol Cell Proteomics 5:1396–1411

Motley A, Bright NA, Seaman MN, Robinson MS (2003) Clathrin-mediated endocytosis in AP-2-depleted cells. J Cell Biol 162:909–918

Müller J, Beck M, Mettbach U, Komis G, Hause G, Menzel D, Samaj J (2010) Arabidopsis MPK6 is involved in cell division plane control during early root development, and localizes to the pre-prophase band, phragmoplast, trans-Golgi network and plasma membrane. Plant J 61:234–248

Murphy JE, Padilla BE, Hasdemir B, Cottrell GS, Bunnett NW (2009) Endosomes: a legitimate platform for the signaling train. Proc Natl Acad Sci USA 106:17615–17622

Nekrasov V, Li J, Batoux M, Roux M, Chu ZH, Lacombe S, Rougon A, Bittel P, Kiss-Papp M, Chinchilla D, van Esse HP, Jorda L, Schwessinger B, Nicaise V, Thomma BP, Molina A, Jones JD, Zipfel C (2009) Control of the pattern-recognition receptor EFR by an ER protein complex in plant immunity. EMBO J 28:3428–3438

Ortiz-Zapater E, Soriano-Ortega E, Marcote MJ, Ortiz-Masia D, Aniento F (2006) Trafficking of the human transferrin receptor in plant cells: effects of tyrphostin A23 and brefeldin A. Plant J 48:757–770

Ovecka M, Berson T, Beck M, Derksen J, Samaj J, Baluska F, Lichtscheidl IK (2010) Structural sterols are involved in both the initiation and tip growth of root hairs in *Arabidopsis thaliana*. Plant Cell 22:2999–23019

Park CJ, Bart R, Chern M, Canlas PE, Bai W, Ronald PC (2010) Overexpression of the endoplasmic reticulum chaperone BiP3 regulates XA21-mediated innate immunity in rice. PLoS ONE 5:e9262

Pérez-Gómez J, Moore I (2007) Plant endocytosis: it is clathrin after all. Curr Biol 17:R217–R219

Reaves BJ, Bright NA, Mullock BM, Luzio JP (1996) The effect of wortmannin on the localisation of lysosomal type I integral membrane glycoproteins suggests a role for phosphoinositide 3-kinase activity in regulating membrane traffic late in the endocytic pathway. J Cell Sci 109:749–762

Rienties IM, Vink J, Borst JW, Russinova E, de Vries SC (2005) The Arabidopsis SERK1 protein interacts with the AAA-ATPase AtCDC48, the 14-3-3 protein GF14lambda and the PP2C phosphatase KAPP. Planta 221:394–405

Robatzek S, Chinchilla D, Boller T (2006) Ligand-induced endocytosis of the pattern recognition receptor FLS2 in Arabidopsis. Genes Dev 20:537–542

Robatzek S, Bittel P, Chinchilla D, Köchner P, Felix G, Shiu SH, Boller T (2007) Molecular identification and characterization of the tomato flagellin receptor LeFLS2, an orthologue of Arabidopsis FLS2 exhibiting characteristically different perception specificities. Plant Mol Biol 64:539–547

Robert S, Chary SN, Drakakaki G, Li S, Yang Z, Raikhel NV, Hicks GR (2008) Endosidin1 defines a compartment involved in endocytosis of the brassinosteroid receptor BRI1 and the auxin transporters PIN2 and AUX1. Proc Natl Acad Sci USA 105:8464–8469

Robinson MS, Bonifacino JS (2001) Adaptor-related proteins. Curr Opin Cell Biol 13:444–453

Robinson D, Hillmer S (1990) Endocytosis in plants. Physiol Plant 79:96–104

Robinson DG, Jiang L, Schumacher K (2008) The endosomal system of plants: charting new and familiar territories. Plant Physiol 147:1482–1492

Rogers S, Wells R, Rechsteiner M (1986) Amino acid sequences common to rapidly degraded proteins: the PEST hypothesis. Science 234:364–368

Russinova E, Borst JW, Kwaaitaal M, Caño-Delgado A, Yin Y, Chory J, de Vries SC (2004) Heterodimerization and endocytosis of Arabidopsis brassinosteroid receptors BRI1 and AtSERK3 (BAK1). Plant Cell 16:3216–3229

Saijo Y, Tintor N, Lu X, Rauf P, Pajerowska-Mukhtar K, Häweker H, Dong X, Robatzek S, Schulze-Lefert P (2009) Receptor quality control in the endoplasmic reticulum for plant innate immunity. EMBO J 28:3439–3449

Seaman MN (2008) Endosome protein sorting: motifs and machinery. Cell Mol Life Sci 65:2842–2858

Shah K, Russinova E, Gadella TW Jr, Willemse J, De Vries SC (2002) The Arabidopsis kinase-associated protein phosphatase controls internalization of the somatic embryogenesis receptor kinase 1. Genes Dev 16:1707–1720

Shiu SH, Bleecker AB (2001) Plant receptor-like kinase gene family: diversity, function, and signaling. Sci STKE 113:re22

Shiu SH, Bleecker AB (2003) Expansion of the receptor-like kinase/Pelle gene family and receptor-like proteins in Arabidopsis. Plant Physiol 132:530–543

Song J, Lee MH, Lee GJ, Yoo CM, Hwang I (2006) Arabidopsis EPSIN1 plays an important role in vacuolar trafficking of soluble cargo proteins in plant cells via interactions with clathrin, AP-1, VTI11, and VSR1. Plant Cell 18:2258–2274

Sorek N, Segev O, Gutman O, Bar E, Richter S, Poraty L, Hirsch JA, Henis YI, Lewinsohn E, Jürgens G, Yalovsky S (2010) An S-acylation switch of conserved G domain cysteines is required for polarity signaling by ROP GTPases. Curr Biol 20:914–920

Stone JM, Collinge MA, Smith RD, Horn MA, Walker JC (1994) Interaction of a protein phosphatase with an Arabidopsis serine-threonine receptor kinase. Science 266:793–795

Takeshita N, Higashitsuji Y, Konzack S, Fischer R (2008) Apical sterol-rich membranes are essential for localizing cell end markers that determine growth directionality in the filamentous fungus *Aspergillus nidulans*. Mol Biol Cell 19:339–351

Tanchak MC, Fowke LC (1987) The morphology of multivesicular bodies in soybean protoplasts and their role in endocytosis. Protoplasma 138:173–182

Taub N, Teis D, Ebner HL, Hess MW, Huber LA (2007) Late endosomal traffic of the epidermal growth factor receptor ensures spatial and temporal fidelity of mitogen-activated protein kinase signaling. Mol Biol Cell 18:4698–4710

Torii KU (2000) Receptor kinase activation and signal transduction in plants: an emerging picture. Curr Opin Plant Biol 3:361–367

Triantafilou M, Morath S, Mackie A, Hartung T, Triantafilou K (2004) Lateral diffusion of Toll-like receptors reveals that they are transiently confined within lipid rafts on the plasma membrane. J Cell Sci 117:4007–4014

Tse YC, Mo B, Hillmer S, Zhao M, Lo SW, Robinson DG, Jiang L (2004) Identification of multivesicular bodies as prevacuolar compartments in *Nicotiana tabacum* BY-2 cells. Plant Cell 16:672–693

Ueda T, Uemura T, Sato MH, Nakano A (2004) Functional differentiation of endosomes in Arabidopsis cells. Plant J 40:783–789

Uemura T, Ueda T, Ohniwa RL, Nakano A, Takeyasu K, Sato MH (2004) Systematic analysis of SNARE molecules in Arabidopsis: dissection of the post-Golgi network in plant cells. Cell Struct Funct 29:49–65

Valdez-Taubas J, Pelham HR (2003) Slow diffusion of proteins in the yeast plasma membrane allows polarity to be maintained by endocytic cycling. Curr Biol 13:1636–1640

Vida TA, Emr SD (1995) A new vital stain for visualizing vacuolar membrane dynamics and endocytosis in yeast. J Cell Biol 128:779–792

Viotti C, Bubeck J, Stierhof YD, Krebs M, Langhans M, van den Berg W, van Dongen W, Richter S, Geldner N, Takano J, Jürgens G, de Vries SC, Robinson DG, Schumacher K (2010) Endocytic and secretory traffic in Arabidopsis merge in the trans-Golgi network/early endosome, an independent and highly dynamic organelle. Plant Cell 22:1344–1357

Voigt B, Timmers AC, Samaj J, Hlavacka A, Ueda T, Preuss M, Nielsen E, Mathur J, Emans N, Stenmark H, Nakano A, Baluska F, Menzel D (2005) Actin-based motility of endosomes is linked to the polar tip growth of root hairs. Eur J Cell Biol 84:609–621

Walker JC, Zhang R (1990) Relationship of a putative receptor protein kinase from maize to the S-locus glycoproteins of Brassica. Nature 345:7436

Wang ZY, Wang Q, Chong K, Wang F, Wang L, Bai M, Jia C (2006) The brassinosteroid signal transduction pathway. Cell Res 16:427–434

Wang J, Cai Y, Miao Y, Lam SK, Jiang L (2009) Wortmannin induces homotypic fusion of plant prevacuolar compartments. J Exp Bot 60:3075–3083

Willemsen V, Friml J, Grebe M, van den Toorn A, Palme K, Scheres B (2003) Cell polarity and PIN protein positioning in Arabidopsis require STEROL METHYLTRANSFERASE1 function. Plant Cell 15:612–625

Zappel NF, Panstruga R (2008) Heterogeneity and lateral compartmentalization of plant plasma membranes. Curr Opin Plant Biol 11:632–640

Zipfel C, Robatzek S, Navarro L, Oakeley EJ, Jones JD, Felix G, Boller T (2004) Bacterial disease resistance in Arabidopsis through flagellin perception. Nature 428:764–767

Zipfel C, Kunze G, Chinchilla D, Caniard A, Jones JD, Boller T, Felix G (2006) Perception of the bacterial PAMP EF-Tu by the receptor EFR restricts Agrobacterium-mediated transformation. Cell 125:749–760

The Protein Quality Control of Plant Receptor-Like Kinases in the Endoplasmic Reticulum

Zhi Hong and Jianming Li

Abstract Plant receptor-like kinases (RLKs) play important roles in regulating plant growth and development and plant–microbe interactions. Like other eukaryotic membrane and secretory proteins, RLKs are cotranslationally inserted into the endoplasmic reticulum (ER) for chaperone-assisted folding to attain their native structures before reaching the plasma membrane to perceive developmental or environmental signals. The ER also houses complex quality control (ERQC) systems that retain incompletely folded proteins for additional folding attempts but eliminate terminally misfolded proteins by ER-associated degradation (ERAD). However, little is known about how the protein folding and ERQC/ERAD events are executed in the plant ER. Recent genetic and biochemical approaches designed to identify regulators of RLK signaling fortuitously discovered various components of the plant ERQC/ERAD systems. These studies have not only dramatically enhanced our understanding of the plant ERQC/ERAD mechanisms that regulate the cell surface expression of RLKs, but have also provided outstanding tools that could identify additional ERQC/ERAD components and uncover novel RLKs involved in plant environment communications.

1 Introduction

Plant receptor-like kinases (RLKs) form a huge monophyletic protein superfamily with ~440 and 790 members in *Arabidopsis* and rice, respectively (Shiu and Bleecker 2001; Lehti-Shiu et al. 2009), and share a common domain organization

Z. Hong
School of Life Sciences, Nanjing University, Nanjing, Jiangsu Province 210093, China
e-mail: zhihong@nju.edu.cn

J. Li (✉)
Department of Molecular, Cellular, and Developmental Biology, University of Michigan, Ann Arbor, MI 48109-1048, USA
e-mail: jian@umich.edu

F. Tax and B. Kemmerling (eds.), *Receptor-like Kinases in Plants*,
Signaling and Communication in Plants 13,
DOI 10.1007/978-3-642-23044-8_14,

consisting of a ligand-binding extracellular domain (ECD), a single-pass transmembrane domain, and an intracellular serine/threonine kinase domain (KD) (Cock et al. 2002). These RLKs can be distinguished into ~20 families based on sequence motifs in their ECDs, such as leucine-rich-repeats (LRRs), lysin-motif (LysM), and lectin-domain (Shiu and Bleecker 2001; Lehti-Shiu et al. 2009). They can also be classified into two general categories depending on their functions. The first group regulates plant growth and development, represented by Brassinosteroid-Insensitive 1 (BRI1) which perceives the plant steroid hormone brassinosteroids (BRs) (Li and Chory 1997) and CLAVATA1 (CLV1), which regulates the development of the shoot apical meristem (SAM) (Clark et al. 1997). The second group participates in plant–microbe interaction, such as Flagellin-Insensitive2 (FLS2), EF-Tu Receptor (EFR), and the chitin elicitor receptor kinase1 (CERK1) that recognizes bacterial flagellin, the bacterial translation elongation factor EF-Tu, and fungal chitin, respectively, which are collectively known as microbe-associated molecular patterns (MAMPs) (Gomez-Gomez and Boller 2000; Zipfel et al. 2006; Miya et al. 2007).

Almost all known plant RLKs are believed to bind their corresponding ligands or perceive uncharacterized developmental or environmental signals on the plasma membrane (Torii 2009). To reach this functional destination, newly synthesized RLKs, like most other eukaryotic membrane and secretory proteins, likely enter the secretory pathway in unfolded states at the endoplasmic reticulum (ER), where they are co-/post-translationally modified by asparagine (Asn or N)-linked glycosylation, form necessary inter-/intra-molecular disulfide bonds, acquire their native conformations, and are sometimes assembled into oligomeric complexes. Because protein functions are largely dictated by their three-dimensional conformations and/or quaternary structures, yet the protein folding is an error-prone process, the folding and assembly states of membrane and secretory proteins are closely monitored by dedicated ER quality control (ERQC) systems, which are composed of molecular chaperones, carbohydrate-binding lectins, and glycan-modifying enzymes. These ERQC systems allow export of only those proteins that have acquired their native conformations or have been successfully assembled into proper oligomeric complexes into the secretory pathways (Ellgaard and Helenius 2003). The ERQC systems also detect and subsequently retain unfolded and improperly assembled proteins for additional rounds of assisted folding and assembly and eventually eliminate terminally misfolded proteins and orphan subunits to prevent their dominant toxic effects on ER physiology by a multistep process known as ER-associated degradation (ERAD) (McCracken and Brodsky 1996; Werner et al. 1996). Most of our knowledge of protein folding and assembly and ERQC/ERAD systems comes from genetic and biochemical studies in yeast and mammalian systems (Burda and Aebi 1999; Ron and Walter 2007; Anelli and Sitia 2008; Hirsch et al. 2009; Braakman and Bulleid 2011). By contrast, our current understanding of similar processes in plants is rather limited (Ceriotti and Roberts 2006; Vitale and Boston 2008; Pattison and Amtmann 2009; Liu and Howell 2010), largely due to a lack of convenient model proteins for genetic studies in model plants. This chapter summarizes and discusses recent fortuitous discoveries from genetic and biochemical studies that were originally designed for identifying regulators of RLK-triggered signaling pathways but instead revealed the involvement of evolutionarily conserved components of eukaryotic ERQC/ERAD systems in assisting RLK folding and in retaining and degrading mutated or misfolded RLKs in the ER.

2 Conserved Eukaryotic ERQC Systems

2.1 N-glycan Biosynthesis

Most membrane and secretory proteins are co-/post-translationally glycosylated at certain Asn residues upon or after entering the ER lumen through the Sec61 translocon (Banerjee et al. 2007). Through interactions with carbohydrate-binding lectins, the constantly modified N-linked glycans regulate protein folding, transport, sorting, degradation, and intracellular signaling (Kato and Kamiya 2007; Molinari 2007). N-glycosylation is catalyzed in the ER by oligosaccharyltransferase (OST), an integral membrane multiprotein complex that transfers a preassembled tetradecasaccharide $Glc_3Man_9GlcNAc_2$ (Glc, Man and GlcNAc for glucose, mannose, and N-acetylglucosamine, respectively) from its membrane-attached dolichylpyrophosphate (Dol-PP) carrier to the Asn residue in the N-X-S/T sequon (X, S, and T represent any nonproline amino acid, serine, and threonine, respectively) on nascent polypeptides (Kelleher and Gilmore 2006). The ordered assembly of Dol-PP-$Glc_3Man_9GlcNAc_2$ starts on the cytoplasmic face of the ER membrane with the addition of GlcNAc(a) to the membrane-embedded dolichylphosphate linker followed by stepwise addition of another GlcNAc and 5 Man residues, generating Dol-PP-$Man_5GlcNAc_2$ (Fig. 1a). This Dol-linked septasaccharide is then flipped over into the ER lumen, where 4 more Man and 3 Glc residues are sequentially added to form Dol-PP-$Glc_3Man_9GlcNAc_2$ containing three dimannose branches (Fig. 1b). The addition of α1,3 Man(h) and α1,2 Man(i) residues, catalyzed by asparagine-linked glycosylation 3 and 9 (ALG3 and ALG9), respectively, forms the middle α1,3- α1,2-dimannose branch (Aebi et al. 1996; Burda et al. 1996); while the addition of α1,6 Man(j) and α1,2 Man(k) residues, catalyzed by ALG12 and ALG9, respectively, generates the upper α1,6-α1,2-dimannose branch (Burda et al. 1999; Frank and Aebi 2005). It is well known that ALG12-catalyzed addition of α1,6 Man(j) occurs only after the ALG3-catalyzed addition of α1,3 Man(h) and the ALG9-catalyzed transfer of α1,2 Man(i), thus ensuring a stepwise assembly of the N-glycan precursor (Burda et al. 1996; Cipollo and Trimble 2000). Adding 3 Glc residues [Glc(l), Glc(m), and Glc(n)] to the lower α1,2-α1,2-dimannose branch, catalyzed consecutively by ALG6, ALG8, and ALG10, is thought to be necessary for the glycan recognition by OST (Burda et al. 1999). It is important to point out that all the cytosolic sugar additions use uridine diphosphate-GlcNAc (UDP-GlcNAc) or guanidine diphosphate-Man (GDP-Man) as donor substrates,

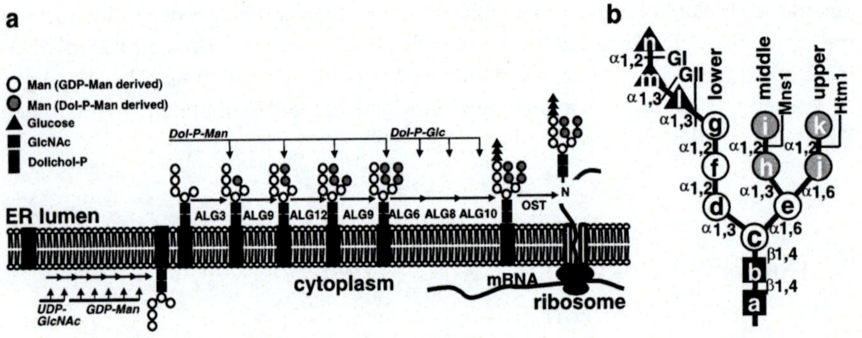

Fig. 1 ([Adapted from Hong et al. 2009)(**a**) Stepwise assembly of the three-branch N-glycan precursor on the ER membrane (**a**). The ordered assembly of Dol-PP-$Glc_3Man_9GlcNAc_2$ starts at the cytosolic face of the ER membrane with the addition of 2 GlcNAc and 5 Man residues from their cytosolic UDP-GlcNAc and GDP-Man donors, respectively, to the membrane-embedded Dol-P linker, generating Dol-PP-$Man_5GlcNAc_2$ that is subsequently flipped over into the ER lumen. Four Man and 3 Glc residues are sequentially added from their respective ER luminal Dol-P-Man and Dol-P-Glc donors to form Dol-PP-$Glc_3Man_9GlcNAc_2$, which is transferred *en mass* to nascent proteins by OST. The dolichol linker, individual sugar residues, their donor substrates, and the ER luminal glycosyltransferases are indicated. The *black bar* indicates the dolichol linker, *square* represents GlcNAc, *circle* denotes Man (*shaded circle* indicating the ER luminal addition), and *triangle* designates Glc. (**b**) The structure of N-linked $Glc_3Man_9GlcNAc_2$ glycan with three dimannose branches (lower, middle, and upper). Lower case letters inside sugar units indicate the order of their addition. The types of sugar linkage and the glycosidases involved in trimming Glc and Man residues (GI, GII, Mns1, and Htm1) are also shown

whereas all the luminal glycosyltransferases use Dol-P-Man or Dol-P-Glc as donor substrates that are synthesized in the cytosol from GDP-Man and UDP-Glc, respectively, and subsequently flipped over into the ER lumen by yet to be characterized flippase(s) (Burda and Aebi 1999).

2.2 N-glycan-Dependent ERQC

Further processing of the N-linked $Glc_3Man_9GlcNAc_2$ inside the ER determines the fate of maturing glycoproteins (Molinari 2007). Immediately after the OST-catalyzed glycan transfer, the terminal α1,2 Glc residue [Glc(n)] is removed by glucosidase I (GI), an integral ER membrane protein, followed by trimming of the middle α1,3 Glc residue [Glc(m)] by glucosidase II (GII), an ER luminal heterodimeric enzyme consisting of one catalytic α-subunit and one regulatory β-subunit (Trombetta 2003) (Fig. 1b). The resulting monoglucosylated N-glycan, $Glc_1Man_9GlcNAc_2$, is specifically recognized by two ER resident lectin-type molecular chaperones (Fig. 2), the ER membrane-anchored calnexin (CNX) and its luminal homolog calreticulin (CRT), which share a similar three-dimensional structure with an N-terminal sugar-binding globular domain plus an extended arm

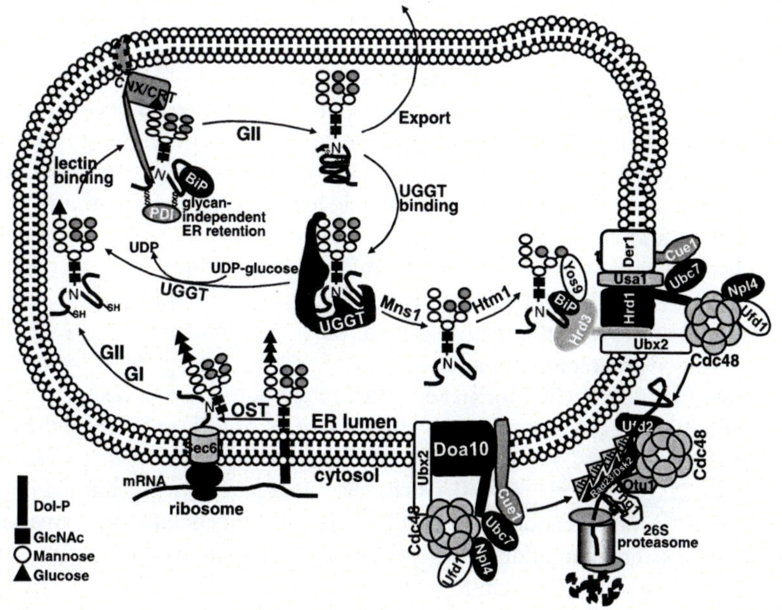

Fig. 2 An overview of the ERQC/ERAD systems in yeast. The rapid trimming of the two Glc residues (*black triangles*) from N-glycans on a nascent polypeptide allows its binding to the ER lectin CNX/CRT. The removal of the last Glc residue liberates a maturing polypeptide from CNX/CRT. A natively folded protein leaves the ER to continue its secretory journey, whereas a misfolded or incompletely folded polypeptide is recognized and reglucosylated by UGGT, thus forcing its reassociation with CNX/CRT for another round of folding. A non native polypeptide can also be detected and retained in the ER by BiP and/or PDI. A glycoprotein that fails to attain its native structure within an allowed time window is removed from the futile folding cycle through sequential trimming of terminal α1,2 Man residues of the middle and upper dimmanose branches by Mns1 and Htm1, respectively, and is subsequently recognized by Hrd3, BiP, and Yos9 that work together to recruit the ERAD client to the membrane-bound Hrd1 complex consisting of Hrd1, Cue1, Ubc7, Usa1, and Der1. The yeast has another membrane-bound E3 ligase complex that contains Doa10, Cue1, and Ubc7. The ubiquitination and subsequent proteasome-mediated degradation of an ERAD substrate requires retrotranslocation through the ER membrane powered by the cytosolic Cdc48-Npl4-Ufd1 complex that is recruited by Ubx2 to the Hrd1 or Doa10 E3 complex. An extracted ERAD client is subsequently delivered to the 26S proteasome for proteolysis through interactions of Cdc48 and the polyubiquitinated ERAD substrate with Ufd2, Otu1, Png1, Rad23, and Dsk2

domain (also known as the P-domain) (Williams 2006). Both CNX and CRT can recruit other ER chaperones or folding catalysts to assist protein folding and disulfide bond formation (Helenius and Aebi 2004), such as luminal binding protein (BiP), an ER-localized member of the heat shock protein 70 (HSP70) family (Otero et al. 2010), and endoplasmic reticulum resident protein 57 (ERp57), a member of the protein disulfide isomerase (PDI) superfamily (Coe and Michalak 2010).

The elimination of the last Glc residue [Glc(I)] by GII liberates maturing glycoproteins from the $Glc_1Man_9GlcNAc_2$-binding lectins. A fully folded glycoprotein interacts with the 53-kDa membrane protein of the ER-Golgi intermediate compartment (ERGIC-53) (Appenzeller et al. 1999) or its homologous mannose-binding lectins, and is subsequently transported from the ER to the Golgi apparatus via transport vesicles for continued glycan modification to produce diverse N-glycan structures (Hauri et al. 2002). By contrast, a misfolded or incompletely folded glycoprotein in higher eukaryotes is recognized by UDP-Glc:glycoprotein glucosyltransferase (UGGT) (Fig. 2), a ~170-kDa ER-resident protein-folding sensor capable of discriminating a natively folded glycoprotein from its non-native conformers (Caramelo and Parodi 2007). UGGT contains a variable client-binding N-terminal domain that probes protein folding states and a highly conserved C-terminal catalytic domain that uses UDP-Glc as the donor substrate to add back a terminal Glc residue to non-native proteins only (Arnold et al. 2000). As a result of this unique reglucosylation activity, the monoglucosylated glycoprotein reassociates with CNX/CRT for an additional round of chaperone-assisted folding. It is generally believed that the alternating activities of GII and UGGT drive the CNX/CRT cycle until the glycoprotein attains its native conformation (Caramelo and Parodi 2008). Interestingly, UGGT only reglucosylates glycoproteins with nearly native conformations but is inactive toward extensively misfolded glycoproteins (Caramelo et al. 2004), suggesting that its physiological function is near the end of a normal protein-folding process.

2.3 Glycan-Independent ERQC

In addition to the N-glycan-dependent GII-UGGT-driven CNX/CRT cycle, the ER houses at least two additional QC systems capable of detecting and retaining unfolded proteins, especially those without any N-linked glycan (Fig. 2). One such system relies on BiP or the 94-kDa glucose-regulated protein (GRP94), the ER-localized member of the HSP90 family in higher eukaryotes (Eletto et al. 2010). BiP is composed of an N-terminal ATP-binding domain and a C-terminal domain that detects and binds hydrophobic patches on improperly folded or incompletely folded proteins in an ATP-dependent manner (Flynn et al. 1991; Blond-Elguindi et al. 1993). GRP94, being much less promiscuous than BiP in client binding (Yang and Li 2005), consists of an N-terminal ATP-binding domain, an acidic linker, a client-recognition middle domain, and a C-terminal dimerization domain (Eletto et al. 2010); however, little is known about its client-recognition mechanism(s). Both BiP and GRP94 bind and release a client protein in an ATP-dependent cycle with their ADP-bound forms exhibiting high substrate affinity and their ATP-bound forms displaying low substrate-binding affinity. The other glycan-independent QC mechanism monitors the oxidation status of one or more cysteine (Cys) residues of proteins regardless of their folding states and retains them in the ER by forming mixed disulfides with PDIs or other ER resident proteins with oxidoreductase activity (Reddy et al. 1996; Anelli et al. 2003, 2007). For instance, the human ER resident protein 44 (ERp44) was a known ER retention factor for two ER resident oxidoreductases, ER oxidoreducin1 α and β (Ero1α and Ero1β), and several unassembled subunits of immunoglobulin M (Anelli et al. 2003).

2.4 ERAD of Terminally Misfolded Proteins

2.4.1 Selection of ERAD Clients

A protein that fails to acquire its native structure within an allowed time window is removed by the ERAD system that involves retrotranslocation through the ER membrane into the cytosol for proteasome-mediated degradation (Vembar and Brodsky 2008). A crucial ERQC decision is to terminate futile chaperone-assisted folding cycles and to divert a terminally misfolded protein into the ERAD pathway. It was previously hypothesized that slow trimming of the α1,2-linked terminal Man residue [Man(i)] from the middle α1,2-α1,3-dimannose branch by the ER-localized α1,2 mannosidase I (ERManI in mammals and Mns1 in yeast) marks a misfolded protein for ERAD (Helenius 1994) because loss-of-function mutations in the yeast Mns1 or pharmacological inhibition of mammalian α1,2 mannosidases by a sugar analog kifunensine (Kif) inhibits ERAD (Jakob et al. 1998; Cabral et al. 2001). The resulting N-linked $Man_8GlcNAc_2$ glycan was thought to be recognized by ER-degradation enhancing α-mannosidase-like proteins [EDEMs, also known as Htm1 for homologous to α-mannosidase I in yeast] to enter the ERAD pathway (Jakob et al. 2001; Oda et al. 2003). However, recent results indicated that while the ERManI/Mns1-catalyzed trimming of the Man(i) from the middle α1,2-α1,3-dimannose branch is a crucial step in ERAD, the actual ERAD signal is an exposed α1,6 Man(j) residue generated through removal of the terminal α1,2 Man(k) residue from the upper α1,2-α1,6-dimmanose branch catalyzed by EDEMs/Htm1 (Quan et al. 2008; Clerc et al. 2009), which specifically bind unfolded proteins in a glycan-independent manner (Cormier et al. 2009) (Fig. 2). The exposed α1,6 Man(j) residue is subsequently recognized by mammalian osteosarcoma 9 (OS9, known as Yos9 in yeast) and XTP3-B for XTP3-transactivated gene B precursor (Hosokawa et al. 2009; Maattanen et al. 2010; Satoh et al. 2010), which are ER luminal lectins with one or two mannose-6-phosphate receptor homology (MRH) domains (Hosokawa et al. 2010). It is worth noting that the other α1,6 Man residue [Man(e)] could also function as the ERAD glycan signal when it becomes exposed by genetic mutations in ALG3 that catalyzes the addition of the α1,3 Man(h) to the α1,6 Man(e) residue (Quan et al. 2008; Clerc et al. 2009).

Selection of ERAD clients depends not only just on the presence of an N-glycan signal but also on their folding states. It was known that Yos9/OS9 physically interacts with the yeast HMG-CoA reductase degradation 3 (Hrd3)/mammalian suppressor enhancer lin12 1-like (Sel1L) protein (Denic et al. 2006; Gauss et al. 2006a, b), a type I membrane protein with a large luminal domain consisting of multiple copies of tetratricopeptide repeat (TPR) motif (Hampton et al. 1996), which recruits unfolded proteins to a membrane-embedded ubiquitin ligase (E3) for subsequent ubiquitination. It is believed that an ERAD client is initially selected by Hrd3/Sel1L and subsequently inspected by the sugar-recognition domain of Yos9/OS9/XTP3-B to ensure degradation of only terminally misfolded glycoproteins but not folding intermediates that lack an α1,6 Man-exposed N-glycan (Denic et al. 2006) (Fig. 2).

2.4.2 Ubiquitination by Membrane-Embedded Ubiquitin Ligase Complexes

Through physical interaction, Hrd3/Sel1L and Yos9/OS9/XTP3-B deliver ERAD substrate to the ER membrane-embedded E3 ubiquitin ligases. Yeast has two such E3 ligases to ubiquitinate different ERAD substrates: the 6-transmembrane-spanning Hrd1 for ERAD clients with folding defects exposed in the ER lumen (ERAD-L substrates) or the ER membrane (ERAD-M substrates) and the 14-transmembrane-spanning protein Doa10 (Degradation of alpha2) for ERAD clients with cytosolic structural errors (ERAD-C substrates) (Kostova et al. 2007) (Fig. 2). Both E3 ligases contain a cytosolic-facing E3-catalytic RING (really-interesting new gene)-finger domain and form multiprotein complexes with other proteins highly conserved between yeast and mammals, including several cytosolic or ER membrane-anchored ubiquitin-conjugating (E2) enzymes and an ER-membrane protein Cue1 (coupling of ubiquitin to ER degradation 1) that recruits cytosolic E2 enzymes to the membrane-bound E3 ligases (Biederer et al. 1997) (Fig. 2). The Hrd1 ligase complex has several additional components including a 4-transmembrane-spanning Der1 (Degradation in the ER 1) (Knop et al. 1996) and a 2-transmembrane-spanning Usa1 (U1 snRNP-associated1) (Carvalho et al. 2006) that is thought to recruit Der1 to Hrd1 or to stabilize the membrane-bound E3 ligase complex (Carvalho et al. 2010; Kanehara et al. 2010) (Fig. 2). Given the sheer complexity of even the simplest multicellular organisms, it is not surprising to know that mammals have a much larger repertoire of membrane-bound E3 ligases involved in ERAD (Kostova et al. 2007), including orthologs of Hrd1 (Kikkert et al. 2004) and Doa10 (Hassink et al. 2005).

2.4.3 Substrate Retrotranslocation and Delivery to Proteasome

An ERAD client has to cross the ER membrane for its ubiquitination due to the cytosolic-facing catalytic domain of Hrd1 or Doa10 and subsequent proteasome-mediated degradation in the cytosol; however, the nature of the retrotranslocation channel (known as retrotranslocon) and its biochemical mechanism remain poorly understood (Bagola et al. 2010). It was previously thought that the Sec61 translocon, which mediates the import of nascent polypeptides into the ER lumen (Rapoport 2007), might also function as the retrotranslocon (Pilon et al. 1997; Plemper et al. 1997). Another potential candidate for the retrotranslocon is the yeast Der1 protein or its three mammalian homologs (derlin1-3) (Hitt and Wolf 2004; Lilley and Ploegh 2004; Ye et al. 2004), which is the 4-transmembrane-spanning component of the Hrd1 E3 complex (Gauss et al. 2006a, b). Recent studies, however, suggested that the retrotranslocation of ERAD-L substrates is likely mediated by the transmembrane domains of the Hrd1 E3 ligase itself (Garza et al. 2009; Carvalho et al. 2010).

The retrotranslocation of a committed ERAD client is powered by a cytosolic Cdc48 (cell-division cycle-48) complex composed of a ring-shaped homohexamer of the AAA-type ATPase Cdc48 [also known as p97 or valosine-containing protein (VCP) in mammals] plus its two cofactors Npl4 (nuclear protein localization 4) and Ufd1 (ubiquitin fusion-degradation 1) (Lord et al. 2002; Raasi and Wolf 2007) (Fig. 2). This cytosolic Cdc48-Npl4-Ufd1 complex is recruited to the Hrd1 or Doa10 E3 ligase complex by a 2-transmembrane-spanning protein known as the ubiquitin regulatory X protein 2 [Ubx2, known as ERASIN in mammals (Liang et al. 2006)] that contains an N-terminal ubiquitin-associated (UBA) domain and a C-terminal Cdc48-binding UBX domain (Schuberth and Buchberger 2005; Wilson et al. 2006). After powering the translocation process, the trimeric Cdc48-Npl4-Ufd1 complex delivers an unfolded protein to the cytosolic proteasome for degradation via its interaction with ubiquitin-binding or ubiquitin-like (UBL) domain-containing proteins, including a polyubiquitin chain-elongating enzyme Ufd2, a deubiquitinase ovarian tumor 1 (Otu1), and two homologous escort factors: radiation sensitive 23 (Rad23) and dominant suppressor of Kar1 (Dsk2) (Raasi and Wolf 2007) (Fig. 2). Both Rad23 and Dsk2 contain a C-terminal UBA domain that binds polyubiquitin chains and an N-terminal UBL domain that delivers polyubiquitinated ERAD substrates to the 26S proteasome (Hartmann-Petersen and Gordon 2004; Medicherla et al. 2004; Richly et al. 2005). It is important to note that Rad23 also interacts with a peptide:N-glycanase (known as Png1 in yeast) (Suzuki et al. 2001), which removes N-glycans from an unfolded ERAD glycoprotein substrate to facilitate its threading through the proteasome channel to reach the central active-site chamber for proteolysis (Yoshida and Tanaka 2010).

2.5 Maintaining ER Homeostasis by Unfolded Protein Response

An overview on ERQC/ERAD would not be complete without discussing the unfolded protein response (UPR), an inter-organelle signaling mechanism that coordinates the protein folding capacity and the protein degradation machinery of the ER to restore ER homeostasis in response to increased abundance of unfolded proteins caused by sudden changes of developmental programs or environmental conditions (Ron and Walter 2007). The competitive titration of BiP from the luminal domain of one or more membrane-bound UPR sensors by unfolded proteins activates the UPR pathways that upregulate the production of molecular chaperones, folding enzymes, and other components of the ERQC/ERAD systems but reduce the general biosynthesis rate of membrane and secretory proteins. Yeast has only one UPR sensor, known as inositol requiring enzyme 1 (Ire1), a transmembrane protein with a stress-sensing luminal domain and a C-terminal domain possessing a serine/threonine kinase activity and an endoribonuclease activity (Cox et al. 1993; Mori et al. 1993), while mammals have at least three UPR sensors that include IRE1, double-stranded-RNA-activated protein kinase-like ER kinase (PERK), and a basic leucine zipper (bZIP)-type transcription factor ATF6 (activating transcription factor 6) (Ron and Walter 2007). Activated Ire1/IRE1 endoribonucleases remove an intron from their specific target mRNAs [*HAC1* (*homologous to ATF/CREB 1*)] for

Ire1 (Cox and Walter 1996) and *Xbp1* (*X-box binding protein 1*) for the mammalian IRE1 (Yoshida et al. 2001; Calfon et al. 2002)], resulting in translation of unconventionally spliced *HAC1/Xbp1* mRNAs into potent bZIP transcriptional activators that bind the upstream DNA UPR elements and activate the promoters of many UPR genes (Ron and Walter 2007). PERK is homologous to IRE1 but lacks the endoribonuclease activity, and upon activation by BiP dissociation and subsequent homodimerization, PERK phosphorylates the α-subunit of the eukaryotic translational initiation factor 2 (eIF2α), thus globally attenuating protein synthesis to reduce the amount of nascent polypeptides into the already stressed folding compartment (Harding et al. 1999; Scheuner et al. 2001). Upon losing BiP, ATF6 travels to the Golgi apparatus for sequential cleavage by two membrane-bound proteases, site-1 protease (S1P) and site-2 protease (S2P), to release its cytosolic portion that contains a transcriptional activation domain, which subsequently moves into the nucleus and binds the so-called ER stress response elements to stimulate the expression of genes encoding ER chaperone and folding catalysts (Haze et al. 1999; Ye et al. 2000; Shen et al. 2002). Interestingly, the *Arabidopsis* genome encodes two IRE homologs (Koizumi et al. 2001), which function similarly to the yeast ire1 or the mammalian IRE1 by targeted splicing of an *Arabidopsis* mRNA encoding AtbZIP60 (Iwata and Koizumi 2005; Deng et al. 2011), and at least 2 ATF6-like bZIP transcriptional factors, AtbZIP17 and AtbZIP28 (Liu et al. 2007a, b; Che et al. 2010) but no single PERK homolog.

3 ER Retention of Mutated BR Receptors

BRI1, a well-studied *Arabidopsis* LRR-RLK, functions as a cell-surface receptor for the plant steroid hormone BRs (Li and Chory 1997). Plants defective in BR biosynthesis or signaling share a characteristic set of growth and developmental defects that include a dwarf stature, delayed flowering, reduced male fertility, and abnormal vascular development (Clouse and Sasse 1998). The BRI1 ECD carries 25 LRRs flanked by disulfide bridge-stabilized N-terminal and C-terminal caps, a 70-amino acid island buried between the 21st and 22nd LRRs, and a total of 14 putative N-glycosylation sites (Li and Chory 1997). A direct ligand-binding assay with *E. coli* expressed fragments of the BRI1 ECD revealed a unique BR-binding domain consisting of the 70-amino-acid island plus the 22nd LRR (Kinoshita et al. 2005), which carries single-amino acid changes in several previously reported *bri1* mutants: *bri1-6*/*bri1-119* [glycine(Gly)644-aspartate(Asp)], *bri1-7* [Gly613-serine(Ser)], *bri1-9* [Ser662-phenylalanine(Phe)], and *bri1-113* [(Gly611-glutamate(Glu)] (Noguchi et al. 1999; Friedrichsen et al. 2000). It was previously thought that these mutations might prevent BR binding or interfere with BRI1 homodimerization or heterodimerization with its coreceptors (Vert et al. 2005). However, a recent genetic screen looking for extragenic suppressors of the *bri1-9* mutant led to a surprising discovery that the Ser662Phe mutation has a marginal effect on the ligand-binding or signaling activity of the mutant BR receptor but likely causes a subtle conformational defect that is detected and consequently prevents bri1-9 from reaching the cell surface by EMS-mutagenized bri1 suppressor1 (EBS1), the sole *Arabidopsis* homolog of the mammalian UGGT (Jin et al. 2007).

It was predicted that EBS1/UGGT adds a terminal Glc residue to N-linked glycans of the mutant BR receptor to allow the monoglucosylated bri1-9 protein to bind the *Arabidopsis* CNX/CRT homologs, thus retaining bri1-9 in the ER (Jin et al. 2007). Indeed, biochemical and confocal microscopic analyses showed that bri1-9 was almost exclusively localized in the ER in an $EBS1^{+}$ background and that *ebs1/uggt* mutations allowed ~ 50% of the total cellular bri1-9 proteins to move out of the ER and reach the plasma membrane where the mutant BR receptor was able to activate the BR signaling pathway to stimulate plant growth (Jin et al. 2007). Thus, the dwarf phenotype of *bri1-9* is attributable to the ER retention of a structurally-defective yet biochemically-competent BR receptor by an overzealous UGGT-based ERQC system in *Arabidopsis*. It is of interest to note that a similar over-vigilant ERQC in human prevents the cell surface localization of the F508-mutated but principally-active ATP-binding cassette-type chloride channel, widely known as cystic fibrosis transmembrane conductance regulator (CFTR). The loss of the Phe508 residue due to deletion of 3 consecutive base pairs is the most common cystic fibrosis mutation and can cause the most severe clinical symptoms of cystic fibrosis (Sanders and Myers 2004).

What could be the lectin molecule that binds the monoglucosylated bri1-9 to retain the mutant BR receptor in the ER? The *Arabidopsis* genome encodes two homologs of CNX, CNX1, and CNX2 (Huang et al. 1993; Boyce et al. 1994), and three homologs of CRT, CRT1-3 that can be grouped into two distinct families: CRT1/2 and CRT3 (Persson et al. 2003). Coimmunoprecipitation experiments showed that bri1-9 interacted with the two CNXs but not with CRT1 or CRT2 and that *ebs1/uggt* mutations blocked the bri1-9-CNX binding; however, neither single *cnx1*/*cnx2* nor the *cnx1 cnx2* double mutation was able to suppress the *bri1-9* mutation (Jin et al. 2009), suggesting that the detected monoglucosylation-dependent bri1-9-CNX interaction plays no major role in retaining bri1-9 in the ER. The genetic identification and subsequent positional cloning of the *Arabidopsis EBS2* gene revealed that it is CRT3 that binds and retains the mutant BR receptor in the ER and that loss-of-function *ebs2*/*crt3* mutations prevent the bri1-9-CRT3 binding, resulting in escape of some bri1-9 receptors from the ER to the cell surface (Jin et al. 2009). A unique ERQC role of CRT3 in retaining bri1-9 was further confirmed by transgenic experiments showing that while overexpression of the *CNX1* or *CRT1* gene failed to complement the *ebs2* mutation in the *bri1-9* background, expression of *CRT3* in an *ebs2 bri1-9* double mutant not only suppressed the suppressor phenotype but also gave rise to a dwarf phenotype much stronger than that of the parental *bri1-9* mutant (Jin et al. 2009), suggesting that CRT3 is a rate-limiting component of the UGGT-based bri1-9 retention system.

What could distinguish CRT3 from CRT1/CRT2 for binding the mutant BR receptor? Sequence alignment revealed that while the three *Arabidopsis* CRTs exhibit high sequence identity in their N-terminal globular domain predicted to bind the monoglucosylated N-glycan and the middle P domain, they differ substantially in their C-terminal sequences. The CRT1/2 C-termini are similar to that of the mammalian CRT rich in acidic residues, which are known to be critical for low-affinity but high capacity Ca^{2+}-binding (Michalak et al. 2009), whereas the CRT3 C-terminus contains clusters of basic residues, implying a crucial role of the C-terminal tail in determining the physiological functions of the plant CRTs in ER Ca^{2+} homeostasis and ERQC. Indeed, while the *Arabidopsis* CRT1 rescued the Ca^{2+}-deficiency of the mouse *CRT*-knockout embryonic fibroblasts (Christensen et al. 2008) and transgenic expression of a maize CRT1/CRT2-member increased ER Ca^{2+} levels (Persson et al. 2001), a chimeric CRT1-3 protein consisting of the N and P domains of CRT1 fused to the CRT3 C-terminus (but not a reciprocally constructed CRT3-1 chimera) was able to complement the *ebs2* mutation when driven by the *CRT1* cognate promoter. Interestingly, while *Arabidopsis CRT1* and *CRT2* were coexpressed under hundreds of experimental conditions with genes encoding known or putative ER chaperones, *EBS2/CRT3* was clustered together with genes predicted to be involved in plant resistance against a wide range of biotic and abiotic environmental stresses (Jin et al. 2009), suggesting that CRT3 might have been evolved to play a specific role in retaining non-native glycoproteins caused by stressful growth conditions.

4 Requirement of the UGGT-CRT3 System for Correctly Folding a Pattern Recognition Receptor

The studies on the bri-9 ER retention provided genetic support for a long-suspected role of UGGT in sensing a subtle conformational change and in working with the CNX/CRT lectins to retain a glycoprotein with a nearly native conformation in the ER. However, the UGGT-driven CNX/CRT cycle is not required for the correct folding of the wild-type BRI1 protein since *ebs1/uggt* or *ebs2/crt3* mutations had little effect on the plant growth and development in a $BRI1^{+}$ background. Nevertheless, loss-of-function *ebs1/uggt* mutations do activate UPR (Jin et al. 2007), suggesting ER accumulation of unfolded proteins and implying that UGGT is required for folding and maturation of many nonessential proteins under standard laboratory growth conditions. This is in sharp contrast to the mammalian UGGT, as a knockout mutation in the mouse UGGT caused an embryo lethal phenotype (Molinari et al. 2005).

An important physiological function of the GII-UGGT-driven CNX/CRT cycle was revealed by two independent genetic screens that aimed to identify key regulators of the plant immune response to the biologically active elf18 peptide derived from the bacterial translation elongation factor EF-Tu (Saijo 2010), which is recognized by EFR, an LRR-containing pattern recognition receptor (PRR) (Zipfel et al. 2006). One screen was based on the activity of elf18 to cause seedling growth inhibition, identifying ~160 *elf18-insensitive* (*elfin*) mutants (Li et al. 2009; Nekrasov et al. 2009), while the other relied on the ability of elf18 to attenuate the stimulatory effect of sucrose on anthocyanin accumulation, isolating >50 "*priority in sweet life*" (*psl*) mutants that still accumulate anthocyanin in the presence of both sucrose and the elf18 peptide (Lu et al. 2009; Saijo et al. 2009). These studies revealed that loss-of-function mutations in either UGGT or CRT3 resulted in misfolding and ER retention of EFR, causing elf18 insensitivity confirmed by several other plant defense response assays. Treatment of the *uggt* or *crt3* mutants with Kif, a well-known inhibitor of α1,2 mannosidases (Elbein et al. 1990), or the widely used proteasome inhibitor MG132, were able to restore the accumulation of EFR but failed to restore the elf18 sensitivity, suggesting that the incompletely folded EFR in the absence of the UGGT-CRT3 system is actively removed by ERAD. It was concluded that the correct folding and complete maturation of EFR absolutely requires the UGGT-CRT3-mediated ERQC system (Saijo 2010).

Surprisingly, neither *uggt* nor *crt3* mutations had any detectable effect on the abundance or subcellular location of FLS2, an EFR-like LRR-RLK that recognizes the bacterial flagellin or its biologically active peptide flg22 to activate a shared signaling pathway with many MAMP-detecting PRRs including EFR (Gomez-Gomez and Boller 2000). Thus, the GII-UGGT-driven CNX/CRT cycle for additional rounds of chaperone-assisted folding is likely needed only for certain inefficiently folded proteins or maturing proteins under stress. Given the limited distribution of EFR-mediated plant defense in the plant kingdom and a relatively more ancient origin of FLS2-trigged plant immunity (Boller and Felix 2009; Zipfel 2009), it was hypothesized that recently evolved LRR-RLKs detecting novel ligands are less likely to fold correctly in their initial folding trial, but require multiple rounds of chaperone-assisted folding (Nekrasov et al. 2009). Consistent with this interpretation, a recent study showed that deglycosylation with tunicamycin, a nucleoside antibiotic that inhibits the 1st addition of GlcNAc in assembling the Dol-linked tetradecasaccharide (Zhu et al. 1992), significantly destabilized nonglycosylated EFR compared to the nonglycosylated FLS2 even though both RLKs were prevented from reaching the plasma membrane (Haweker et al. 2010). As expected from the differential requirement of the CNX/CRT cycle-assisted structural maturation for EFR and FLS2, mutations in Staurosporin and temperature sensitive–3A) STT3A, one of the two catalytic components of the OST complex (Koiwa et al. 2003; Kelleher and Gilmore 2006), affect the protein abundance and biological functions of EFR but not FLS2 (Nekrasov et al. 2009; Saijo et al. 2009; Haweker et al. 2010). Similarly, mutations in the catalytic α-subunit [also known as RSW3 for RADIAL SWELLING 3 (Burn et al. 2002)] and the ER-retention β subunit of GII, which generates the initial set of monoglucosylated N-glycans and works together with UGGT to drive the CNX/CRT cycle, inhibit the accumulation, plasma membrane localization, and receptor function of EFR but not that of FLS2 (Lu et al. 2009; von Numers et al. 2010).

A requirement of CRTs to mediate the folding and maturation of RLKs seems to be a common feature in plants. It was recently shown that RNA-interference (RNAi)-mediated silencing of two tobacco (*Nicotiana benthamiana*) CRTs, NbCRT2 and NbCRT3, led to a significant reduction of the protein abundance of the tobacco Induced Receptor Kinase (IRK), a 5-LRR-containing RLK required for the tobacco immune receptor N to trigger a hypersensitive response and other plant defense responses against *Tobacco mosaic virus* (Caplan et al. 2009).

5 Involvement of Other ER Chaperones in the Folding and QC of Plant RLKs

5.1 Retention of bri1-5 by Three Independent Mechanisms

As discussed earlier, the mammalian ER houses several other ERQC systems capable of detecting misfolded or incompletely folded polypeptides, such as the BiP-mediated detection of hydrophobic patches on unfolded proteins and a thiol-mediated retention of proteins with unpaired Cys residue(s) (Anelli and Sitia 2008). Studies of another *bri1* dwarf mutant revealed the presence of similar CNX/CRT-independent retention systems in plant cells. A biochemical assay analyzing glycoforms of previously reported bri1 mutant proteins with extracellular mutations identified bri1-5, carrying a Cys69-tyrosine(Tyr) mutation in the N-terminal cap of the BRI1 ECD (Noguchi et al. 1999), as another ER-retained mutant BR receptor, which was further confirmed by confocal microscopic analysis of a fluorescently tagged bri1-5 fusion protein (Hong et al. 2008). Surprisingly, *ebs1/uggt* mutations failed to suppress but instead enhanced the dwarf phenotype of *bri1-5* despite the fact that bri1-5 interacted with CNXs in a monoglucosylation-dependent manner. It was predicted that bri1-5 is retained in the ER by additional retention systems whose components are likely overproduced due to the *uggt* mutation-induced UPR (Jin et al. 2007).

This prediction was consistent with sequence analysis and molecular modeling of the Cys69Tyr mutation (Hong et al. 2008). The Cys69 is a highly conserved residue among many LRR-RLKs and is generally believed to form a disulfide bond with another highly conserved Cys residue, Cys62 (van der Hoorn et al. 2005; Kolade et al. 2006). This presumed disulfide bridge is thought to be critical for maintaining the correct N-terminal cap structure to shield the N-terminal edge of the hydrophobic core of the solenoid LRR structure (Di Matteo et al. 2003; Choe et al. 2005). Thus, the Cys69Tyr mutation not only destroys the Cys62-Cys69 disulfide bridge, generating an orphan Cys residue that could form a mixed disulfide linkage with a PDI-like protein similar to the mammalian ERp44 (Anelli et al. 2003), but also alters the N-terminal cap structure, likely exposing the hydrophobic residues of the 1st BRI1 LRR detectable by UGGT and/or BiP.

Biochemical and transgenic experiments seemed to support these predictions. Coimmunoprecipitation experiments demonstrated that bri1-5 interacted quite strongly with BiP, while RNAi-mediated silencing of three *Arabidopsis BiP* genes somewhat suppressed the *bri1-5* dwarfism and partially restored its

BR-sensitivity (Hong et al. 2008), suggesting a major role for BiP in retaining bri1-5 in the ER. Because the *Arabidopsis* genome lacks an *ERp44* homologous gene but encodes a total of 22 PDI-like proteins (Houston et al. 2005), it was not practical to investigate a direct involvement of a PDI-like protein in retaining the Cys69Tyr-mutated BR receptor. Instead, a transgenic experiment was performed in the *bri1-5* background to show that while the Cys62Tyr-bri1-5 protein was still retained in the ER, the elimination of the orphan Cys62 residue by changing it to a Tyr residue did enable export of a small percentage of the doubly mutated bri1 to the plasma membrane, thus increasing the activity of the transgenically expressed Cys62Tyr-bri1-5 (compared to transgenically expressed bri1-5 or Cys62Tyr-mutated bri1) to rescue the *bri1-5* mutation (Hong et al. 2008).

A recent genetic study discovered a very interesting intragenic suppressor of *bri1-5*, known as *bri1-5R1* (Belkhadir et al. 2010). Sequence analysis identified a Gly87Glu mutation located just 18 amino acids downstream of the Cys69 residue. Immunoblot analysis revealed partial accumulation and dual localization (ER and plasma membrane) of the Gly87Glu-bri1-5, indicating that the Gly87Glu mutation in the 1st LRR somewhat compensates for the folding abnormality caused by the Cys69Tyr mutation. Interestingly, while Gly87 is highly conserved among many *Arabidopsis* LRR-RLKs, including the two BR-binding BRI1 homologs [AtBRL1 and AtBRL3 (Cano-Delgado et al. 2004)], BAK1 [a 5-LRR-containing BRI1 coreceptor (Li et al. 2002; Nam and Li 2002)], FLS2, and CLV1, Glu or its smaller analog Asp, is known to occupy the position of Gly87 in the BRI1 homologs of other plant species (Belkhadir et al. 2010). It remains to be determined whether the Gly87Glu mutation helps to seal the edge of the hydrophobic core of the solenoid LRR structure or introduces a steric hindrance that blocks disulfide bond formation between the orphan Cys62 with a Cys residue of an ERp44-like protein.

5.2 Assisted Folding of EFR by the BiP Chaperone System

As the major chaperone of the ER, BiP plays at least three independent roles in protein folding and ERQC/ERAD by detecting and binding hydrophobic patches on protein surface to import nascent polypeptides, to retain incompletely folded proteins in the ER, and to select terminally misfolded polypeptides for ERAD. Such multifunctionality of BiP is believed to be the result of BiP association with various ER resident J-domain-containing HSP40 co-chaperones known as ERdjs (Otero et al. 2010). Yeast has 4 ERdjs, including Sec63, Scj1 (*Saccharomyces cerevisiae* DnaJ homologue 1), Jem1 (DnaJ-like protein of the ER membrane 1), and Erj5 (ER-localized J domain-containing protein 5) (Carla Fama et al. 2007); mammals have a total of 7 ERdj chaperones [ERdj1-7 (Otero et al. 2010)]; and the *Arabidopsis* genome encodes a total of 6 ERdjs that include two Sec63 homologs (AtERdj2A and AtERdj2B), two Scj1 homologs (AtERdj3A and AtERdj3B), a single Jem1 homolog (AtP58^{IPK}), and a potential Erj5 homolog (At1g61770)

(Yamamoto et al. 2008). The binding of Sec63-type ERdjs helps BiP import nascent polypeptides through the Sec61 translocon into the ER lumen (Rudiger et al. 1997; Matlack et al. 1999). Recent studies suggest that ERdj3 and ERdj6 help in recruiting incompletely folded clients to BiP to facilitate their folding, whereas ERdj4 and ERdj5 are important in recognizing terminally misfolded proteins for ERAD (Maattanen et al. 2010).

The genetic screen for *elfin* mutants revealed a crucial role of the BiP chaperone system in the correct folding and maturation of EFR through identification of an additional *ELFIN* gene that encodes a 218-amino-acid protein AtSDF2, homologous to the mammalian stromal cell-derived factor 2 (SDF2) containing three repeats of the MIR domain (Nekrasov et al. 2009), which was named after the three founding members of the MIR-containing family: mannosyltransferase, inositol-3-phosphate receptor, and ryanodine receptor (Ponting 2000). In mammalian cells, SDF2 is a UPR-induced protein (Fukuda et al. 2001) and a component of a large multi-chaperon complex that contains BiP, ERdj3, PDI, GRP94, cyclophilin B, and UGGT but conspicuously lacks CNX/CRT (Meunier et al. 2002). Consistent with the mammalian discovery, AtSDF2 was found to be induced by UPR (Schott et al. 2010) and to bind BiP and AtERdj3B but not its close homolog AtERdj3A (Nekrasov et al. 2009). It is of interest to note that three independent EMS-generated *erdj3b* alleles were recovered from the *elfin* mutant collection (Nekrasov et al. 2009).

Loss-of-function mutations in AtSDF2 or AtERdj3B led to a significant reduction of EFR abundance, apparently by affecting the folding and maturation of the elf18-binding LRR-RLK because all detectable transgenically expressed EFR proteins in a loss-of-function *sdf2* mutant were localized to the ER (Nekrasov et al. 2009). Interestingly, this loss-of-function *sdf2* mutation also affected the folding efficiency of FLS2 since ~50% of all cellular FLS2 proteins were retained in the ER in the *sdf2* mutant and might also interfere with the maturation and trafficking of other MAMP-recognition PRRs as the *sdf2* mutant was more susceptible to some bacterial and fungal pathogens than single *efr* or *fls2* mutants (Nekrasov et al. 2009). However, AtSDF2 is apparently not required for the correct folding and maturation of CERK1, a LysM-type RLK that detects fungal chitin (Miya et al. 2007), since the *sdf2* mutant retained the same immunity response to the chitin as the wild-type control (Nekrasov et al. 2009).

The observations that loss-of-function mutations affecting the UGGT-GII-driven CNX/CRT cycle prevented the folding and maturation of EFR but not FLS2, whereas *sdf2* mutations affected the accumulation and plasma membrane localization of both EFR and FLS2 (albeit to a lesser extent) suggest that the SDF2-ERdj3-BiP chaperone system likely functions upstream of the CNX/CRT cycle known to act near the end of a protein folding process. This hypothesis is consistent with the presence of UGGT but not CNX/CRT in the mammalian SDF2-ERdj3-BiP-containing multi-chaperone complex, although EBS1/UGGT has not yet been detected in the *Arabidopsis* SDF2-ERdj3B-BiP complex. The presence of this folding sensor in a BiP chaperone complex allows transfer of a maturing polypeptide with a near-native conformation to the CNX/CRT cycle for an additional round of chaperone-assisted folding, thus permitting sequential actions of the two ERQC systems to fold and mature polypeptides with inherent low folding efficiency. Interestingly, such a "hand-off" mechanism between the two chaperone systems was previously identified for the correct folding and maturation of the vesicular stomatitis virus glycoprotein G (Hammond and Helenius 1994) and the type I membrane glycoprotein tyrosinase (Wang et al. 2005a, b) in mammalian cells.

5.3 *A Crucial Role of GRP94 for the Biological Functions of CLV Proteins and/or Complexes*

Unlike BiP that has hundreds of known client proteins and many co-chaperones, the ER-localized HSP90 protein GRP94 has a rather short client list and no known cochaperone (Eletto et al. 2010). One of the potential GRP94 clients is the Arabidopsis CLV1, a 21-LRR-containing RLK that regulates SAM development (Clark et al. 1997), CLV2, a kinase-lacking LRR-containing receptor-like protein that forms a CLV1-independent signaling receptor complex (Jeong et al. 1999; Gish and Clark 2011), or CLV3 that produces a peptide ligand for the CLV1 and/or CLV2-containing receptor complexes (Fletcher et al. 1999). A T-DNA insertional mutation in the *Arabidopsis* GRP94 homolog, known as SHEPHERD or SHD (Ishiguro et al. 2002), resulted in enlarged SAM and floral meristem (FM), similar to that of the *Arabidopsis clv* (*clv1*, *clv2*, and *clv3*) mutants (Clark et al. 1993, 1995; Kayes and Clark 1998). Subsequent genetic studies revealed that SHD works in the same genetic pathway with the three CLV proteins to control SAM/FM development, suggesting that the CLV signaling functions require the activity of SHD (Ishiguro et al. 2002). In line with this genetic data, it was subsequently found that while overexpression of a *CLV3* transgene caused mis-specification of meristematic cells and a premature SAM/FM termination in a *SHD*$^+$ background, similar CLV3 overproduction had no detectable effect on the SAM/FM defects of the *shd* mutant. However, it remains unknown whether the *shd* mutation inhibits the folding and maturation of individual CLV proteins, the assembly of CLV1- and/or CLV2-containing receptor complexes, or the secretion of the CLV3 peptide ligand. In mammals, GRP94 is required for the folding and maturation of all known LRR-containing Toll-like immune receptors as they were largely retained in the ER when GRP94 was eliminated in mice (Yang et al. 2007), and it also provides a necessary chaperone function that aids the secretion of insulin-like growth factor because GRP94-deficient mouse embryonic stem cells failed to produce the peptide growth factor in response to stresses (Wanderling et al. 2007; Ostrovsky et al. 2009, 2010). It is worth noting that the *shd* mutation also causes a disorganized root apical meristem (RAM) phenotype and inhibits pollen tube elongation causing male sterility (Ishiguro et al. 2002), suggesting the potential requirement of GRP94 in the folding of additional RLKs or other membrane and secretory proteins involved in RAM development and pollen tube elongation.

5.4 A Potential Role of a Cytosolic HSP90 Complex in the Maturation of OsCERK

Because RLKs contain both ECD, which is translocated into the ER lumen through the Sec61 translocon, and a cytosolic KD, their correct folding or maturation likely involves cytosolic chaperones as well. One prominent cytosolic chaperone system is HSP90 that hydrolyzes ATP to drive folding and maturation of a wide range of protein clients involved in protein trafficking, signal transduction, and immunity. Unlike its ER counterpart, cytosolic HSP90 interacts with many different co-chaperones that link HSP90 with other protein chaperones, regulate its ATPase activity, or recruit specific clients (Taipale et al. 2010). One such co-chaperone is Hsp organizer protein (Hop) [also known as stress-inducible (Sti) in yeast], a TRP domain-containing protein that shuttles client proteins from HSP70 to HSP90 (Frydman and Hohfeld 1997). A recent study showed that a rice Hop/Sti1 homolog, known as Hop/Sti1a, interacted with a rice HSP90 and OsCERK (Chen et al. 2010), a rice RLK carrying the LysM domain that functions as a PRR for fungal chitin (Shimizu et al. 2011). Importantly, RNAi-mediated silencing of Hop/Sti1a reduced the efficiency of the plasma membrane targeting of OsCERK and consequently the plant responses to the chitin (Chen et al. 2010), suggesting that the folding and maturation of this chitin-binding RLK requires the cytosolic HSP90 chaperone system. The HSP90-Hop/Sti chaperone-cochaperone pair is therefore mechanistically similar to the HSP90-SGT1 (Suppressor of G-two allele of Skp1, also a TPR-containing cochaperone) chaperone system that regulates the folding and/or stability of plant NBS (nucleotide-binding site)-LRR proteins (Shirasu 2009), which are intracellular immune sensors capable of detecting pathogenic effector molecules to mount the so-called "effector-triggered immunity" (DeYoung and Innes 2006). Interestingly, the HSP90-Hop/Sti1a complex also interacted with the rice homologs of FLS2 and BAK1 (Chen et al. 2010); however it remains to be tested whether this interaction is of physiological significance in the maturation and/or assembly of the FLS2-BAK1 receptor complex in rice. Although an earlier study showed that mutations in AtSDF2 had little effect on the chitin sensitivity in *Arabidopsis* (Nekrasov et al. 2009), correct folding and maturation of the LysM-containing OsCERK and its *Arabidopsis* homolog CERK1 likely involves not only the cytosolic HSP90 but also ER-localized chaperones and QC systems independent of the SDF2-ERdj3-BiP complex. The requirement of both ER and cytosolic chaperones for protein folding and maturation was previously known for the human insulin receptor, a disulfide-linked tetrameric receptor tyrosine kinase that requires both the ER luminal CRT and the cytoslic HSP90 to stabilize its cell surface expression (Ramos et al. 2007).

6 ERAD of the Mutated and Misfolded Plant RLKs

6.1 *Requirement of Fully Assembled Glycans to Generate an ERAD Glycan Signal*

The facts that the abundance of bri1-5, bri1-9, and the incompletely folded EFR in the absence of proper chaperones could be increased by treatment with Kif (Hong et al. 2008, 2009; Nekrasov et al. 2009; Saijo et al. 2009), a widely used inhibitor of α1,2 mannosidase (Elbein et al. 1990), not only confirmed the existence of similar ERAD processes in plant cells for removing mutated or misfolded RLKs but also suggested involvement of Man-trimming steps in generating a necessary ERAD glycan signal. Interestingly, Kif treatment was able to suppress the dwarf phenotype of both *bri1* mutants but had little effect on the elf18 sensitivity of the *uggt/crt3* mutants (Hong et al. 2008, 2009; Nekrasov et al. 2009; Saijo et al. 2009). This is likely caused by saturation of the ERQC system by overaccumulation of bri1-5 or bri1-9 that are structurally imperfect yet biochemically functional. Consistent with this explanation, overproduction of bri1-5 and bri1-9 in the *bri1-5* and *bri1-9* mutants, respectively, resulted in expression of a considerable amount of the mutant BR receptors on the plasma membrane (Hong et al. 2008, 2009). By contrast, the abundance of EFR in Kif-treated *uggt/crt3* mutants might not be high enough to saturate non-UGGT-based ERQC systems responsible for retaining the incompletely folded EFR in the ER. Alternatively, the recognition of elf18 is less tolerant to subtle variations from the native EFR conformation compared to the perception of the plant steroid hormones by the two mutant BR receptors.

Consistent with the ability of Kif to rescue the dwarf phenotype of both *bri1-5* and *bri1-9* mutants (Hong et al. 2008, 2009), and the demonstrated inhibitory effect of the yeast *Δalg12* mutation on ERAD of misfolded glycoproteins (Jakob et al. 1998), genetic screens for suppressors of both *bri1* mutations led to the identification of the *EBS4* gene, which encodes the *Arabidopsis* ortholog of the yeast and human ALG12 catalyzing the addition of the α1,6 Man(j) to the other α1,6 Man residue [Man(e)] (Fig. 1) (Hong et al. 2009). In line with recent studies on *alg3* mutants with T-DNA insertional mutations in the *Arabidopsis* homolog of the yeast and mammalian ALG3 catalyzing the ER luminal addition of the α1,3 Man(h) to the α1,6 Man(e) residue (Fig. 1) (Henquet et al. 2008; Kajiura et al. 2010), it was found that the incompletely assembled glycans, which were predicted to lack the entire upper α1,2-α1,6-dimannose branch of $Glc_3Man_9GlcNAc_2$ in the *ebs4/alg12* mutants, were efficiently transferred from their dolichol linkers to bri1-5, bri1-9, BRI1, and other glycoproteins (Hong et al. 2009). The truncated N-glycans in both *alg3* and *ebs4/alg12* mutants were properly processed in the Golgi apparatus to form mature complex N-glycans commonly observed in wild-type *Arabidopsis* plants, thus causing no detectable effect on plant growth or development under normal conditions (Henquet et al. 2008; Hong et al. 2009; Kajiura et al. 2010). However, the shortened N-glycans on bri1-5, bri1-9, and other ER-retained glycoproteins lack the α1,6 Man(j) residue that was recently shown to be the ERAD glycan signal (Quan et al. 2008; Clerc et al. 2009). This is likely the reason why *ebs4/alg12* mutations inhibit the ERAD of bri1-5, bri1-9, and possibly many other terminally misfolded unknown glycoproteins, causing activation of UPR (Hong et al. 2009). Although the predicted $Man_6GlcNAc_2$ N-glycan on bri1-5 or bri1-9 in *ebs4/alg12* mutants does contain another α1.6 Man residue [Man(e)] that could be recognized by an OS9/Yos9-like ERAD lectin (Quan et al. 2008; Clerc et al. 2009), the ER likely lacks the necessary mannosidases to remove the entire middle α1,2-α1,3-dimannose branch in order to expose this secondary α1,6 Man residue. That an exposed α1,6 Man residue is the necessary glycan signal to mark an ERAD client in *Arabidopsis* needs further confirmatory genetic and biochemical studies.

6.2 *Conserved ERAD Machinery in* Arabidopsis *to Remove Mutated or Misfolded RLKs*

While the identities of a plant ERAD glycan signal and its binding lectin(s) await to be discovered, it was recently shown that successful ERAD of both bri1-5 and bri1-9 requires EBS5, the *Arabidopsis* homolog of the yeast Hrd3/mammalian Sel1L (Su et al. 2011). As discussed previously, Hrd3/Sel1L is a client recognition factor that recognizes and recruits misfolded or incompletely folded proteins to the membrane-bound Hrd1 E3 ligase complex. The *Arabidopsis* genome encodes two Hrd3/Sel1L homologs, Sel1A/Hrd3A and Sel1B/Hrd3B; however, only one of them works in the ERAD process, whereas the other is likely a pseudogene that fails to produce a functional protein (Liu et al. 2011; Su et al. 2011). Loss-of-function mutations in EBS5/Sel1A/Hrd3A but not Sel1B/Hrd3B block the ERAD of bri1-5 and bri1-9, resulting in accumulation of the two mutant BR receptors and the rescue of their corresponding dwarf phenotypes (Liu et al. 2011; Su et al. 2011). EBS5/Sel1A/Hrd3A interacted with bri1-5 or bri1-9 but not with the wild-type BRI1 and complemented an ERAD defect of the yeast *Δhrd3* mutant (Su et al. 2011). Thus, the degradation of the two mutant BR receptors likely involves the so-called "bipartite signal": an exposed α1,6 Man residue and a non-native conformation with exposed hydrophobic patches on the surface of the two mutant LRR-RLKs. Similar to what was observed in *ebs4* mutants, loss-of-function mutations of *ebs5/sel1a/hrd3a* had little effect on plant development under standard laboratory growth conditions; however, the *ebs5/sel1a/hrd3a* mutations did activate the UPR and somehow inhibited a normal signaling process that regulates plant salt tolerance (Liu et al. 2011).

In yeast, Hrd3, Yos9, and Kar2 (karyogamy2, the yeast ortholog of BiP) form a luminal surveillance complex that selects misfolded glycoproteins for ERAD (Denic et al. 2006) and mutations in Kar2/BiP significantly slow down an ERAD process (Plemper et al. 1997; Nishikawa et al. 2001). It remains unknown whether BiP is also a component of the plant ERAD client-recognition complex or required to maintain the solubility of a terminally misfolded ERAD client for retrotranslocation. Although a single *Arabidopsis bip2* mutant was previously shown to be compromised in rapid secretion of plant pathogenesis-related proteins necessary to establish systemic acquired resistance in plants (Wang et al. 2005a, b), the functional redundancy between multiple BiP homologs (Hong et al. 2008) and the multifunctionality of BiP in protein import, folding, ER retention, and ERAD (Maattanen et al. 2010) present a technical challenge to use a loss-of-function genetic approach to investigate its biochemical functions in the folding, maturation, and ERQC/ERAD of plant RLKs (Nekrasov et al. 2009). Interestingly, a recent transgenic study suggested that BiP likely participates in an ERAD process to remove a rice MAMP-recognition LRR-RLK, XA21 (Park et al. 2010), which binds a bacterial sulfated peptide activator of XA21-mediated immunity (Ax21) (Song et al. 1995; Lee et al. 2009). The BiP-XA21 relationship was initially revealed by a proteomic study that aimed to identify components of the XA21-mediated plant immunity (Park et al. 2010). Consistent with the detected binding between XA21 and BIP3 (the most abundant member of the 5 rice BiP homologs), a fluorescently tagged XA21 protein was found largely not only in the ER but also on the plasma membrane when expression was driven by the *XA21* native promoter in transgenic rice. Given an earlier demonstration of the plasma membrane-mediated recognition of Ax21 by XA21 (Lee et al. 2006), this dual localization pattern suggested that XA21 might be an inefficiently folded protein that is largely retained by one or more ERQC systems in the ER. Subsequent studies showed that overproduction of BIP3 did not affect normal growth or general plant defenses but specifically compromised the XA21-mediated immunity, likely by reducing the abundance of XA21. Therefore, BIP3 is not involved in promoting the correct folding and maturation of the Ax21 receptor but instead stimulates the degradation of XA21 likely by an ERAD process. Further biochemical and transgenic studies will be needed to determine the exact biochemical function of the rice BiP3 in promoting XA21 degradation.

In both yeast and mammalian cells, Hrd3/Sel1L form a tight complex with Hrd1 and is thus considered as a cofactor for the 6-transmembrane-spanning E3 ligase. The *Arabidopsis* genome encodes two homlogs, of the yeast and mammalian Hrd1 E3 ligases: Hrd1A and Hrd1B. The *Arabidopsis bri1-5* and *bri1-9* dwarf mutants provided nice systems to investigate whether the two Hrd1 homologs play a role in a plant ERAD process. While T-DNA insertional mutations in Hrd1A or Hrd1B had little effect on the ERAD of bri1-5 and bri1-9 or the morphology of the corresponding *bri1* mutants, simultaneous elimination of the two Hrd1 homologs blocked the degradation of the two mutant BR receptors and partially suppressed their dwarf phenotypes (Su et al. 2011). It was therefore concluded that both Hrd1A and Hrd1B function redundantly in the ERAD of the ER-retained bri1-5 and bri1-9. However, it remains to be determined whether the two mutant BR receptors are indeed ubiquitinated before their retrotranslocation from the ER membrane into the cytosol for degradation. A recent study suggested that retrotranslocation of an ERAD client through the ER membrane might not require ubiquitination because a lysine-free form of the catalytic A subunit of ricin (RTA for ricin toxin A chain), a ribosome-inactivating toxin from castor bean, could still be retrotranslocated from the ER into the cytosol in tobacco leaf epidermal cells (Di Cola et al. 2005).

It was well known that the retrotranslocation of ERAD clients is powered by the cytosolic Cdc48-Ufd1-Npl4 trimeric complex (Raasi and Wolf 2007). The *Arabidopsis* genome encodes three Cdc48-like proteins, Cdc48A, Cdc48B, and Cdc48C with Cdc48A being capable of complementing a yeast *cdc48* mutant (Feiler et al. 1995; Rancour et al. 2002). While it remains unknown whether ERAD of the two mutant BR receptors or an incompletely folded EFR requires such an AAA-ATPase complex, previous studies showed that the plant Cdc48A was needed for degrading several mutated forms of the barley 7-transmembrane-spanning powdery mildew resistance (MLO) protein and a mutant form of an *Arabidopsis* vacuolar carboxypeptidase in *Arabidopsis* protoplasts (Muller et al. 2005; Yamamoto et al. 2010) and for retrotranslocation of RTA in tobacco leaf protoplasts (Marshall et al. 2008).

6.3 *Proteasome-Dependent/-Independent ERAD Pathways*

Despite requirements of similar ERAD components for degradation, bri1-5, bri1-9, and incompletely folded EFR might use distinct degradation machinery for their cytosolic proteolysis because treatment with the proteasome inhibitor MG132 inhibited the degradation of bri1-9 and incompletely folded EFR but had little effect on the ERAD of bri1-5 (Hong et al. 2008; Saijo et al. 2009). Further studies are needed to understand the exact biochemical mechanism by which bri1-5 is degraded in *Arabidopsis*. Studies in mammalian and yeast cells have shown that there are several proteasome-independent ERAD processes, such as protease-mediated degradation in the ER or cytosol and a lysosomal or autophagic degradation pathway (Schmitz and Herzog 2004). Previous investigations suggested that plant cells might also use lytic vacuoles for degrading ER resident proteins such as BiP and other artificial ERAD substrates (Pedrazzini et al. 1997; Brandizzi et al. 2003; Tamura et al. 2004; Pimpl et al. 2006). Because chemical or genetic inhibition of bri1-5 ERAD results in a significant phenotypic suppression of the *bri1-5* mutant, screening for second site mutations that restore the wild-type morphology to the *bri1-5* dwarf mutant could lead to identification of novel ERAD components and eventual elucidation of the biochemical mechanism by which bri1-5 is degraded.

7 Conclusions and Future Perspectives

These unexpected yet exciting discoveries coming from biochemical and genetic screens for regulators of RLK-mediated signaling and subsequent investigations have significantly expanded our understanding of folding, maturation, and QC of

RLKs in plant cells. These studies have demonstrated that *Arabidopsis* is a superb multicellular genetically tractable model organism to study ER-mediated protein folding and ERQC because loss-of-function mutations in conserved ERQC/ERAD components often cause embryo lethality in mammals but have no detectable effect on plant growth and development under standard laboratory growth conditions. Given the types of genetic screens that were originally designed, it is anticipated that positional cloning of additional *ELFIN/PLS* genes will lead to identification of more components of the ER chaperone networks necessary for folding and maturation of EFR, while cloning of more *EBS* genes will uncover more participants of the ERQC/ERAD systems. The availability of multiple RLKs as convenient model glycoproteins and their associated morphological and physiological phenotypes will enable functional studies of putative chaperones, lectins, folding catalysts, and glycan-modifying enzymes that were previously uncovered through microarray approaches (Martinez and Chrispeels 2003; Noh et al. 2003; Kamauchi et al. 2005; Iwata et al. 2010).

On the other hand, careful physiological analyses of the previously characterized mutants and/or transgenic lines should reveal novel physiological functions for the identified components of ERQC/ERAD machinery and N-glycan metabolic pathways. Consistent with this prediction, recent studies demonstrated a causal relationship between mutations affecting N-glycan biosynthesis and ERQC/ERAD and plant resistance to various biotic or abiotic stresses (Koiwa et al. 2003; Kang et al. 2008; Qin et al. 2008; Zhang et al. 2009; Saijo 2010; Ceriotti 2011; Liu et al. 2011). It is anticipated that further biochemical and proteomic studies of those *Arabidopsis* mutants and transgenic lines could lead to discovery of novel RLKs that require ERQC/ERAD to control their expression at the plasma membrane and play important roles in plant immunity and stress tolerance, thus also significantly increasing our knowledge of plant RLKs.

Acknowledgements Works in the authors' laboratories were supported in part by a grant from National Natural Science Foundation of China (31070246) to ZH and a grant from National Institutes of Health (GM060519) to JL.

References

Aebi M, Gassenhuber J, Domdey H, te Heesen S (1996) Cloning and characterization of the *ALG3* gene of *Saccharomyces cerevisiae*. Glycobiology 6:439–444

Anelli T, Sitia R (2008) Protein quality control in the early secretory pathway. EMBO J 27:315–327

Anelli T, Alessio M, Bachi A, Bergamelli L, Bertoli G, Camerini S, Mezghrani A, Ruffato E, Simmen T, Sitia R (2003) Thiol-mediated protein retention in the endoplasmic reticulum: the role of ERp44. EMBO J 22:5015–5022

Anelli T, Ceppi S, Bergamelli L, Cortini M, Masciarelli S, Valetti C, Sitia R (2007) Sequential steps and checkpoints in the early exocytic compartment during secretory IgM biogenesis. EMBO J 26:4177–4188

Appenzeller C, Andersson H, Kappeler F, Hauri HP (1999) The lectin ERGIC-53 is a cargo transport receptor for glycoproteins. Nat Cell Biol 1:330–334

Arnold SM, Fessler LI, Fessler JH, Kaufman RJ (2000) Two homologues encoding human UDP-glucose:glycoprotein glucosyltransferase differ in mRNA expression and enzymatic activity. Biochemistry 39:2149–2163

Bagola K, Mehnert M, Jarosch E, Sommer T (2010) Protein dislocation from the ER. Biochim Biophys Acta 1808:925–936

Banerjee S, Vishwanath P, Cui J, Kelleher DJ, Gilmore R, Robbins PW, Samuelson J (2007) The evolution of N-glycan-dependent endoplasmic reticulum quality control factors for glycoprotein folding and degradation. Proc Natl Acad Sci USA 104:11676–11681

Belkhadir Y, Durbak A, Wierzba M, Schmitz RJ, Aguirre A, Michel R, Rowe S, Fujioka S, Tax FE (2010) Intragenic suppression of a trafficking-defective brassinosteroid receptor mutant in *Arabidopsis*. Genetics 185:1283–1296

Biederer T, Volkwein C, Sommer T (1997) Role of Cue1p in ubiquitination and degradation at the ER surface. Science 278:1806–1809

Blond-Elguindi S, Cwirla SE, Dower WJ, Lipshutz RJ, Sprang SR, Sambrook JF, Gething MJ (1993) Affinity panning of a library of peptides displayed on bacteriophages reveals the binding specificity of BiP. Cell 75:717–728

Boller T, Felix G (2009) A renaissance of elicitors: perception of microbe-associated molecular patterns and danger signals by pattern-recognition receptors. Annu Rev Plant Biol 60:379–406

Boyce JM, Coates D, Fricker MD, Evans DE (1994) Genomic sequence of a calnexin homolog from *Arabidopsis thaliana*. Plant Physiol 106:1691

Braakman I, Bulleid NJ (2011) Protein folding and modification in the mammalian endoplasmic reticulum. Annu Rev Biochem 80:71–99

Brandizzi F, Hanton S, DaSilva LL, Boevink P, Evans D, Oparka K, Denecke J, Hawes C (2003) ER quality control can lead to retrograde transport from the ER lumen to the cytosol and the nucleoplasm in plants. Plant J 34:269–281

Burda P, Aebi M (1999) The dolichol pathway of N-linked glycosylation. Biochim Biophys Acta 1426:239–257

Burda P, te Heesen S, Brachat A, Wach A, Dusterhoft A, Aebi M (1996) Stepwise assembly of the lipid-linked oligosaccharide in the endoplasmic reticulum of *Saccharomyces cerevisiae*: identification of the *ALG9* gene encoding a putative mannosyl transferase. Proc Natl Acad Sci USA 93:7160–7165

Burda P, Jakob CA, Beinhauer J, Hegemann JH, Aebi M (1999) Ordered assembly of the asymmetrically branched lipid-linked oligosaccharide in the endoplasmic reticulum is ensured by the substrate specificity of the individual glycosyltransferases. Glycobiology 9:617–625

Burn JE, Hurley UA, Birch RJ, Arioli T, Cork A, Williamson RE (2002) The cellulose-deficient *Arabidopsis* mutant *rsw3* is defective in a gene encoding a putative glucosidase II, an enzyme processing N-glycans during ER quality control. Plant J 32:949–960

Cabral CM, Liu Y, Sifers RN (2001) Dissecting glycoprotein quality control in the secretory pathway. Trends Biochem Sci 26:619–624

Calfon M, Zeng H, Urano F, Till JH, Hubbard SR, Harding HP, Clark SG, Ron D (2002) IRE1 couples endoplasmic reticulum load to secretory capacity by processing the *XBP-1* mRNA. Nature 415:92–96

Cano-Delgado A, Yin Y, Yu C, Vafeados D, Mora-Garcia S, Cheng JC, Nam KH, Li J, Chory J (2004) BRL1 and BRL3 are novel brassinosteroid receptors that function in vascular differentiation in *Arabidopsis*. Development 131:5341–5351

Caplan JL, Zhu X, Mamillapalli P, Marathe R, Anandalakshmi R, Dinesh-Kumar SP (2009) Induced ER chaperones regulate a receptor-like kinase to mediate antiviral innate immune response in plants. Cell Host Microbe 6:457–469

Caramelo JJ, Parodi AJ (2007) How sugars convey information on protein conformation in the endoplasmic reticulum. Semin Cell Dev Biol 18:732–742

Caramelo JJ, Parodi AJ (2008) Getting in and out from calnexin/calreticulin cycles. J Biol Chem 283:10221–10225

Caramelo JJ, Castro OA, de Prat-Gay G, Parodi AJ (2004) The endoplasmic reticulum glucosyltransferase recognizes nearly native glycoprotein folding intermediates. J Biol Chem 279:46280–46285

Carla Fama M, Raden D, Zacchi N, Lemos DR, Robinson AS, Silberstein S (2007) The *Saccharomyces cerevisiae YFR041C/ERJ5* gene encoding a type I membrane protein with a J domain is required to preserve the folding capacity of the endoplasmic reticulum. Biochim Biophys Acta 1773:232–242

Carvalho P, Goder V, Rapoport TA (2006) Distinct ubiquitin-ligase complexes define convergent pathways for the degradation of ER proteins. Cell 126:361–373

Carvalho P, Stanley AM, Rapoport TA (2010) Retrotranslocation of a misfolded luminal ER protein by the ubiquitin-ligase Hrd1p. Cell 143:579–591

Ceriotti A (2011) Waste disposal in the endoplasmic reticulum, ROS production and plant salt stress response. Cell Res 21:555–557

Ceriotti A, Roberts LM (2006) Endoplasmic reticulum-associated protein degradation in plant cells. In: Robinson DG (ed) The plant endoplasmic reticulum. Springer, Berlin, pp 75–98

Che P, Bussell JD, Zhou W, Estavillo GM, Pogson BJ, Smith SM (2010) Signaling from the endoplasmic reticulum activates brassinosteroid signaling and promotes acclimation to stress in *Arabidopsis*. Sci Signal 3:ra69

Chen L, Hamada S, Fujiwara M, Zhu T, Thao NP, Wong HL, Krishna P, Ueda T, Kaku H, Shibuya N, Kawasaki T, Shimamoto K (2010) The Hop/Sti1-Hsp90 chaperone complex facilitates the maturation and transport of a PAMP receptor in rice innate immunity. Cell Host Microbe 7:185–196

Choe J, Kelker MS, Wilson IA (2005) Crystal structure of human toll-like receptor 3 (TLR3) ectodomain. Science 309:581–585

Christensen A, Svensson K, Persson S, Jung J, Michalak M, Widell S, Sommarin M (2008) Functional characterization of *Arabidopsis* calreticulin1a: a key alleviator of endoplasmic reticulum stress. Plant Cell Physiol 49:912–924

Cipollo JF, Trimble RB (2000) The accumulation of Man(6)GlcNAc(2)-PP-dolichol in the *Saccharomyces cerevisiae Δalg9* mutant reveals a regulatory role for the Alg3p α1,3-Man middle-arm addition in downstream oligosaccharide-lipid and glycoprotein glycan processing. J Biol Chem 275:4267–4277

Clark SE, Running MP, Meyerowitz EM (1993) *CLAVATA1*, a regulator of meristem and flower development in *Arabidopsis*. Development 119:397–418

Clark SE, Running MP, Meyerowitz EM (1995) *CLAVATA3* is a specific regulator of shoot and floral meristem development affecting the same processes as *CLAVATA1*. Development 121:2057–2067

Clark SE, Williams RW, Meyerowitz EM (1997) The *CLAVATA1* gene encodes a putative receptor kinase that controls shoot and floral meristem size in Arabidopsis. Cell 89:575–585

Clerc S, Hirsch C, Oggier DM, Deprez P, Jakob C, Sommer T, Aebi M (2009) Htm1 protein generates the N-glycan signal for glycoprotein degradation in the endoplasmic reticulum. J Cell Biol 184:159–172

Clouse SD, Sasse JM (1998) BRASSINOSTEROIDS: essential regulators of plant growth and development. Annu Rev Plant Physiol Plant Mol Biol 49:427–451

Cock JM, Vanoosthuyse V, Gaude T (2002) Receptor kinase signalling in plants and animals: distinct molecular systems with mechanistic similarities. Curr Opin Cell Biol 14:230–236

Coe H, Michalak M (2010) ERp57, a multifunctional endoplasmic reticulum resident oxidoreductase. Int J Biochem Cell Biol 42:796–799

Cormier JH, Tamura T, Sunryd JC, Hebert DN (2009) EDEM1 recognition and delivery of misfolded proteins to the SEL1L-containing ERAD complex. Mol Cell 34:627–633

Cox JS, Walter P (1996) A novel mechanism for regulating activity of a transcription factor that controls the unfolded protein response. Cell 87:391–404

Cox JS, Shamu CE, Walter P (1993) Transcriptional induction of genes encoding endoplasmic reticulum resident proteins requires a transmembrane protein kinase. Cell 73:1197–1206

Deng Y, Humbert S, Liu JX, Srivastava R, Rothstein SJ, Howell SH (2011) Heat induces the splicing by IRE1 of a mRNA encoding a transcription factor involved in the unfolded protein response in Arabidopsis. Proc Natl Acad Sci USA 108:7247–7252

Denic V, Quan EM, Weissman JS (2006) A luminal surveillance complex that selects misfolded glycoproteins for ER-associated degradation. Cell 126:349–359

DeYoung BJ, Innes RW (2006) Plant NBS-LRR proteins in pathogen sensing and host defense. Nat Immunol 7:1243–1249

Di Cola A, Frigerio L, Lord JM, Roberts LM, Ceriotti A (2005) Endoplasmic reticulum-associated degradation of ricin A chain has unique and plant-specific features. Plant Physiol 137:287–296

Di Matteo A, Federici L, Mattei B, Salvi G, Johnson KA, Savino C, De Lorenzo G, Tsernoglou D, Cervone F (2003) The crystal structure of polygalacturonase-inhibiting protein (PGIP), a leucine-rich repeat protein involved in plant defense. Proc Natl Acad Sci USA 100:10124–10128

Elbein AD, Tropea JE, Mitchell M, Kaushal GP (1990) Kifunensine, a potent inhibitor of the glycoprotein processing mannosidase I. J Biol Chem 265:15599–15605

Eletto D, Dersh D, Argon Y (2010) GRP94 in ER quality control and stress responses. Semin Cell Dev Biol 21:479–485

Ellgaard L, Helenius A (2003) Quality control in the endoplasmic reticulum. Nat Rev Mol Cell Biol 4:181–191

Feiler HS, Desprez T, Santoni V, Kronenberger J, Caboche M, Traas J (1995) The higher plant *Arabidopsis thaliana* encodes a functional CDC48 homologue which is highly expressed in dividing and expanding cells. EMBO J 14:5626–5637

Fletcher JC, Brand U, Running MP, Simon R, Meyerowitz EM (1999) Signaling of cell fate decisions by *CLAVATA3* in *Arabidopsis* shoot meristems. Science 283:1911–1914

Flynn GC, Pohl J, Flocco MT, Rothman JE (1991) Peptide-binding specificity of the molecular chaperone BiP. Nature 353:726–730

Frank CG, Aebi M (2005) ALG9 mannosyltransferase is involved in two different steps of lipid-linked oligosaccharide biosynthesis. Glycobiology 15:1156–1163

Friedrichsen DM, Joazeiro CA, Li J, Hunter T, Chory J (2000) Brassinosteroid-insensitive1 is a ubiquitously expressed leucine-rich repeat receptor serine/threonine kinase. Plant Physiol 123:1247–1256

Frydman J, Hohfeld J (1997) Chaperones get in touch: the Hip-Hop connection. Trends Biochem Sci 22:87–92

Fukuda S, Sumii M, Masuda Y, Takahashi M, Koike N, Teishima J, Yasumoto H, Itamoto T, Asahara T, Dohi K, Kamiya K (2001) Murine and human SDF2L1 is an endoplasmic reticulum stress-inducible gene and encodes a new member of the Pmt/rt protein family. Biochem Biophys Res Commun 280:407–414

Garza RM, Sato BK, Hampton RY (2009) In vitro analysis of Hrd1p-mediated retrotranslocation of its multispanning membrane substrate 3-hydroxy-3-methylglutaryl (HMG)-CoA reductase. J Biol Chem 284:14710–14722

Gauss R, Jarosch E, Sommer T, Hirsch C (2006a) A complex of Yos9p and the HRD ligase integrates endoplasmic reticulum quality control into the degradation machinery. Nat Cell Biol 8:849–854

Gauss R, Sommer T, Jarosch E (2006b) The Hrd1p ligase complex forms a linchpin between ER-lumenal substrate selection and Cdc48p recruitment. EMBO J 25:1827–1835

Gish LA, Clark SE (2011) The RLK/Pelle family of kinases. Plant J 66:117–127

Gomez-Gomez L, Boller T (2000) FLS2: an LRR receptor-like kinase involved in the perception of the bacterial elicitor flagellin in *Arabidopsis*. Mol Cell 5:1003–1011

Hammond C, Helenius A (1994) Quality control in the secretory pathway: retention of a misfolded viral membrane glycoprotein involves cycling between the ER, intermediate compartment, and Golgi apparatus. J Cell Biol 126:41–52

Hampton RY, Gardner RG, Rine J (1996) Role of 26S proteasome and HRD genes in the degradation of 3-hydroxy-3-methylglutaryl-CoA reductase, an integral endoplasmic reticulum membrane protein. Mol Biol Cell 7:2029–2044

Harding HP, Zhang Y, Ron D (1999) Protein translation and folding are coupled by an endoplasmic-reticulum-resident kinase. Nature 397:271–274

Hartmann-Petersen R, Gordon C (2004) Protein degradation: recognition of ubiquitinylated substrates. Curr Biol 14:R754–R756

Hassink G, Kikkert M, van Voorden S, Lee SJ, Spaapen R, van Laar T, Coleman CS, Bartee E, Fruh K, Chau V, Wiertz E (2005) TEB4 is a C4HC3 RING finger-containing ubiquitin ligase of the endoplasmic reticulum. Biochem J 388:647–655

Hauri HP, Nufer O, Breuza L, Tekaya HB, Liang L (2002) Lectins and protein traffic early in the secretory pathway. Biochem Soc Symp 73–82

Haweker H, Rips S, Koiwa H, Salomon S, Saijo Y, Chinchilla D, Robatzek S, von Schaewen A (2010) Pattern recognition receptors require N-glycosylation to mediate plant immunity. J Biol Chem 285:4629–4636

Haze K, Yoshida H, Yanagi H, Yura T, Mori K (1999) Mammalian transcription factor ATF6 is synthesized as a transmembrane protein and activated by proteolysis in response to endoplasmic reticulum stress. Mol Biol Cell 10:3787–3799

Helenius A (1994) How N-linked oligosaccharides affect glycoprotein folding in the endoplasmic reticulum. Mol Biol Cell 5:253–265

Helenius A, Aebi M (2004) Roles of N-linked glycans in the endoplasmic reticulum. Annu Rev Biochem 73:1019–1049

Henquet M, Lehle L, Schreuder M, Rouwendal G, Molthoff J, Helsper J, van der Krol S, Bosch D (2008) Identification of the gene encoding the α1,3-mannosyltransferase (ALG3) in *Arabidopsis* and characterization of downstream N-glycan processing. Plant Cell 20:1652–1664

Hirsch C, Gauss R, Horn SC, Neuber O, Sommer T (2009) The ubiquitylation machinery of the endoplasmic reticulum. Nature 458:453–460

Hitt R, Wolf DH (2004) Der1p, a protein required for degradation of malfolded soluble proteins of the endoplasmic reticulum: topology and Der1-like proteins. FEMS Yeast Res 4:721–729

Hong Z, Jin H, Tzfira T, Li J (2008) Multiple mechanism-mediated retention of a defective brassinosteroid receptor in the endoplasmic reticulum of *Arabidopsis*. Plant Cell 20:3418–3429

Hong Z, Jin H, Fitchette AC, Xia Y, Monk AM, Faye L, Li J (2009) Mutations of an α1,6 mannosyltransferase inhibit endoplasmic reticulum-associated degradation of defective brassinosteroid receptors in *Arabidopsis*. Plant Cell 21:3792–3802

Hosokawa N, Kamiya Y, Kamiya D, Kato K, Nagata K (2009) Human OS-9, a lectin required for glycoprotein endoplasmic reticulum-associated degradation, recognizes mannose-trimmed N-glycans. J Biol Chem 284:17061–17068

Hosokawa N, Kamiya Y, Kato K (2010) The role of MRH domain-containing lectins in ERAD. Glycobiology 20:651–660

Houston NL, Fan C, Xiang JQ, Schulze JM, Jung R, Boston RS (2005) Phylogenetic analyses identify 10 classes of the protein disulfide isomerase family in plants, including single-domain protein disulfide isomerase-related proteins. Plant Physiol 137:762–778

Huang L, Franklin AE, Hoffman NE (1993) Primary structure and characterization of an *Arabidopsis thaliana* calnexin-like protein. J Biol Chem 268:6560–6566

Ishiguro S, Watanabe Y, Ito N, Nonaka H, Takeda N, Sakai T, Kanaya H, Okada K (2002) SHEPHERD is the *Arabidopsis* GRP94 responsible for the formation of functional CLAVATA proteins. EMBO J 21:898–908

Iwata Y, Koizumi N (2005) An *Arabidopsis* transcription factor, AtbZIP60, regulates the endoplasmic reticulum stress response in a manner unique to plants. Proc Natl Acad Sci USA 102:5280–5285

Iwata Y, Sakiyama M, Lee M-H, Koizumi N (2010) Transcriptomic response of *Arabidopsis thaliana* to tunicamycin-induced endoplasmic reticulum stress. Plant Biotechnol 27:161–171

Jakob CA, Burda P, Roth J, Aebi M (1998) Degradation of misfolded endoplasmic reticulum glycoproteins in *Saccharomyces cerevisiae* is determined by a specific oligosaccharide structure. J Cell Biol 142:1223–1233

Jakob CA, Bodmer D, Spirig U, Battig P, Marcil A, Dignard D, Bergeron JJ, Thomas DY, Aebi M (2001) Htm1p, a mannosidase-like protein, is involved in glycoprotein degradation in yeast. EMBO Rep 2:423–430

Jeong S, Trotochaud AE, Clark SE (1999) The *Arabidopsis CLAVATA2* gene encodes a receptor-like protein required for the stability of the CLAVATA1 receptor-like kinase. Plant Cell 11:1925–1934

Jin H, Yan Z, Nam KH, Li J (2007) Allele-specific suppression of a defective brassinosteroid receptor reveals a physiological role of UGGT in ER quality control. Mol Cell 26:821–830

Jin H, Hong Z, Su W, Li J (2009) A plant-specific calreticulin is a key retention factor for a defective brassinosteroid receptor in the endoplasmic reticulum. Proc Natl Acad Sci USA 106:13612–13617

Kajiura H, Seki T, Fujiyama K (2010) *Arabidopsis thaliana ALG3* mutant synthesizes immature oligosaccharides in the ER and accumulates unique N-glycans. Glycobiology 20:736–751

Kamauchi S, Nakatani H, Nakano C, Urade R (2005) Gene expression in response to endoplasmic reticulum stress in *Arabidopsis thaliana*. FEBS J 272:3461–3476

Kanehara K, Xie W, Ng DT (2010) Modularity of the Hrd1 ERAD complex underlies its diverse client range. J Cell Biol 188:707–716

Kang JS, Frank J, Kang CH, Kajiura H, Vikram M, Ueda A, Kim S, Bahk JD, Triplett B, Fujiyama K, Lee SY, von Schaewen A, Koiwa H (2008) Salt tolerance of *Arabidopsis thaliana* requires maturation of N-glycosylated proteins in the Golgi apparatus. Proc Natl Acad Sci USA 105:5933–5938

Kato K, Kamiya Y (2007) Structural views of glycoprotein-fate determination in cells. Glycobiology 17:1031–1044

Kayes JM, Clark SE (1998) *CLAVATA2*, a regulator of meristem and organ development in *Arabidopsis*. Development 125:3843–3851

Kelleher DJ, Gilmore R (2006) An evolving view of the eukaryotic oligosaccharyltransferase. Glycobiology 16:47R–62R

Kikkert M, Doolman R, Dai M, Avner R, Hassink G, van Voorden S, Thanedar S, Roitelman J, Chau V, Wiertz E (2004) Human HRD1 is an E3 ubiquitin ligase involved in degradation of proteins from the endoplasmic reticulum. J Biol Chem 279:3525–3534

Kinoshita T, Cano-Delgado A, Seto H, Hiranuma S, Fujioka S, Yoshida S, Chory J (2005) Binding of brassinosteroids to the extracellular domain of plant receptor kinase BRI1. Nature 433:167–171

Knop M, Finger A, Braun T, Hellmuth K, Wolf DH (1996) Der1, a novel protein specifically required for endoplasmic reticulum degradation in yeast. EMBO J 15:753–763

Koiwa H, Li F, McCully MG, Mendoza I, Koizumi N, Manabe Y, Nakagawa Y, Zhu J, Rus A, Pardo JM, Bressan RA, Hasegawa PM (2003) The STT3a subunit isoform of the Arabidopsis oligosaccharyltransferase controls adaptive responses to salt/osmotic stress. Plant Cell 15:2273–2284

Koizumi N, Martinez IM, Kimata Y, Kohno K, Sano H, Chrispeels MJ (2001) Molecular characterization of two Arabidopsis Ire1 homologs, endoplasmic reticulum-located transmembrane protein kinases. Plant Physiol 127:949–962

Kolade OO, Bamford VA, Ancillo Anton G, Jones JD, Vera P, Hemmings AM (2006) In vitro characterization of the cysteine-rich capping domains in a plant leucine rich repeat protein. Biochim Biophys Acta 1764:1043–1053

Kostova Z, Tsai YC, Weissman AM (2007) Ubiquitin ligases, critical mediators of endoplasmic reticulum-associated degradation. Semin Cell Dev Biol 18:770–779

Lee SW, Han SW, Bartley LE, Ronald PC (2006) From the Academy: Colloquium review. Unique characteristics of *Xanthomonas oryzae* pv. *oryzae* AvrXa21 and implications for plant innate immunity. Proc Natl Acad Sci USA 103:18395–18400

Lee SW, Han SW, Sririyanum M, Park CJ, Seo YS, Ronald PC (2009) A type I-secreted, sulfated peptide triggers XA21-mediated innate immunity. Science 326:850–853

Lehti-Shiu MD, Zou C, Hanada K, Shiu SH (2009) Evolutionary history and stress regulation of plant receptor-like kinase/pelle genes. Plant Physiol 150:12–26

Li J, Chory J (1997) A putative leucine-rich repeat receptor kinase involved in brassinosteroid signal transduction. Cell 90:929–938

Li J, Wen J, Lease KA, Doke JT, Tax FE, Walker JC (2002) BAK1, an *Arabidopsis* LRR receptor-like protein kinase, interacts with BRI1 and modulates brassinosteroid signaling. Cell 110:213–222

Li J, Zhao-Hui C, Batoux M, Nekrasov V, Roux M, Chinchilla D, Zipfel C, Jones JD (2009) Specific ER quality control components required for biogenesis of the plant innate immune receptor EFR. Proc Natl Acad Sci USA 106:15973–15978

Liang J, Yin C, Doong H, Fang S, Peterhoff C, Nixon RA, Monteiro MJ (2006) Characterization of erasin (UBXD2): a new ER protein that promotes ER-associated protein degradation. J Cell Sci 119:4011–4024

Lilley BN, Ploegh HL (2004) A membrane protein required for dislocation of misfolded proteins from the ER. Nature 429:834–840

Liu L, Cui F, Li Q, Yin B, Zhang H, Lin B, Wu Y, Xia R, Tang S, Xie Q (2011) The endoplasmic reticulum-associated degradation is necessary for plant salt tolerance. Cell Res. doi:10.1038/cr.2010.181

Liu JX, Howell SH (2010) Endoplasmic reticulum protein quality control and its relationship to environmental stress responses in plants. Plant Cell 22:2930–2942

Liu JX, Srivastava R, Che P, Howell SH (2007a) An endoplasmic reticulum stress response in *Arabidopsis* is mediated by proteolytic processing and nuclear relocation of a membrane-associated transcription factor, bZIP28. Plant Cell 19:4111–4119

Liu JX, Srivastava R, Che P, Howell SH (2007b) Salt stress responses in Arabidopsis utilize a signal transduction pathway related to endoplasmic reticulum stress signaling. Plant J 51:897–909

Lord JM, Ceriotti A, Roberts LM (2002) ER dislocation: Cdc48p/p97 gets into the AAAct. Curr Biol 12:R182–R184

Lu X, Tintor N, Mentzel T, Kombrink E, Boller T, Robatzek S, Schulze-Lefert P, Saijo Y (2009) Uncoupling of sustained MAMP receptor signaling from early outputs in an Arabidopsis endoplasmic reticulum glucosidase II allele. Proc Natl Acad Sci USA 106:22522–22527

Maattanen P, Gehring K, Bergeron JJ, Thomas DY (2010) Protein quality control in the ER: the recognition of misfolded proteins. Semin Cell Dev Biol 21:500–511

Marshall RS, Jolliffe NA, Ceriotti A, Snowden CJ, Lord JM, Frigerio L, Roberts LM (2008) The role of CDC48 in the retro-translocation of non-ubiquitinated toxin substrates in plant cells. J Biol Chem 283:15869–15877

Martinez IM, Chrispeels MJ (2003) Genomic analysis of the unfolded protein response in Arabidopsis shows its connection to important cellular processes. Plant Cell 15:561–576

Matlack KE, Misselwitz B, Plath K, Rapoport TA (1999) BiP acts as a molecular ratchet during posttranslational transport of prepro-a factor across the ER membrane. Cell 97:553–564

McCracken AA, Brodsky JL (1996) Assembly of ER-associated protein degradation in vitro: dependence on cytosol, calnexin, and ATP. J Cell Biol 132:291–298

Medicherla B, Kostova Z, Schaefer A, Wolf DH (2004) A genomic screen identifies Dsk2p and Rad23p as essential components of ER-associated degradation. EMBO Rep 5:692–697

Meunier L, Usherwood YK, Chung KT, Hendershot LM (2002) A subset of chaperones and folding enzymes form multiprotein complexes in endoplasmic reticulum to bind nascent proteins. Mol Biol Cell 13:4456–4469

Michalak M, Groenendyk J, Szabo E, Gold LI, Opas M (2009) Calreticulin, a multi-process calcium-buffering chaperone of the endoplasmic reticulum. Biochem J 417:651–666

Miya A, Albert P, Shinya T, Desaki Y, Ichimura K, Shirasu K, Narusaka Y, Kawakami N, Kaku H, Shibuya N (2007) CERK1, a LysM receptor kinase, is essential for chitin elicitor signaling in *Arabidopsis*. Proc Natl Acad Sci USA 104:19613–19618

Molinari M (2007) N-glycan structure dictates extension of protein folding or onset of disposal. Nat Chem Biol 3:313–320

Molinari M, Galli C, Vanoni O, Arnold SM, Kaufman RJ (2005) Persistent glycoprotein misfolding activates the glucosidase II/UGT1-driven calnexin cycle to delay aggregation and loss of folding competence. Mol Cell 20:503–512

Mori K, Ma W, Gething MJ, Sambrook J (1993) A transmembrane protein with a cdc2^{+}/CDC28-related kinase activity is required for signaling from the ER to the nucleus. Cell 74:743–756

Muller J, Piffanelli P, Devoto A, Miklis M, Elliott C, Ortmann B, Schulze-Lefert P, Panstruga R (2005) Conserved ERAD-like quality control of a plant polytopic membrane protein. Plant Cell 17:149–163

Nam KH, Li J (2002) BRI1/BAK1, a receptor kinase pair mediating brassinosteroid signaling. Cell 110:203–212

Nekrasov V, Li J, Batoux M, Roux M, Chu ZH, Lacombe S, Rougon A, Bittel P, Kiss-Papp M, Chinchilla D, van Esse HP, Jorda L, Schwessinger B, Nicaise V, Thomma BP, Molina A, Jones JD, Zipfel C (2009) Control of the pattern-recognition receptor EFR by an ER protein complex in plant immunity. EMBO J 28:3428–3438

Nishikawa SI, Fewell SW, Kato Y, Brodsky JL, Endo T (2001) Molecular chaperones in the yeast endoplasmic reticulum maintain the solubility of proteins for retrotranslocation and degradation. J Cell Biol 153:1061–1070

Noguchi T, Fujioka S, Choe S, Takatsuto S, Yoshida S, Yuan H, Feldmann KA, Tax FE (1999) Brassinosteroid-insensitive dwarf mutants of Arabidopsis accumulate brassinosteroids. Plant Physiol 121:743–752

Noh SJ, Kwon CS, Oh DH, Moon JS, Chung WI (2003) Expression of an evolutionarily distinct novel BiP gene during the unfolded protein response in *Arabidopsis thaliana*. Gene 311:81–91

Oda Y, Hosokawa N, Wada I, Nagata K (2003) EDEM as an acceptor of terminally misfolded glycoproteins released from calnexin. Science 299:1394–1397

Ostrovsky O, Ahmed NT, Argon Y (2009) The chaperone activity of GRP94 toward insulin-like growth factor II is necessary for the stress response to serum deprivation. Mol Biol Cell 20:1855–1864

Ostrovsky O, Eletto D, Makarewich C, Barton ER, Argon Y (2010) Glucose regulated protein 94 is required for muscle differentiation through its control of the autocrine production of insulin-like growth factors. Biochim Biophys Acta 1803:333–341

Otero JH, Lizak B, Hendershot LM (2010) Life and death of a BiP substrate. Semin Cell Dev Biol 21:472–478

Park CJ, Bart R, Chern M, Canlas PE, Bai W, Ronald PC (2010) Overexpression of the endoplasmic reticulum chaperone BiP3 regulates XA21-mediated innate immunity in rice. PLoS One 5: e9262

Pattison RJ, Amtmann A (2009) N-glycan production in the endoplasmic reticulum of plants. Trends Plant Sci 14:92–99

Pedrazzini E, Giovinazzo G, Bielli A, de Virgilio M, Frigerio L, Pesca M, Faoro F, Bollini R, Ceriotti A, Vitale A (1997) Protein quality control along the route to the plant vacuole. Plant Cell 9:1869–1880

Persson S, Wyatt SE, Love J, Thompson WF, Robertson D, Boss WF (2001) The Ca^{2+} status of the endoplasmic reticulum is altered by induction of calreticulin expression in transgenic plants. Plant Physiol 126:1092–1104

Persson S, Rosenquist M, Svensson K, Galvao R, Boss WF, Sommarin M (2003) Phylogenetic analyses and expression studies reveal two distinct groups of calreticulin isoforms in higher plants. Plant Physiol 133:1385–1396

Pilon M, Schekman R, Romisch K (1997) Sec61p mediates export of a misfolded secretory protein from the endoplasmic reticulum to the cytosol for degradation. EMBO J 16:4540–4548

Pimpl P, Taylor JP, Snowden C, Hillmer S, Robinson DG, Denecke J (2006) Golgi-mediated vacuolar sorting of the endoplasmic reticulum chaperone BiP may play an active role in quality control within the secretory pathway. Plant Cell 18:198–211

Plemper RK, Bohmler S, Bordallo J, Sommer T, Wolf DH (1997) Mutant analysis links the translocon and BiP to retrograde protein transport for ER degradation. Nature 388:891–895

Ponting CP (2000) Novel repeats in ryanodine and IP3 receptors and protein O-mannosyltransferases. Trends Biochem Sci 25:48–50

Qin C, Qian W, Wang W, Wu Y, Yu C, Jiang X, Wang D, Wu P (2008) GDP-mannose pyrophosphorylase is a genetic determinant of ammonium sensitivity in *Arabidopsis thaliana*. Proc Natl Acad Sci USA 105:18308–18313

Quan EM, Kamiya Y, Kamiya D, Denic V, Weibezahn J, Kato K, Weissman JS (2008) Defining the glycan destruction signal for endoplasmic reticulum-associated degradation. Mol Cell 32:870–877

Raasi S, Wolf DH (2007) Ubiquitin receptors and ERAD: a network of pathways to the proteasome. Semin Cell Dev Biol 18:780–791

Ramos RR, Swanson AJ, Bass J (2007) Calreticulin and Hsp90 stabilize the human insulin receptor and promote its mobility in the endoplasmic reticulum. Proc Natl Acad Sci USA 104:10470–10475

Rancour DM, Dickey CE, Park S, Bednarek SY (2002) Characterization of AtCDC48. Evidence for multiple membrane fusion mechanisms at the plane of cell division in plants. Plant Physiol 130:1241–1253

Rapoport TA (2007) Protein translocation across the eukaryotic endoplasmic reticulum and bacterial plasma membranes. Nature 450:663–669

Reddy P, Sparvoli A, Fagioli C, Fassina G, Sitia R (1996) Formation of reversible disulfide bonds with the protein matrix of the endoplasmic reticulum correlates with the retention of unassembled Ig light chains. EMBO J 15:2077–2085

Richly H, Rape M, Braun S, Rumpf S, Hoege C, Jentsch S (2005) A series of ubiquitin binding factors connects CDC48/p97 to substrate multiubiquitylation and proteasomal targeting. Cell 120:73–84

Ron D, Walter P (2007) Signal integration in the endoplasmic reticulum unfolded protein response. Nat Rev Mol Cell Biol 8:519–529

Rudiger S, Buchberger A, Bukau B (1997) Interaction of Hsp70 chaperones with substrates. Nat Struct Biol 4:342–349

Saijo Y (2010) ER quality control of immune receptors and regulators in plants. Cell Microbiol 12:716–724

Saijo Y, Tintor N, Lu X, Rauf P, Pajerowska-Mukhtar K, Haweker H, Dong X, Robatzek S, Schulze-Lefert P (2009) Receptor quality control in the endoplasmic reticulum for plant innate immunity. EMBO J 28:3439–3449

Sanders CR, Myers JK (2004) Disease-related misassembly of membrane proteins. Annu Rev Biophys Biomol Struct 33:25–51

Satoh T, Chen Y, Hu D, Hanashima S, Yamamoto K, Yamaguchi Y (2010) Structural basis for oligosaccharide recognition of misfolded glycoproteins by OS-9 in ER-associated degradation. Mol Cell 40:905–916

Scheuner D, Song B, McEwen E, Liu C, Laybutt R, Gillespie P, Saunders T, Bonner-Weir S, Kaufman RJ (2001) Translational control is required for the unfolded protein response and in vivo glucose homeostasis. Mol Cell 7:1165–1176

Schmitz A, Herzog V (2004) Endoplasmic reticulum-associated degradation: exceptions to the rule. Eur J Cell Biol 83:501–509

Schott A, Ravaud S, Keller S, Radzimanowski J, Viotti C, Hillmer S, Sinning I, Strahl S (2010) *Arabidopsis* stromal-derived Factor2 (SDF2) is a crucial target of the unfolded protein response in the endoplasmic reticulum. J Biol Chem 285:18113–18121

Schuberth C, Buchberger A (2005) Membrane-bound Ubx2 recruits Cdc48 to ubiquitin ligases and their substrates to ensure efficient ER-associated protein degradation. Nat Cell Biol 7:999–1006

Shen J, Chen X, Hendershot L, Prywes R (2002) ER stress regulation of ATF6 localization by dissociation of BiP/GRP78 binding and unmasking of Golgi localization signals. Dev Cell 3:99–111

Shimizu T, Nakano T, Takamizawa D, Desaki Y, Ishii-Minami N, Nishizawa Y, Minami E, Okada K, Yamane H, Kaku H, Shibuya N (2011) Two LysM receptor molecules, CEBiP and OsCERK1, cooperatively regulate chitin elicitor signaling in rice. Plant J 64:204–214

Shirasu K (2009) The HSP90-SGT1 chaperone complex for NLR immune sensors. Annu Rev Plant Biol 60:139–164

Shiu SH, Bleecker AB (2001) Receptor-like kinases from *Arabidopsis* form a monophyletic gene family related to animal receptor kinases. Proc Natl Acad Sci USA 98:10763–10768

Song WY, Wang GL, Chen LL, Kim HS, Pi LY, Holsten T, Gardner J, Wang B, Zhai WX, Zhu LH, Fauquet C, Ronald P (1995) A receptor kinase-like protein encoded by the rice disease resistance gene, *Xa21*. Science 270:1804–1806

Su W, Liu Y, Xia Y, Hong Z, Li J (2011) Conserved endoplasmic reticulum-associated degradation system to eliminate mutated receptor-like kinases in Arabidopsis. Proc Natl Acad Sci USA 108:870–875

Suzuki T, Park H, Kwofie MA, Lennarz WJ (2001) Rad23 provides a link between the Png1 deglycosylating enzyme and the 26S proteasome in yeast. J Biol Chem 276:21601–21607

Taipale M, Jarosz DF, Lindquist S (2010) HSP90 at the hub of protein homeostasis: emerging mechanistic insights. Nat Rev Mol Cell Biol 11:515–528

Tamura K, Yamada K, Shimada T, Hara-Nishimura I (2004) Endoplasmic reticulum-resident proteins are constitutively transported to vacuoles for degradation. Plant J 39:393–402

Torii KU (2009) Transmembrane receptors in plants: receptor kinases and their ligands. Annu Plant Rev 33:1–29

Trombetta ES (2003) The contribution of N-glycans and their processing in the endoplasmic reticulum to glycoprotein biosynthesis. Glycobiology 13:77R–91R

van der Hoorn RA, Wulff BB, Rivas S, Durrant MC, van der Ploeg A, de Wit PJ, Jones JD (2005) Structure-function analysis of cf-9, a receptor-like protein with extracytoplasmic leucine-rich repeats. Plant Cell 17:1000–1015

Vembar SS, Brodsky JL (2008) One step at a time: endoplasmic reticulum-associated degradation. Nat Rev Mol Cell Biol 9:944–957

Vert G, Nemhauser JL, Geldner N, Hong F, Chory J (2005) Molecular mechanisms of steroid hormone signaling in plants. Annu Rev Cell Dev Biol 21:177–201

Vitale A, Boston RS (2008) Endoplasmic reticulum quality control and the unfolded protein response: insights from plants. Traffic 9:1581–1588

von Numers N, Survila M, Aalto M, Batoux M, Heino P, Palva ET, Li J (2010) Requirement of a homolog of glucosidase II b-subunit for EFR-mediated defense signaling in *Arabidopsis thaliana*. Mol Plant 3:740–750

Wanderling S, Simen BB, Ostrovsky O, Ahmed NT, Vogen SM, Gidalevitz T, Argon Y (2007) GRP94 is essential for mesoderm induction and muscle development because it regulates insulin-like growth factor secretion. Mol Biol Cell 18:3764–3775

Wang D, Weaver ND, Kesarwani M, Dong X (2005a) Induction of protein secretory pathway is required for systemic acquired resistance. Science 308:1036–1040

Wang N, Daniels R, Hebert DN (2005b) The cotranslational maturation of the type I membrane glycoprotein tyrosinase: the heat shock protein 70 system hands off to the lectin-based chaperone system. Mol Biol Cell 16:3740–3752

Werner ED, Brodsky JL, McCracken AA (1996) Proteasome-dependent endoplasmic reticulum-associated protein degradation: an unconventional route to a familiar fate. Proc Natl Acad Sci USA 93:13797–13801

Williams DB (2006) Beyond lectins: the calnexin/calreticulin chaperone system of the endoplasmic reticulum. J Cell Sci 119:615–623

Wilson JD, Liu Y, Bentivoglio CM, Barlowe C (2006) Sel1p/Ubx2p participates in a distinct Cdc48p-dependent endoplasmic reticulum-associated degradation pathway. Traffic 7:1213–1223

Yamamoto M, Maruyama D, Endo T, Nishikawa S (2008) *Arabidopsis thaliana* has a set of J proteins in the endoplasmic reticulum that are conserved from yeast to animals and plants. Plant Cell Physiol 49:1547–1562

Yamamoto M, Kawanabe M, Hayashi Y, Endo T, Nishikawa S (2010) A vacuolar carboxypeptidase mutant of *Arabidopsis thaliana* is degraded by the ERAD pathway independently of its N-glycan. Biochem Biophys Res Commun 393:384–389

Yang Y, Li Z (2005) Roles of heat shock protein gp96 in the ER quality control: redundant or unique function? Mol Cells 20:173–182

Yang Y, Liu B, Dai J, Srivastava PK, Zammit DJ, Lefrancois L, Li Z (2007) Heat shock protein gp96 is a master chaperone for toll-like receptors and is important in the innate function of macrophages. Immunity 26:215–226

Ye J, Rawson RB, Komuro R, Chen X, Dave UP, Prywes R, Brown MS, Goldstein JL (2000) ER stress induces cleavage of membrane-bound ATF6 by the same proteases that process SREBPs. Mol Cell 6:1355–1364

Ye Y, Shibata Y, Yun C, Ron D, Rapoport TA (2004) A membrane protein complex mediates retro-translocation from the ER lumen into the cytosol. Nature 429:841–847

Yoshida Y, Tanaka K (2010) Lectin-like ERAD players in ER and cytosol. Biochim Biophys Acta 1800:172–180

Yoshida H, Matsui T, Yamamoto A, Okada T, Mori K (2001) *XBP1* mRNA is induced by ATF6 and spliced by IRE1 in response to ER stress to produce a highly active transcription factor. Cell 107:881–891

Zhang M, Henquet M, Chen Z, Zhang H, Zhang Y, Ren X, van der Krol S, Gonneau M, Bosch D, Gong Z (2009) *LEW3*, encoding a putative α1,2-mannosyltransferase (ALG11) in N-linked glycoprotein, plays vital roles in cell-wall biosynthesis and the abiotic stress response in *Arabidopsis thaliana*. Plant J 60:983–999

Zhu X, Zeng Y, Lehrman MA (1992) Evidence that the hamster tunicamycin resistance gene encodes UDP-GlcNAc:dolichol phosphate N-acetylglucosamine-1-phosphate transferase. J Biol Chem 267:8895–8902

Zipfel C (2009) Early molecular events in PAMP-triggered immunity. Curr Opin Plant Biol 12:414–420

Zipfel C, Kunze G, Chinchilla D, Caniard A, Jones JD, Boller T, Felix G (2006) Perception of the bacterial PAMP EF-Tu by the receptor EFR restricts *Agrobacterium*-mediated transformation. Cell 125:749–760

Index

F. Tax and B. Kemmerling (eds.), *Receptor-like Kinases in Plants*,
Signaling and Communication in Plants 13,
DOI 10.1007/978-3-642-23044-8, © Springer-Verlag Berlin Heidelberg 2012